Vernacular Architecture
Support for Territorial Development

International colloquium on the theme:
«Vernacular architecture, a support for territorial development»
National School of Architecture of Rabat, 4-5 October 2024,
Rabat, Morocco

Editors

Hassane Kharmich[1], Khalid El Harrouni[1], Khadija Baba[2], Khadija Karibi[1]

[1]National School of Architecture in Rabat, BP 6372, Rabat Institutes, Rabat, Morocco

[2]Higher School of Technology, Sale, BP 227, Sale Medina, Morocco

Peer review statement

All papers published in this volume of "Materials Research Proceedings" have been peer reviewed. The process of peer review was initiated and overseen by the above proceedings editors. All reviews were conducted by expert referees in accordance to Materials Research Forum LLC high standards.

Published under License by **Materials Research Forum LLC**
Millersville, PA 17551, USA

Published as part of the proceedings series
Materials Research Proceedings
Volume 47 (2025)

ISSN 2474-3941 (Print)
ISSN 2474-395X (Online)

ISBN 978-1-64490-338-4 (Print)
ISBN 978-1-64490-339-1 (eBook)

Distributed worldwide by

Materials Research Forum LLC
105 Springdale Lane
Millersville, PA 17551
USA
https://mrforum.com

Manufactured in the United State of America
10 9 8 7 6 5 4 3 2 1

Table of Contents

Preface

In its formal diversity and ecological adaptations, Vernacular architecture embodies an exceptional mastery of the environment. It reveals a collective heritage, embodying the diversity and richness of natural, human, and socio-cultural contexts. This architecture is the result of ancestral knowledge in the use of local materials and pure craft techniques. It is an authentic expression of the environment's potential, testifying to the deep bond between communities and their natural habitat.

The variability of vernacular architecture is evident from region to region, influenced by the physical characteristics of the land (plains, mountains, desert, etc.), lifestyles (sedentary, transhumant, etc.), settlement patterns, and the local availability of materials (stone, earth, wood, etc.). The construction techniques employed contribute to shaping extremely diverse and rich architectural landscapes, ranging from simple nomadic tents to solid sedentary homes, all the while adapting to the challenges of the terrain, the rigors of climate, and the scarcity of natural resources.

Internationally recognized for its particularity and uniqueness, vernacular architecture has been the subject of numerous international charters and has captured the interest of various authors and researchers. It has inspired architectural movements and mobilized many players in the field of territorial development and marketing. This craze has enriched our understanding of vernacular architecture, particularly its characteristics and formal typologies, as well as the approaches and methods used to analyze its productions. These studies address both the material and immaterial, natural and human, technical and aesthetic dimensions of this field.

Vernacular architecture, characterized by its virtuosity, ingenuity, technicality, and flexibility, has regained interest in its advantages in terms of thermal comfort, resource savings, and harmony with the environment while meeting contemporary requirements. Emblematic architects of the modern movement such as Frank Lloyd Wright, Le Corbusier, and Alvar Aalto, as well as more recent figures such as Hassan Fathy and Luis Barragán, have drawn inspiration from this architecture to create a "new vernacular architecture". This revival, based on artisanal production, demonstrates the continuity of traditions adapted to today's needs.

Vernacular architecture is attracted by its adaptability and continuous evolution. It presents itself as a passive solution, maximizing the use of sunlight and natural ventilation while respecting site constraints and local know-how. It symbolizes identity and a culture while demonstrating resilience in the face of various environmental and climatic risks. Despite these qualities, the vulnerability of these structures to disasters such as earthquakes and floods underscores the intrinsic fragility of this heritage, which, while resilient, remains exposed to major risks.

In the face of threats of vulnerability, abandonment, and destructive transformation, vernacular architecture in rural areas continues to symbolize a strong local identity. Its bioclimatic and ecological virtues remain a source of inspiration, despite the increasing standardization and globalization of architectural models. Taking an interest in this heritage also means contributing to the preservation and enhancement of the rural environment, fostering sustainable economic and human development. This heritage, rich in lessons on the rational use of traditional building materials and techniques, calls for restoration and rehabilitation strategies that aim not only for conservation, but also for the active integration of these buildings into the dynamics of local development.

From this argument stem four main areas of focus in which the multiple interventions should fit:

Focus 1: Vernacular architecture: Richness and diversity of forms of expression

Focus 2: Vernacular architecture: restoration issues and revitalization challenges

Focus 3: Vernacular architecture: Natural risks and resilience

Focus 4: Vernacular architecture: A source of inspiration and a lever for development

Colloquium expectations

This scientific Colloquium, organized by the National School of Architecture of Rabat (ENA Rabat), in partnership with its Doctoral Studies Center "Architecture and related Disciplines", its research structure LabHAUT (Habitat, Architecture, and Urbanization of Territories Laboratory) and its UNESCO Chair "Education and Research in Sustainable Bioclimatic Urban Planning and Architecture", and in collaboration with the CNRST (National Center for Scientific and Technical Research), and the University of Le Havre Normandie, proposes to examine, on the one hand, the architectural production described as vernacular, appreciated for its diversity and richness, its specificities and singularities adapted to various environments and contexts. On the other hand, it will look at local building skills and techniques, analyzing their organization, use, adaptability, evolution, and durability. By extension, this symposium explores the benefits and virtues of vernacular architecture in the field of eco-responsible practices and comfort, aiming to make it an invaluable source of inspiration for new construction.

It is also proposed to examine national and international experiences in protecting, restoring, and rehabilitating this essential architectural heritage. Depending on the context, these rehabilitation initiatives are interpreted through various prisms, including a) the regulatory and institutional aspect, b) intervention and governance strategies and techniques, and c) the social perception of this type of architecture. These interventions seek to preserve the authenticity of the structures while adapting the original architectural forms to contemporary requirements, thus balancing conservation and modernization.

Finally, the aim is to reveal the potential of vernacular architecture as a driving force for revitalization and territorial promotion. The transformation of these buildings into tourist attractions, places of natural beauty, and spaces of communion between man and the environment illustrates their growing appeal. These conversions, manifested in restaurants, bed & breakfasts, or exhibition areas for local produce, testify to their successful integration into the local economy. This revival is sparking debate on the relationship between heritage and tourism, as well as on the interaction between terroir and globalization, highlighting the importance of these structures in contemporary development dynamics and providing inspiration for new forms of urbanization.

These are the major objectives behind the present international symposium on: Vernacular Architecture, a Support for Territorial Development.

Committees

Coordination Committee

- M. Ibnyahya Abdelhadi, Director of ENA Rabat
- Prof. Kharmich Hassane, Scientific Coordinator of Colloquium and LabHAUT
- Prof. El Harrouni Khalid, Deputy Director of Research, CEDoc Director
- Prof. Karibi Khadija, Scientific Coordinator of Doctoral Studies
- M. Zerrad Youssef, SG of ENA Rabat
- Mme Cherkaoui Ghizlane, Head of Communication, ENA Rabat
- Mme Fars Sara; Phd Student, ENA Rabat
- Mme Majidi Salma, Phd Student, ENA Rabat

Scientific Committee

- Prof. Hassan Kharmich, ENA Rabat
- Prof. Khadija Karibi, ENA Rabat
- Prof. Khalid El Harrouni, ENA Rabat
- Prof. Ouafa Messous, ENA Rabat
- Prof. Mouna M'hammedi, ENA Rabat
- Prof. Assia Lamzah, ENA Rabat
- Prof. Hicham Mouloudi, ENA Rabat
- Prof. Sanae Aljem, ENA Rabat
- Prof. Youssef Talibi, ENA Rabat
- Prof. Iman Benkirane, ENA Rabat
- Prof. Mouhcine Ben Aïcha, ENA Rabat
- Dr. Karima Berdouz, ENA Rabat
- Prof. Rachid Alillouch, ENA Fès
- Prof. Majid Mansour, ENA Rabat
- Prof. Soummaya Samadi, ENA Rabat
- M. Hakim Cherkaoui, ENA Rabat
- Dr. Nezha Tlemçani Mekkaoui, ENA Rabat
- Prof. Tarik Harroud, INAU, Rabat
- Prof. Mohamed Tamim, INAU, Rabat
- Dr. Mohammed Akazzaf , ENA Rabat
- M. Aziz Ouahabi, ENA Rabat
- Prof. Samuel Deprez, Le Havre Normandie University
- Prof. Tariq Ouahbi, Le Havre Normandie University
- Prof. Mohamed Azrour, Faculty of Science and Technology, Errachidia
- Prof. Khadija Baba, Superior School of Technology, Salé
- Dr. Nadya Rouizem Labied, Sup. National School of Architecture, Paris Val de Seine
- Prof. Brahim Al Fasskaoui, Moulay Ismail University, FLSH Meknés
- Prof. Abdelaziz Adidi, INAU, Rabat
- Prof. Mohammed Hanzaz, INAU, Rabat
- Prof. Nabil Mekaoui, Mohammadia School of Engineering, Rabat

Organizing Committee

Doctoral students of the CEDoc Architecture and associated disciplines, ENA Rabat

Vernacular architecture: (Re) Visiting the concept

LAMZAH Assia

National School of Architecture, BP 6372, Avenue Allal Al Fassi, Rabat. Morocco

a.lamzah@enarabat.ac.ma

Keywords: Architecture, Vernacular, Postcolonial, Historiography

Abstract. Vernacular architecture refers to buildings produced using local materials and construction techniques that reflect specific social and cultural contexts and environments. It is widely described as an architecture that is the product of the geographical, social, economic, and cultural contexts and is mostly presented as eco-friendly, functional, and community-driven. Moreover, vernacular architecture is constructed with the empirical accumulated knowledge of the building masters using local resources and bears important lessons about the cultural, social, and technical contexts that produced it. Such buildings and landscapes are immovable cultural heritage assets that reflect the lifestyle of the region where they are located. However, in our postmodern world, where our understanding of the local has changed, and the spatial constructs of culture are shifting under the pressure of globalization and global networks --both physical and virtual-- the concept of vernacular architecture is far from being unproblematic. The meaning, scale, and symbolism of vernacular architecture are not universally agreed. Recent literature shed light on some of its problematic aspects, such as its multiple and shifting meanings, and more importantly its exclusion from the mainstream discourse of architecture and its history. Vernacular architecture has been widely understood as the architectural language of the people; the non-experts. To some extent, Architecture started to be seen not only as different from vernacular but as superior.

The present paper provides a critical reading of the content and evolution of the concept of vernacular architecture since the rise of Western modernism. It analyses two main aspects of the concept of vernacular architecture. The first aspect is its meaning and its evolution over time: does it mean local? traditional? authentic? has this meaning changed over time? How? It argues that its meaning is dynamic and that the vernacular is more about the process than the final architectural result. The second aspect is the place of vernacular architecture in the mainstream of the history of architecture. Knowing the power of conventions and their system of authority, the paper critically analyses the vernacular as other than official, professional, real Architecture. Finally, the paper provides new methodologies and perspectives on the definition, understanding, and analysis of vernacular architecture.

Introduction

Vernacular architecture refers to buildings produced using local materials and construction techniques that reflect specific social and cultural contexts and environments. It is widely described as an architecture that is the product of the geographical, social, economic, and cultural contexts and is mostly presented as eco-friendly, functional, and community-driven. Moreover, vernacular architecture is constructed with the empirical accumulated knowledge of the building masters using local resources and bears important lessons about the cultural, social, and technical contexts that produced it. Such buildings and dwellings, shaped according to the culture, beliefs, and values of the societies, are immovable cultural heritage assets that reflect the lifestyle of the region where they are located. However, in a context where our understanding of the local has changed, and the spatial constructs of culture are shifting as the phenomenon of globalization/global networks both physical and virtual, the concept of vernacular architecture is far from being unproblematic. The meaning, scale, and symbolism of vernacular architecture are not universally agreed upon: neighborhoods, markets, and community equipment. Recent literature shed light on some of its problematic aspects, such as its multiple and shifting meanings, and its exclusion from the mainstream discourse of architecture and its history.

The present paper addresses questions such as: how can we define vernacular architecture in relationship to other related concepts? Traditional, local, authentic…etc.? How did these definitions develop over time? How was vernacular architecture presented and represented over time by architectural historians? Canonical discourse? Did vernacular architecture's meaning change over time? What is the status of vernacular architecture in opposition to official, professional, high-tech, high culture? Are these dualities true? Useful? If not what new readings, and methodologies to use? (re)the interpretation we may have of vernacular architecture?

The paper aims to provide a critical reading of the content and evolution of the concept of vernacular architecture. It analyses two aspects of the concept of vernacular architecture. The first is its meaning over time. For this, it presents a brief historiography of the evolution of the concept. It argues that vernacular architecture had different meanings over time and that it has not always been "a legitimate" component of what real, legitimate, professional architecture is/should be. It also points out that the vernacular is more about the process than the final result of architecture. The second aspect is the place of vernacular architecture in the mainstream of the history of architecture, knowing the power of conventions and their system of authority. In the end, the paper proposes new methodologies for a new (re) reading and interpretations of vernacular architecture. Finally, the results of the paper are to demystify new expectations of vernacular architecture, as it is not limited to its physical and timely aspects, but also its anthropological and political ones.

Short historiography of the concept

Since its first apparition of the concept of vernacular architecture in the 1830s, a literature review of the evolution of the concept shows three main phases. In the first phase, since the rise of modernism at the end of the 19th Century, architectural writers have tended to admire what they called "traditional buildings" for the immediate relationship between form and function. The vernacular was mainly associated with the traditional, premodern. It also evolved to be defined as "folk" / popular. As an example, Olivier [1] presents the Folklorist movement of the early 20th Century, where vernacular architecture was associated with functional shelters for local people. Vernacular architecture was equivalent to what was built to meet the needs; what was built from the available [2].

The second phase in the historiography of the concept of vernacular architecture corresponds to the end of the 20th century, when vernacular architecture started to be seen according to social and cultural aspects, and no longer as an example of modernist buildings that illustrate the modernist principle of form follows function. Vernacular architecture starts to be defined and analyzed as complex social and cultural relations spatially constructed [3].

The third phase of the historiography of the concept concurs with the conceptual and doctrinal shift of the postmodern era. During this period, and in addition to functional, social, and cultural aspects of vernacular architecture, its political aspect starts to be implemented. For example, governments started to use architectural typologies, defined as vernacular, to establish and reinforce national/subnational identities [4].

Under the light of critical studies, postcolonial theory, and critical regionalism, several critics and scholars have criticized the way vernacular architecture is defined, by whom? For whom? They also analyzed its place within the education curricula. Indeed, the study of "everyday buildings" of vernacular architecture has been marginalized by architectural historians and cultural theorists until the late 80s. Indeed, for scholars, such as De Certau and Lefebvre, vernacular architecture was equivalent to "Anonymous" or "Unknown", what Robert Venturi described as "the Ugly" and "the Ordinary." [5]

Venturi [6] explores the conventional, the vernacular, and the undersigned. [7] take as an object of study the contemporary, "fallen" ordinary buildings and landscapes. They oppose the every day "is" to the ideal "ought" of modernism, asserting that before dictating to it, architecture must learn from its environment

Until the late 80s vernacular architecture was seen as different from "Personal Gestures of Architects", as anonymous; as non-Place.

Some scholars, following Venturi, extended the vernacular to new architectural and landscape typologies such as vacant lots, the backs of buildings, and the underside of bridges, appropriated by urban street art, graffiti artists, skateboarders… etc [8]). These new urban spaces, landscapes, and architectural typologies are interesting because they challenge the classical definition and understanding of vernacular architecture. They also help construct a more dynamic contextualization of vernacular architecture in its relationship to the social, cultural, economic, and political aspects. Other scholars extended vernacular architecture to include Oil refineries, old industrial equipment, power stations, grain silos… etc [9]. The Vernacular ended up being a mixture of unrelated buildings and places. But still, the common point between these elements is that they are all placed outside what architects and architecture historians consider "Architecture". The vernacular is defined as the architecturally other.

The Vernacular as Other

A critical reading of the previously presented historiography, and the ever-growing number of studies on vernacular architecture during the last decades, shows that one of the main problematic aspects is that the vernacular has always been removed from the mainstream discourse and the practice of Architecture. For example, it has been removed, for a long period, from the analysis and study of the history of architecture, as a source of inspiration and learning and as a witness of the relationship between humans and their built environments and landscapes. It has been considered different from practical, professional architecture. Another possible reason for this exclusion is that it was seen as a threat or unworthy of attention.

As a result, the vernacular as local, started to be presented in juxtaposition with the official. It started to become the Other: anti-quoted, archaic, traditional (fixed past), low culture, and low professional. It was therefore constructed and perceived as opposed to contemporary/modern, innovative architecture. Some scholars have even begun to question whether the vernacular can be considered architecture. Indeed, Architecture then is not only different from the Vernacular, but it is superior

From aesthetic theory to the history of art and architecture, it is interesting to analyze what is architecture. What is art? What is not? The place of the vernacular was not always present. The same argument can be applied to architecture education; the role and place of vernacular architecture in architecture curricula and if present, how is it constructed and (re) presented.

This marginalization of the vernacular needs to be contextualized (western/nonwestern, regional, local…etc) to be better understood within the fields of history, ethnography, cultural anthropology, and architecture.

The Vernacular is more about the process than the final architectural result. It may be traditional and contemporary. The false dualities traditional/modern, expert/non-expert, and local/global are not only false but may limit our understanding of the vernacular. They may lead to essentialism, which is the "Belief in the true essence of things, the invariable and fixed properties which define the witness of a given entity" [10]. These dualities are also reductive. Applied to vernacular architecture, they ignore individuality, complexity, and shift of knowledge of this specific architecture. Moreover, such dualities make the depth, complexity, and range of vernacular architecture and its meaning within their specific contexts inaccessible.

New Perspectives of reading and (re)interpretation

To overcome the previously discussed challenges, be it the shifting meaning of the vernacular and the danger of applying false dichotomies, vernacular architecture needs new methodologies. It should be analyzed and presented as a dynamic process of architectural production and landscape construction. This architecture is not necessarily an architecture "without architects" or an architecture for ordinary people, but the result of an accumulation of constructive design decisions of many anonymous actors who contributed to the creation of very localized and contextual architectural models and typologies, tested over time.

Additionally, these new methodologies proposed may be defined according to an integrated approach that combines the perspectives and methodologies of different academic disciplines and professions, and that attempts to deal with the dynamic interrelationships between building traditions, cultural identities, and environments on an ethnographic and descriptive, as well as a comparative and theoretical level [11].

Conclusion

Vernacular Architecture refers to buildings produced using local materials and construction techniques that reflect specific social and cultural contexts and environments. Indeed, this architecture constructed with the empirical accumulated knowledge of the building masters using local resources may bear important lessons. Such buildings and dwellings, shaped according to the culture, beliefs, and values of the societies, are immovable cultural heritage assets that reflect the lifestyle of the region where they are located.

Morocco has a rich vernacular architecture that witnesses its history and local identity. It testifies to traditional design wisdom, great capacity of adaptability to the context, and a great genius and know-how that developed over centuries in different regions and territories of the country. But, some of these buildings are now abandoned due to the change of social life and needs, suffering degradation and deterioration. We need definitely to save these buildings as they are important architectural heritage, to be preserved for future generations. Additionally, in Morocco and during the last decades, most parts of vernacular architecture, both urban and rural, suffered great deterioration that resulted in some cases in the complete demolition of vernacular buildings. Also, local people tend recently to value more concrete architecture, because they associate it with modernity and development and tend to ignore and even abandon these old traditional buildings, be it for habitat or community equipment. The various fortified and collective granaries, for example, present a large diversity, in their typology, their construction techniques (stone and rammed earth), and their geographic location. Moreover, in the architectural heritage preservation agenda in Morocco, vernacular architecture did not benefit from high interest, in comparison to other types of architecture. Therefore, vernacular architecture still needs to be documented and evaluated to be well understood.

But, to save this architecture, we need a critical reading of its content and evolution both on a theoretical and practical level. The concept of vernacular architecture has been considerably analyzed in the literature during the last two decades. However, we may encounter different ways of defining vernacular architecture due to different points of view, the literature used, and different periods. Additionally, given the fundamental transformations of what the local means and the political agendas of architecture education, theory, and practice, the vernacular needs to be revisited. In the postmodern global era, the world has changed. In the contemporary context, mainly under the shed of critical regionalism and postcolonial theory, the concept of vernacular architecture and the relevance of its understanding need to be critically studied.

The present paper presents a historiography of the concept of vernacular architecture and critically analyses its contents and categorization as other than normal/official.

The results of the present study are to demystify new expectations of vernacular architecture, as it is not limited to its physical and timely aspects, but also its anthropological and political ones.

Contextualizing and critically analyzing the historiography of the concept of vernacular architecture, and defining it more as a process than a final architectural object, constitute the first steps toward its understanding and therefore its preservation and rehabilitation.

References

[1] Olivier, Paul, Built to Meet Needs: Cultural Issues in Vernacular Architecture, Routledge, London, 2006. https://doi.org/10.4324/9780080476308

[2] Olivier, Paul, Built to Meet Needs: Cultural Issues in Vernacular Architecture, Routledge, London, 2006, and Al Sayyad, 2004. https://doi.org/10.4324/9780080476308

[3] Jackson, John, Discovering the Vernacular Landscape, Yale University Press, Massachusetts, 1989.

[4] Brown, Robert and Daniel Maudlin, "Concepts of Vernacular Architecture", in Greig Crysler, Stephen Cairns and Hilde Heynen (Eds), The Sage Handbook of Architectural Theory, London: SAGE Publications Ltd, 2012, pp 340-355. https://doi.org/10.4135/9781446201756.n21

[5] Venturi, Denise Scott Brown, and Steven Izenour, Learning from Las Vegas: The Forgotten Symbolism of Architectural Form, Cambridge, MIT, 1972; rev. ed. Cambridge, MA: MIT Press, 1977.

[6] Robert Venturi, Complexity and Contradiction in Architecture, New York: Museum of Modern Art, 1966.

[7] Venturi, Denise Scott Brown, and Steven Izenour, Learning from Las Vegas: The Forgotten Symbolism of Architectural Form, Cambridge, MIT, 1972; rev. ed. Cambridge, MA: MIT Press, 1977.

[8] Augé, Marc, Non-places: Introduction to an Anthropology of Supermodernity, New York, Verso,1995. And Chase et al, 1999 in Greig Crysler, et al (Eds), The Sage Handbook of Architectural Theory, London: SAGE Publications Ltd, 2012, pp 347.

[9] Brown, Robert and Daniel Maudlin, "Concepts of Vernacular Architecture", in Greig Crysler, Stephen Cairns and Hilde Heynen (Eds), The Sage Handbook of Architectural Theory, London: SAGE Publications Ltd, 2012, pp 340-355. https://doi.org/10.4135/9781446201756.n21

[10] Fuss, 1989, cited in Brown, Robert and Daniel Maudlin, "Concepts of Vernacular Architecture", in Greig Crysler, et al (Eds), The Sage Handbook of Architectural Theory, London: SAGE Publications Ltd, 2012, pp 345

[11] Rapoport, 1969, in Victor Buchli, An Anthropology od Architecture, London, Bloomsbury, and Lawrence, R, Learning from colonial houses and lifestyles', in M. Turan (ed.) Vernacular Architecture: Paradigms of Environmental Response, Aldershot: Avebury, 1990, and Asquith, L and M. Vellinga (eds), Vernacular Architecture in the Twenty-First Century: Theory, Education and Practice, London, Taylor & Francis, 1982.

The heritage of the Bulgarian revival in the central Balkans region: Vernacular architecture - restoration issues and revitalization challenges

Tzvetelina DASKALOVA[1,a *]

[1]Hristo Smirnenski bul, Sofia, Bulgaria

[a] cdask_far@uacg.bg

Keywords: Bulgarian Revival, Vernacular Architecture, Restoration, Central Balkans, Historical Heritage, Revitalization

Abstract. Over the past century, the concept of "obsolescence" has gained prominence, highlighting the ephemeral nature of architecture. The rapid advancement of building systems and technologies has accelerated the depreciation and aging of structures, prompting a shift in our understanding of a building's "lifespan." However, with evolving mindsets, it is essential to reconsider the value of vernacular buildings even not classified as Immovable Cultural Heritage. Some can be saved and revitalized, offering significant meaning for future development while preserving the cultural identity of cities and their inhabitants, as well as the layered historical memory they represent.

Introduction

The challenge lies in making complex adaptations of the vernacular buildings to meet modern requirements. This involves employing technologies that enhance living comfort while safeguarding historical value. An objective assessment must take into account the various factors affecting buildings and their environments, guiding a rational redesign of historical heritage in today's economic and social context. This approach seeks to update, maintain, and enhance interior spaces while maximizing the adaptation and enrichment of the exteriors.

Sustainable development is especially relevant for industrialized and post-industrial societies, where resource efficiency is critical. Buildings that face restoration challenges can often find a "second life" by adapting to new functions. For these vernacular buildings, some of which represent the architectural heritage of our country, some of them not, they need to be preserved and appreciated over time, but meeting the evolving criteria. These buildings often carry historical, national, and cultural significance that must be preserved to maintain the heritage spirit to a greater extent than their material essence.

The study of the typology and construction specifics of vernacular buildings from the Revival period in Bulgaria's Central Balkan region is crucial for preserving cultural and historical heritage. The dissemination of this knowledge will complement and expand upon existing scientific research in this field, particularly for other regions.

Historical Context

The architectural achievements of the Bulgarian Revival period represent the culmination in Bulgaria's national architectural heritage. This period is often referred to as our "Renaissance," but due to historical circumstances, it manifested later than in Western Europe—between the mid and late 19th century. The Bulgarian Revival house, in particular, encapsulates the architectural genius of the old masters.

The period from 1780 to 1878, known as the "Mature Revival," was marked by a remarkable surge in construction, both in terms of quantity and quality. This growth was driven by improved economic conditions, an increasing demand for public buildings, and a rising cultural interest and national self-awareness among Bulgarians.

As architect Chavdar Angelov points out, although the final years of Ottoman rule (1860-1878) are considered the beginning of modern times, this era saw the emergence of large masonry houses, prominent balconies, and solid cornices, with early Western European influences [1]. These developments, however, were mainly confined to major cities, while traditional building practices continued to dominate in rural areas long after Bulgaria's liberation.

The construction techniques of this period were based on earlier local traditions, but with improved craftsmanship, attention to detail, and refinements in certain technologies. While some structural and stylistic elements were influenced by Asia Minor and Tsarigrad (Istanbul) houses, as well as Aegean architecture, the architectural achievements of this period are predominantly the original creations of Bulgarian master builders. The purest examples, largely free from foreign influence, are found in the architecture of northern Stara Planina (the Balkans), which are the focus of this study.

The Central Balkans region, located in the heart of Bulgaria, encompasses much of the Pre-Balkan area and Stara Planina, excluding its easternmost sections. Stara Planina, also known as the Balkan Mountains, spans a total length of 530 km and is divided into three parts: Western, Central, and Eastern, separated by the Zlatishki and Vratnik passes.

This region holds significant tourism potential, offering a unique blend of cultural, historical, and architectural heritage along with diverse natural attractions. The region showcases authentic Bulgarian traditions, crafts, folk celebrations, and festivals. It also offers numerous opportunities for mountain and cultural tourism, as well as activities such as hiking, recreational tourism, adventure, ecotourism, rural tourism, religious pilgrimage, and skiing. In addition, the Central Balkans National Park, which has been on UNESCO's World Heritage Tentative List since 2011, holds the PAN parks certificate, attesting to the preservation of its biodiversity and the development of sustainable tourism, which provides income for the local population. The park safeguards forest massifs, plant and animal species, and historical monuments of global importance for both science and culture. Visitors can enjoy the unique landscapes, rock formations, and waterfalls, with the highest peak, Mount Botev, standing at 2,376 meters.

Building Typologies and Construction Techniques

A defining feature of the Bulgarian Revival architecture is its focus on tectonics and functionality, where every structural element serves a purpose and is functionally determined. This period saw a harmonious unity of structure, materials, and form. Even decorative elements were subordinated to the tectonic logic of the building.

Initially, the designs were not theoretically underpinned but based on the practical experience of master builders. It was only after Bulgaria's liberation that builders and architects who had studied abroad began applying theoretical knowledge to local construction techniques. Formal architectural education in Bulgaria began in 1942 with the founding of the Higher Technical School, which later became the University of Architecture, Civil Engineering, and Geodesy. The first Bulgarian professor of architecture, Hrabar Popov, founder and organizer of architectural education in Bulgaria, creating curricula, and establishing the foundations of the field [2].

In the traditional houses of the Balkan region, locally sourced materials such as stone, wood, brick, and rammed earth were commonly used. For security reasons, homes were often enclosed by high stone walls supported by horizontal beams, adapting to the terrain's irregularities. A large wooden gate would open into a courtyard, with the house and its outbuildings occupying the space.

Most Revival-period houses had stone foundations and a ground floor, with one or more upper floors reserved for living spaces. The materials and construction technologies used during this time reflected the region's socioeconomic conditions and building traditions.

Materials Research Proceedings 47 (2025) 6-14 https://doi.org/10.21741/9781644903391-2

Key structural systems of the period include:

- Solid structures with load-bearing walls: Brick, stone, composite, or adobe.
- Timber-framed structures: These included post-and-plank walls, half-timbered walls with infills of wattle and daub, adobe, laths and plaster, stone, or brick.

Fig.1 Kurtpashova kula – solid stone structure, 17th Cen., source: Wikipedia

Fig.2 Stone house from 19th Cen. source: own photo

In the Central Balkans, mixed structures were the most common. Typically, the ground floor was constructed of stone masonry, while the upper floors featured post-and-plank walls. One of the critical factors for assessing the structural performance of these buildings is their behavior under vertical and horizontal loads. Their resilience, particularly to seismic activity, is crucial for evaluating their future use and potential for adaptation.

Fig.3 Half-timbered walls, 18th Cen., source: own photo

Fig.4 Half-timbered walls, 18th Cen., source: own photo

The post-and-plank walls were designed for lateral resistance, as the planks were fitted into grooves in the posts. This structural configuration made these buildings better suited to withstand seismic forces. The primary components of post-and-plank timber walls include vertical posts, horizontal beams, and planks, all connected by traditional joints and wooden pegs, without the use of metal nails or screws. The post and plank systems are rare outside of Bulgaria, where they can be found in mountainous regions with forests that provide adequate wood.

Fig.5 Benkovski house, Koprivshtitsa, Post and plank structure, built in 1831, source: Wikipedia

Fig.6 Lafchieva house, Dryanovo, Post and plank structure, built in 1840, source: own photo

Socio-Cultural Value of Vernacular Buildings

Understanding a building's socio-cultural value is essential in its evaluation. This value encompasses not only its historical and scientific significance but also its emotional resonance with communities. The preservation of authentic materials and traditional construction methods plays a pivotal role in maintaining these values.

The architectural monument's significance lies not only in its appearance but also in the integrity of its structural components, which reflect the construction technologies specific to its time. Preservation efforts must strive to retain the original design, materials, and techniques to ensure the building's authenticity and allow future generations to recognize its historical context.

As Hrabar Popov noted in Tectonics and Structures in the Architectural Heritage of Bulgaria (1972), the value of an architectural object is not only in its external form but also in the architectural construction elements and overall structure. Furthermore, national peculiarities can be found in both unique monuments and ordinary residential buildings, which must be protected as part of Bulgaria's heritage [3].

Legal Aspects of Heritage Conservation

In Bulgaria, architectural and historical values are protected by a comprehensive legal framework. This framework regulates the national system of conservation and outlines the roles of state and municipal bodies. The National Institute for Immovable Cultural Heritage (NIICH-NINKN) plays a significant role in this process, supporting the Ministry of Culture in the preservation and promotion of real estate cultural heritage.

Despite these protections, many Revival-period buildings are privately owned, and the responsibility for their maintenance often falls on the owners. Unfortunately, many lack the financial resources to properly preserve these historic structures. Additionally, many valuable buildings have not been officially classified as immovable cultural heritage, making them ineligible for legal and financial support.

Development Potential and Restoration Challenges

The value of Bulgarian Revival architecture is indisputable, yet many restoration projects in recent years have failed to respect the original work of master builders. The inappropriate use of modern materials such as concrete and polystyrene foam has resulted in significant changes to the buildings' structural behaviour and microclimate, which threatens their historical authenticity.

Fig.7 Examples of bad intervention with inappropriate materials (new bricks and cement) source: own photos

Restoration efforts must begin with a thorough analysis of the building's architectural and structural features. This analysis will determine which elements are original and which are the result of inappropriate repairs. To ensure historical integrity, restoration must focus on using materials with characteristics similar to the originals. Recycled materials from other similar buildings slated for demolition may also be considered.

Key principles for restoration of such buildings must be much close to those for classified ones and include:

1. Dismantling non-original materials.
2. Restoring compromised elements using materials that match the original as closely as possible.
3. Respecting the building's original design and ensuring that any new elements are distinguishable from the historic ones.
4. Preserving the building's aesthetic and historical essence.

The choice of restoration methods depends on the degree of structural authenticity. For buildings that have retained their original structure, minimal intervention is necessary. However, for buildings where the supporting structure has been completely replaced, restoration is much more difficult and may require innovative solutions that prioritize the preservation of the building's integrity.

If structural reinforcement is needed, any changes should be aligned with the building's historical features, ensuring minimal disruption to its spatial interactions and overall design.

Case study from Leshnitza village

A restored house in the village of Leshnitsa will be presented as an example of good practice.

In this specific case - the house is not classified as an immovable cultural heritage and this gives wider possibilities for reconstruction and adaptation and even the use of modern building materials, but the owners desire to preserve the integrity of the house as much as possible.

Although the relatively well-preserved structure, some of the elements are compromised and in need of replacement. It is a typical mixed structure with a ground floor of stone masonry and a half-timbered upper-floor walls with infills of wattle and daub, and plaster. The roof is traditional from local slate tiles.

Fig.8 Leshnitza house, before restoration, Source: Lilya Tsankova

As much as possible the existing members are retained. Aged building parts from other similar structures are used where necessary. The whole reconstruction of the structure is made using similar craft methods and tools as the original. The infills – originally by wattle and daub – are replaced with composite panels with thermal insulation with characteristics close to the original. The refurbishment takes into account the microclimate in the premises, namely the temperature-humidity performance.

Preserving the original residential function of this vernacular house after its restoration is not only a practical solution but also a powerful means of maintaining its cultural, historical, and architectural continuity. Living in a house that was originally built for residential purposes allows modern inhabitants to engage directly with the past. Modern interventions, such as upgrading insulation, or adding eco-friendly utilities, can be made in a way that enhances the building's functionality while maintaining its originality.

Fig.9 Leshnitza house, During and after restoration works. Source: Lilya Tsankova

A case study from the town of Kotel

The second example refers to the town of Kotel - the Kozichkovi houses are dwellings in the heart of the historical old part of the town, classified as immovable cultural heritage.

Fig.10 Kozitchkovi dwellings, Kotel, second half of the 19th century, source: own photos

In 2021 an international training course took place in Kotel and was organized by Meshtra – Traditional Knowledge and Crafts in collaboration with European Heritage Volunteers, the University of Architecture, Civil Engineering, and Geodesy in Sofia, and the Municipality of Kotel [4]. Both international and Bulgarian students participated in the project and were introduced to the process of architectural documentation of the specific features of vernacular architecture and also the methods for 3D laser scanning of buildings and the use of photogrammetry in such cases. The municipality aimed to restore the houses to their original state and convert them into active workshop spaces for local artisans. As a result, the detailed documentation of the current condition was a crucial part of the project's initial stage and served as the foundation for planning future emergency repairs and conservation interventions.

The dwellings are an example of "post-and-plank" structure. It consists of vertical pillars (posts) and horizontal beams, made of hard timber. The spaces between the posts are not filled with adobe or some other infill, as in timber-framed structures, but specific timber planks are inserted into the posts with tongue-and-groove joints that improve their structural behavior. All the elements work together, and the walls act in a load-bearing manner.

Fig. Kozitchkovi dwellings, Kotel, wall and floor structure, source: own photos

The main issues identified are due to neglect and water damage, particularly seen in the decaying wood and collapsing structures. Damage to slate roof tiles from weathering, deterioration of the mud binding the tiles, and their weight have allowed rainwater to seep in, causing the wooden elements to rot.

To address these concerns, a restoration strategy is proposed based on the following key principles:

Preservation of Historical Authenticity: Restoration efforts must retain the original appearance and heritage significance of the buildings, using traditional methods and techniques whenever possible.

Retention of Original Materials: The goal is to preserve the original materials as much as possible. When replacements are necessary, they should be made with locally sourced materials that are easily distinguishable from the originals. In the case of wood, the new materials must be of the same type, with similar age and moisture content, to match the original wood.

Reversibility of Interventions: All restoration work should be reversible to allow for future preservation, adhering to the conservation philosophy outlined by ICOMOS (1999b) [5].

The project must consider not only the physical and architectural value of the dwellings but also their intangible cultural heritage, such as traditional craftsmanship and the cultural significance of these homes to the local community (ICOMOS 1999a).

Restoring the timer structures should enable the community to continue engaging with them, reflecting their cultural and historical importance [6].

The restored homes must be suitable for contemporary living standards, addressing micro-climate control and essential facilities, while preserving the traditional character.

The suggested new function, as a Cultural and Educational space, is a great opportunity to maintain the traditional crafts, where skills like local carpentry for example can be taught.

This restoration plan seeks to balance the need for practical use with the importance of preserving the historical and cultural heritage of the Kozichkovi houses, ensuring they remain both functional and a vital part of the local community's identity.

Fig. 1 South façade, part of the first architectural sketches, made by students

Conclusion

The vernacular architecture of the Bulgarian Revival period in the Central Balkans holds immense cultural, historical, and architectural value. These buildings represent the craftsmanship and ingenuity of Bulgarian builders, whose methods and materials have endured for centuries. Preservation efforts must strike a balance between maintaining historical authenticity and adapting buildings to modern needs.

Restoration of these architectural treasures requires a multidisciplinary approach that integrates historical knowledge, architectural analysis, sustainable practices, and legal protections. By embracing these principles, we can ensure that Bulgaria's rich architectural heritage continues to thrive for future generations.

References

[1] Angelov Ch., Building construction in Bulgaria, vol 2., Marin Drinov Publishing, 2012, ISBN: 978-954-322-492-0

[2] Popov Hr., Tectonics and structures of the architectural heritage in Bulgaria, Tehnika Publishing, 1972,

[3] Popov Hr., Tectonics and structures of the architectural heritage in Bulgaria, Tehnika Publishing, 1972,

[4] Information on https://meshtrango.com/

[5] ICOMOS, Charter on the built vernacular heritage (1999) Ratified by the ICOMOS 12th General Assembly, in Mexico, October 1999.

[6] ICOMOS, Principles for the conservation of wooden built heritage, adopted by the 19th ICOMOS General Assembly, New Delhi, India, 15 December 2017

Materials Research Proceedings 47 (2025) 15-29 https://doi.org/10.21741/9781644903391-1

Erasing the architectural and built identity of the Marais Vernier: A chronicle of the suburban dream as an expression of consumer society

Samuel DEPREZ

Université Le Havre Normandie - Équipe IDEES Le Havre - UMR 6266 CNRS, 25, rue Philippe Lebon - BP.1123 - 76063 Le Havre Cedex France

samuel.deprez@univ-lehavre.fr

Keywords: Natural Environment and Landscape Organization, Architectural and Built Identity, Land Use Change, Sururbanisation, Consumer Society

Abstract. At the heart of an ancient meander of the Seine, the Marais Vernier micro-region (France - Normandy) is renowned for its vast wetlands, an ecological sanctuary protected by multiple French and European schemes. It is also an exceptional area in terms of architecture and buildings, the fruit of local determinisms that have dictated the nature and form of the habitat and land. While the "chaumière", a thatched, half-timbered house, is the land's defining feature, its place in the landscape is gradually disappearing under the weight of the urbanization movement initiated in the early 1960s and the opening up of the area. Since then, in a random arrangement and denial of the existing, singularity and richness, the various forms of built-up areas that have marked each new phase of individual construction in France since the post-war period have been sedimented. The result has been a long process of erasing built, land, and architectural identity, driven by the realization of the suburban dream and the myth of the single-family home as the material manifestation of consumer society.

Introduction

Nestled in an ancient meander of the Seine, some thirty kilometers from Le Havre (Normandy, France), the Marais Vernier micro-region is today primarily known and recognized for its vast wetlands, an ecological sanctuary for countless protected species, enhanced and preserved by multiple national and European classification measures. The area's uniqueness and richness stem from an ancient hydrogeological process that over time has dictated its shape, topography and soil structure. During the great ice ages of the Quaternary, the erosive action of water on the chalky plateaux of the Caux to the north of the Seine's present bed and the Roumois to the south formed numerous meanders, giving the river its original sinuous course. During the last ice age, the river was redesigned and shifted to a northerly position, abandoning the meander it formed at the Marais Vernier as it is today. Enclosed by a barrier beach, the area became a tidal mudflat, where 4,500 hectares of one of France's largest peat bogs gradually formed. In the southern part of the area, a dense forest borders the summit of the hillside, which closes off the area and marks the boundary of this vast semi-circular depression, where the first identified traces of human settlement date back to the 11th century.

These original witnesses mark the starting point of a complex and singular process of anthropization that took place over a long period to create the conditions for human life in a natural environment that was hostile in many respects. Based on the dual principles of adaptation and appropriation, a habitat and land use with original attributes were gradually designed and structured, based on making the most of existing resources and dealing with physical constraints, such as marked topography and the presence of water in the soil, which posed the challenge of settling and living in these places. The inventiveness and genius of the local people have given rise

Materials Research Proceedings 47 (2025) 15-29 https://doi.org/10.21741/9781644903391-1

to several distinctive features, built forms, and spatial organization, the traces of which can still be seen today, and for which the area is renowned as a heritage site and tourist attraction.

Long protected from the influences of modern society by its landlocked location, the Marais Vernier has nevertheless undergone profound upheaval since the 1960s and the commissioning in 1959 of the Pont de Tancarville, which provides a continuous crossing of the river and opens up the area to Le Havre, its conurbation and the industrial and petrochemical complex of the Seine estuary. This entry into the age of modernity, of which the automobile is one of the most emblematic symbols, marks the beginning of a new phase in the process of anthropization of this micro-territory, which calls into question the built identity and original landscape of the Marais Vernier and threatens to erase its future. This is the starting point for our reflection on the changes that have been at work for almost seven decades now in the architectural and built landscape of the village of Marais Vernier, from which this micro-territory, comprising six communes, takes its name by extension. We'll be taking a critical look at these trajectories, between the disappearance of a built identity and the erasure of the foundations that enabled these places to “make territory”.

1 - From the original natural environment to a living space built *ex-nihilo*: the Marais Vernier or the genesis of a territory as a process

The first contact with the Marais Vernier micro-region confronts the visitor with the singularity of its morphology and topography. From the wooded hillside a hundred meters above sea level, an immense amphitheater unfolds before the visitor's eyes, all the way to the chalky ledge 4.5 kilometers away, at the foot of which flows the Seine, now stabilized in its dyked bed.

Fig.1 - The Marais Vernier from the hillside panorama. Photo: S. DEPREZ - 09.2024

The beauty of this remarkable landscape, the serenity it exudes, and the feeling of suspended time, like a privileged interlude away from the noise and bustle of the industrial zone on the horizon on the other bank of the river. Before taking the road that will soon take him through the forest to the commune of Marais Vernier, however, he is certainly far from suspecting that the components of this landscape have all, and each at its level, both determined man's ability to settle and dictated the modalities of his presence, foremost among which is his habitat.

1.1- Buildings as an expression of territorial anchorage: exploiting and enhancing the resource environment

If the Marais Vernier has one well-known attribute, it's the original buildings, of which there are still vivid traces today. Despite more or less skillful work, these traces help us to understand the consubstantial link between the *thatched cottage* and the environment in which it was built. Each element expresses the ability of the people of Marais to transform their surroundings into a resource for building their dwelling, according to similar principles depending on the status of the

Materials Research Proceedings 47 (2025) 15-29 https://doi.org/10.21741/9781644903391-1

occupants - farm owners or workers, farmers or laborers. From the woodlands on the hillside, he extracted the *timber-framed* structure made of a complex assembly of different oak pieces (Fig. 2), which would ensure the rigidity of the whole and its durability over time. From the subsoil, he extracts blocks of chalk and flint, which he cuts, superimposes, arranges, and combines with a natural lime-based mortar to form the bedrock that supports the dwelling while protecting it from the dampness of the soil. Further on, in the reed bed at the heart of the marsh, he cuts and gathers reeds, which he assembles into bunches to form the roof, at the top of which the clay ridge fitted to the house and planted with iris, a marsh plant, ensures watertightness (Fig. 3). Over time, slate has supplanted thatch roofing, transforming the *thatched cottage* into a *Norman house, but* in the Marais Vernier there are still a few examples of this traditional know-how, which is now threatened because local resources are limited and the thatcher's trade no longer appeals.

Photography / Production : S.DEPREZ - ULHN - UMR IDEES 66266 CNRS - 09.2024

Fig.2 - The thatched cottage, a traditional half-timbered house

As you wander around, you'll notice how the gaps between the timbers are filled in to close off the walls, another way of making the most of what's available locally to build your home. On a horizontal wooden lath made from split branches or hazel twigs, *torchis* is projected by hand, a mixture of clayey loam, sometimes animal matter, hay and water, smoothed to facilitate sliding and the evacuation of rainwater (Fig. 3). In this way, the various resources of the environment are exploited and valorized by drawing from each of these horizons, from the soil to the subsoil, from the wooded hillside to the wet marsh. Beyond the general matrix, *thatched cottages* take on singular attributes depending on the social status of their dwellings. Thatched cottages with narrow half-timbering or high foundations, because they require more wood and materials for their construction, as well as more openings, bear witness to the wealth and affluence of their owners, whereas those of day laborers or workers are much smaller, with just one door and a single window.

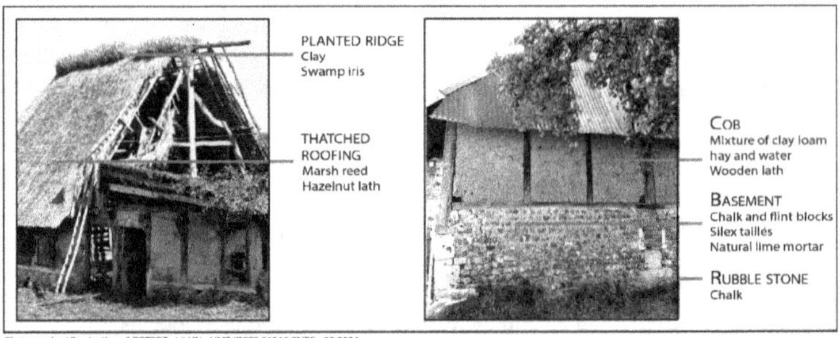

Photography / Production : S.DEPREZ - ULHN - UMR IDEES 66266 CNRS - 09.2024

Fig.3 - Construction diagram of a traditional thatched cottage

This link to the environment as a determining factor in the structure of the building is expressed with equal acuity in its organization into three distinct levels, made possible by the use of each material. The ground floor (R-1), as we would call it today, houses the *cafoutin*, a semi-buried storeroom where fruit and vegetables are kept in the dark, particularly those collected from the apple-tree planted courtyard surrounding the dwelling. The living quarters are located on the first floor, with a common room, a storeroom, and one or two bedrooms, depending on the number of generations living under the same roof. On the same level, a stable and/or cowshed sometimes extended the building. They can also be found in the form of isolated buildings close to the house. The second floor of the house can be used as an attic if the structure can support the weight of sacks of grain, sheaves of hay, or straw. In recent decades, it has been transformed into a living space, often housing bedrooms and offices lit by dormer windows or skylights.

Photography / Production : S.DEPREZ - ULHN - UMR IDEES 66266 CNRS - 09.2024

Fig.4 - Organization of the traditional thatched cottage

However, the *thatched cottage* as an expression of human adaptation to inhabit an environment has developed into much more than a simple architectural singularity. It has given rise, without being formalized as such, to a veritable territorial project which, in an overall logic, articulates buildings and land dedicated to livestock rearing, cultivation, and food production. This anthropization has taken place in a context constrained by a marked topography and the presence of water, both of which have dictated the conditions of human presence, determined the organization of space, and presupposed inventiveness on the part of the inhabitants.

Materials Research Proceedings 47 (2025) 15-29 https://doi.org/10.21741/9781644903391-1

1.2 - Constrained architecture and urbanization.

Marais Vernier is structured around the main road at the foot of the hillside. From this axis, housing developed in two different ways: along the road, over a width of a few dozen meters, particularly in the heart of the village; perpendicular to the road and in the direction of the slope beyond this first curtain of buildings (fig. 5). Numerous surviving buildings bear witness to this topographically dictated urbanization, creating a singular landscape. On the hillside, dwellings with asymmetrical facades can be seen, where one end, sometimes less than a meter high, seems to be buried in the ground, while the other rests on high timbers to reach the same roof level (fig. 6). Beyond the road, where the differences in levels are less marked, the same pattern applies, producing a dwelling with lower ceilings on the first floor and smaller cellars and storerooms below.

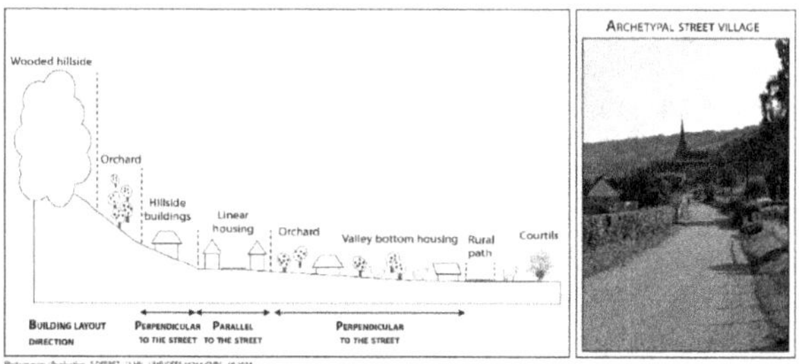

Fig. 5 - Functional structuring of the Marais-Vernier space

Photography / Production : S.DEPREZ - ULHN - UMR IDEES 66266 CNRS - 09.2024

Fig. 6 - Building on slopes: two expressions of building adaptation

The determinism of topography can be seen just as clearly in the way land use is allocated and organized. Between the wooded hillside and the first houses built along the slope are apple orchards grazed by small herds of sheep and cattle. They extend beyond the road and its built-up front to the rural road that separates the inhabited part of the commune from the wet marshland. This organization of space is based on a singular plot of land. On either side of the street are small, narrow strips of land on which the buildings and surrounding ornamental gardens are built. On the second front, the plots develop vertically in a single block: rising up to the wooded edge in the southern part; joining the rural road in the northern part. Beyond this, the *courtils (small fields)* take their place, *in* the form of long, narrow strips, sometimes over a kilometer in length. In the Marais Vernier commune, the most extensive of these "measure just 650 meters before giving way to communal property" [1]. The most common rarely exceed a few dozen meters in width. Their origins date back to the Middle Ages, and their organization is defined in response to the needs of the inhabitants and the constraints of the environment. "Vegetable gardens, which gave their name to the courtils, developed in the less humid areas. Towards the marsh, they extend into meadows and pastures. Then, in the heart of the marsh, the common pastures, a vast area with very imprecise boundaries that is often flooded. To the north, along the sea or the Seine, the coastline contains a few ploughed areas". [2]. The *courtyards* are delimited by narrow ditches lined with pollarded willows and alders, which serve as drainage systems and fences to keep the animals in. From the pruning of these trees, regularly carried out to give them and maintain their specific shape, the inhabitants extract firewood, precious in this environment subject to high humidity. They also use the finest branches to make baskets and other basketry used in daily farming activities.

In this way, the unique architectural and land identity of the Marais-Vernier, for which it is renowned today, was first and foremost an expression of the inhabitants' appropriation of the area, laying the foundations for an original territorial system in which each stratum corresponds to a use and responds to a need, enabling it to function in complete autonomy and creating the conditions for a subsistence economy as an obligatory response to the closed and landlocked situation of the Marais Vernier micro-region. Here again, topography played a decisive role, which we can now read, with a little hindsight, and by reinscribing it in a multiscalar perspective under the dual register of constraint and boon. The site's location at the bottom of a valley, far from the first towns and small centralities, forced the inhabitants to think in terms of self-organization and to find in the environment the conditions for their presence and survival. We owe this heritage to an environment shaped by man, the product of know-how and social organization that have shaped this "landscape of identity" [3], understood as both a collective construct and a singular materiality deeply rooted in a place to the point of "making territory" in the sense of a space "lived in, perceived, appropriated" [4]. In this sense, the landscape of identity defines and characterizes only this "territory", in these "places alone". This underscores the importance of a typo-morphological analysis of buildings and land, such as the one we are carrying out, in order to establish a frame of reference from which to analyze the evolution of this landscape and beyond.

2 - The future of an original territorial identity in a changing society

This exercise is the subject of our second part, in which, on the basis of our immersion *in situ* and the repositioning of the Marais Vernier in the wider Seine estuary, we set out to explain the changes that have taken place in the original framework, the determining factors behind them and their consequences for the very identity of the commune as we have outlined it. In addition to factual observation of the changes at work, our aim is to extend our analysis of the link between society, territory and landscape, using buildings and land as entry points to express and reveal an individual and collective relationship with places and territories. Through the physical and organizational markers visible in the built and urbanized fabric, a certain number of changes can be seen as illustrations of the more contemporary links established between the people of Marais and the territory in which they live, and which contribute to the very rewriting of the village's identity. Beyond the relationship between residents and the place where they live, we see these changes as an expression of the profound changes taking place in society, and of societal aspirations for a way of life and living that the Marais Vernier now bears the scars of, and which are reexamining its very nature and its future.

2.1 - The Marais Vernier, a chronicle of territories under the influence

If the Marais Vernier has undergone one profound change in recent decades, it is undoubtedly in its links with the surrounding area, which it has gradually opened up. The inauguration of the Pont de Tancarville and the creation of the A131 freeway that accompanied it in the early 1960s initiated a profound change in the territorial situation. From a situation of functional enclavement posed by topography, which for the built system had forced us to think about "ensured(d) its balance, cohesion and solidity" [5], the micro-region suddenly found itself projected into an enlarged territorial system driven by the presence of a permanent and continuous service between the two banks of the Seine (Fig. 7).

This new openness has brought new dynamics to the area, which are rapidly gaining momentum in a two-pronged movement from and towards the Marais Vernier. On the first front, a large number of Marquais were recruited by petrochemical and automotive companies in the industrial zone of the Seine estuary, near Le Havre, and Port-Jérôme to the west, served by a ferry across the river. Gradually, salaried employment grew among the community's working population, who turned away from traditional trades, particularly livestock breeding and crafts. Some left the village to settle in modern housing in the greater Le Havre area. Since 1968, each census has shown more people leaving than arriving, and household sizes have steadily shrunk. Another indicator of these dynamics is the steady increase in the number of vacant homes, which has doubled between 2010 and 2021. It now stands at 10% of the residential stock, with 266 units [6]. In the village, the decline in agricultural activity and the disaffection of some of the traditional buildings is evident. Unused farm buildings are maintained at a lower cost by protecting them from the elements with metal roofing, which is both less expensive than thatch roofing and easier to install. Beyond the roof, there's no need for renovation work or preventive action to prevent rapid deterioration. As far as housing is concerned, many abandoned houses with closed shutters, similar to those shown above (Fig.6), can now be seen along the roadside. Mirroring these trajectories, vast modern farm sheds are taking their place in the landscape on the initiative of farmers still in business, giving it resolutely contemporary attributes.

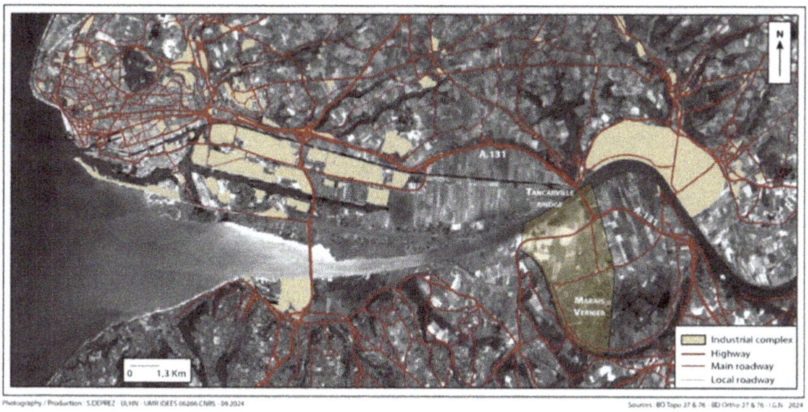

Fig. 7: The Marais-Vernier, an area under the influence of economic and urban centers

Photography / Production : S.DEPREZ - ULHN - UMR IDEES 66266 CNRS - 09.2024

Fig.8 - The difficulty of maintaining agricultural and residential buildings over time

But while some people are leaving the Marais Vernier, others, attracted by the calm of the area and the beauty of the landscape away from the hustle and bustle of urban life, are also looking to settle here, just as some native residents wish to stay. These two groups of people have a dual dynamic when it comes to building. By acquiring existing properties, they contribute to maintaining the original habitat, despite taking a few liberties with its foundations and random interventions. However, the constraints of renovation, the high cost of the work, and the difficulty of reintegrating traditional buildings into the codes and expectations of modernity in terms of comfort and amenities lead many of them to opt for the construction of a new property, with the question of available building land as a prerequisite. Here, the question is posed in very specific terms, since the organization of land plots was a structuring element of daily life and the village, but also because the topography leaves few opportunities. For some years now, we've been witnessing a rewriting of this original layout, driven by a logic of supply and demand between owners of vast plots of land whose original use is no longer valid, and buyers wishing to establish themselves. For the former, the motivations are mainly financial. Often on low incomes or modest pensions, they see the sale of part of their land as a welcome capital opportunity. Some also opt to subdivide their land to build a more comfortable home for their retirement. For the latter, it's an opportunity to buy a building plot at a reasonable price (around 45 euros per square meter, i.e. two to four times cheaper than in the Le Havre conurbation), some thirty kilometers from this urban center, and its jobs. While almost half of the principal residences (91) were built before 1919, according to the inventory, 69 have been built since 1970, including 21 between 2006 and 2018. And our field observations have revealed further construction since then.

In terms of landscape and land organization, this real estate dynamic has brought about major changes: in terms of buildings, the houses erected over the last five decades represent a sharp break with the original morphology, in terms of both the shape of the buildings and their placement within the plot; in terms of plots, the subdivision into small lots is erasing the native matrix and, with it, the founding principles of land structuring (Fig.9).

Fig.9 - The difficulty of maintaining agricultural and residential buildings over the long term

The result of this modern urbanization movement is continuous urban sprawl, which makes the original building an increasingly discrete reality, almost invisible when the traditional thatched cottage is hemmed in between modern pavilions which, because of the topography from the heart of the town, dominate and erase it (Fig. 10). In this way, the Marais Vernier becomes a privileged setting for neo-urbanites and native-born villagers alike to make their dream of a new home come true, illustrating perhaps more forcefully than in other areas the weight of suburbanization as it has been manifesting itself in France for several decades now. In our view, it is a dazzling expression of the consumer society that has been the matrix of both society and planning [7] for some sixty years now, placing the residential function at a distance from the workplace, with the automobile as the medium between these two time-spaces of daily life - another attribute if ever there was one of consumerist thinking and planning.

The Marais Vernier is a reflection of this interdependent relationship between (consumer) society, landscape, and territory, with its many strands, which is leading to a profound rethinking of suburban spaces and their identity: the arrival of populations with no previous ties to the territory, the transformation of traditional agricultural land into residential land, the rewriting of original land parcels, the introduction of new built forms (including housing estates, from which our study area still escapes), the regular appearance of new architectural styles as building systems and materials evolve...), and the introduction of exogenous attributes (exotic plants for hedges, verandas and pergolas, shelters for camper vans...). In the Marais Vernier commune, this heterogeneity is now a major trend, apparently irremediable since the dynamic remains and certainly irrevocable since what has been erased will no longer be and what has recently been built is destined to remain. Put into perspective with the national context, it's part of the order of things as the territory evolves according to these principles. Perhaps, as the political discourse legitimizes, this is necessary to bring about a generational renewal, maintain a village school, preserve one last shop, and avoid the risk of abandonment and even greater abandonment. However, the argument

Materials Research Proceedings 47 (2025) 15-29 | https://doi.org/10.21741/9781644903391-1

merits a more detailed reading, based on an analysis of the discourse and practices of residents, to validate or invalidate the correlation highlighted between new urbanization and communal (re)dynamization...

From our current reading angles, the question remains that of the modalities of this new time of urbanization, which *de facto* induces the question of the place of the individual and the collective in the maintenance, preservation, and evolution of territorial identity, to avoid a dilution of original attributes in a disparate aggregate of constructions and land developments, whose sole logic responds to the conception of what each person makes of his or her habitat and property, which, from our point of view, is prejudicial to the very notion of territory.

Photography / Production : S.DEPREZ - ULHN - UMR IDEES 66266 CNRS - 09.2024

Fig.10 - Changes in housing, its forms, and surroundings

Materials Research Proceedings 47 (2025) 15-29 https://doi.org/10.21741/9781644903391-1

2.2 - From a unified, "collectively-built" landscape to a heterogeneous, "individually-built" landscape

A quick tour of the commune reveals the extent of the shift from a unified landscape - the visible expression of a singular territoriality in which productive activities and housing, land, and buildings were conceived as a coherent whole, where a sense of belonging and strong social interactions were expressed in the same space - to a support environment chosen for building one's residence because it offers amenities sought after in choosing one's living environment. Contrary to the original plan, the different types of buildings on view express a relationship of individual consumption with the landscape, where each person builds, renovates, modernizes, and embellishes his or her own home, without looking at the surrounding area or taking into account what, on a village scale, cements its identity, expresses its specific features, gives coherence to the whole and, through the materiality of the buildings, land and natural elements, contributes to "making territory".

Careful observation of old and new buildings has produced an enlightening typology of the built mosaic that characterizes the Marais Vernier today (Fig.11). There are two opposing approaches to traditional housing: on the one hand, the original Norman thatched cottage is being maintained and preserved, although only a handful of units remain; on the other, traditional dwellings are being modernized, with the original features of their facades disappearing under thick rendering and their roofs being replaced by natural or artificial slate. At the same time, the pavilion dream is expressed in the search for continuity with the existing, where the house is adorned with faux half-timbering based on the timber-framed structure. More frequently, it takes the form of radical breaks, with breeze-block and concrete construction opening up new perspectives in terms of volumetry and architectural composition. While the double-sloped roof can be seen as a last reminder of the buildings of yesteryear, the wide openings and bay windows break with the reduced openings and small-paned windows. The ease of terracing now makes it possible to disregard topographical constraints for construction, and the original layout is overlaid by a haphazard, almost anarchic distribution of buildings, another expression of the individual's contribution to the material recomposition of the landscape. Even more audaciously, the architects have reinterpreted the wooden house in a size and volume that have nothing in common with what the Marais Vernier has given us to see and look at, regardless of when it was built. Unique yet decisive in its imposing size, and even more at odds with the village's visual identity than its predecessor, this architectural UFO has taken its place at the foot of the hillside, as a blatant expression of a consummate landscape, where the choice of plot overlooking the inhabited areas is dictated by the desire to enjoy a peripheral view of the entire marsh from the house's living rooms.

Fig.11 - A mosaic of built forms: from unity to heterogeneity

There's nothing to unify these different expressions of urbanization, all the result of the personal wishes and choices of their owners. This blurring of the built and land frameworks is equally evident in the natural part of the landscape, where high conifer hedges, veritable plant walls, close off plots of land from outside view, in a radical departure from the open landscape of orchards and low hedges of local species. This natural enclosure is often accompanied by the use of rigid fences, which have become the norm for private property boundaries. Here again, a new landscape is being created, reinforced by the installation of modern gates as new expressions of what, through the offerings of the major DIY chains, consumer society is both proposing and installing. Its influence on the evolution of buildings and, through them, of landscapes, can be seen with the same vivacity in the addition of new attributes around the home, in the register of amenities (verandas and conservatories, pergolas), comfort (roller shutters) or energy conversion (photovoltaic panels), to the installation of which each proceeds (Fig.12). These individual choices give rise to marked trajectories that leave their mark on landscapes and can profoundly alter their nature and original features in a sometimes very short space of time.

Fig.12 - Modern attributes as new landscape markers

This process, illustrated by the singular example of the Marais Vernier, is a manifest reality throughout France, and almost no commune escapes this movement. It highlights civil society's capacity to use the habitat as a medium to promote singular forms of territoriality, understood as "a modality of action by which individuals collectively compose a common good and experience it through the relationships they maintain together with the outside world, in ways of knowing and valuing space that is specific to them and that they share" [8], and then rediscussing them through individual initiatives that, juxtaposed, nourish an overall trajectory of erasing primary territoriality, without necessarily replacing it with another. It therefore seems legitimate to re-examine the role and weight of politics and the instruments at its disposal as guarantors of the evolution of territories and the territorialities that originally qualified them. In terms of urban planning and architecture, part of the answer lies in planning documents and their regulations. In the commune of Marais Vernier, the National Town Planning Regulations (RNU) apply in the absence of a Carte Communale (CC) or a Plan Local d'Urbanisme (PLU), which would have defined strong prescriptive principles for future urbanization within a coherent village-wide framework. In the absence of such a plan, land development is governed by the general principles of urban planning in force throughout France. The result is a case-by-case management of requests for building permits or plot subdivisions, with no territorial background or prescriptions on building, its forms, recommended or non-recommended materials, chromatic ranges to be respected, authorized fences or vegetation... as may be set out in urban planning documents drawn up by local authorities. While the French urban planning code (Art. R.111-21) sets the preservation of the interest and character of the surrounding area as the objective of the RNU, the numerous examples we have presented demonstrate that many authorizations have been issued when the project was detrimental to the identity of the commune. Although the illustration is striking, it should be pointed out that in many areas governed by town planning documents, residential development based around the detached house and its various architectural and stylistic variations has been authorized, producing unmarked landscapes in contact with traditional buildings, as in the commune of Saint-Aubin-sur-Quillebeuf, at the western end of the Marais-Vernier micro-region. This is where the subtlety of the rule is expressed in its very essence, where over and above its respect and application, the question of its content and objectives arises, which brings us back to the very notion of territorialities, those of the past and those we wish to promote, in continuity, rupture or complementarity. This also raises the question of the rule's ability to keep pace with the changes and influences brought about by the consumer society as a planning matrix and to provide the necessary safeguards. This is also where the relationship between individuals and the rule comes into play, where between distrust and rejection of principles seen as obstacles to the freedom to enjoy one's property, the development of private property, both built and unbuilt, and its evolution over time are often written in denial of the rule and against its principles, with all the consequences for local identity.

Conclusion

The Marais Vernier micro-region, and the commune to which it owes its name, vividly illustrate man's ability to appropriate an inhospitable environment, to mobilize its elements to create the resources necessary for his settlement, to make his mark on it to create the conditions for his subsistence, to organize it materially through buildings and land, and to develop a social life based on solidarity and traditions around activities (farming, livestock breeding...), and thus "to make territory". They certainly owe their uniqueness to a landlocked situation that has created the conditions for this social and territorial microcosm and enabled it to be maintained over time, along with the singular, unchanging landscape, until recently, when everything accelerated.

This territory is also a chronicle of the short term, where in the space of a few decades, what seemed intangible has been re-discussed as it has been opened up, opening it up to external influences in the same way that consumer society, transformed into a matrix of urban planning and development, has imposed its own on it, through the behaviour of each individual. Over the past sixty years or so, the Marais Vernier's territorial foundations have been, if not erased, at least nuanced, and the movement that began it is still going strong today. Political awareness of these far-reaching changes is now emerging, with the DREAL (Direction Régionale de l'Environnement, de l'Aménagement et du Logement) filing an application for "site" classification in 2023. The purpose of this national provision is to protect classified sites, with the aim of ensuring the preservation of natural monuments and sites whose artistic, historical, scientific, legendary or picturesque character is of general interest. Classification as a site constitutes a public easement which is binding on all owners. While this project may be seen as a welcome opportunity to influence the dynamics of urbanization, many elected representatives are fiercely opposed to it, seeing in this possible classification additional procedures and constraints, particularly in terms of town planning. For if the procedure were to be completed by 2025, as envisaged, "all projects falling wholly or partly within the perimeter of the listed site will be subject to the procedure for authorizing works in a listed site" [9], as well as to the reasoned opinion of the relevant government departments, the departmental commission for nature, landscapes and sites, and the Architecte des Bâtiments de France.

Only time will tell, but perhaps the outlines of a coherent urban planning policy will finally take shape here, based on an affirmed recognition of the architectural, built and landscape identity of the Marais Vernier and the need to maintain the last material witnesses of an original territoriality which, in reality, will never be the same again.

References

[1] T. Lecomte, Le Marais-Vernier, un territoire entre passé et modernité. in: Études Normandes, 1e année, n°1,2017,. pp. 12-19. https://doi.org/10.3406/etnor.2017.3499

[2] B. Penna, S. Mitaut, M. Valette. Le Marais vernier. Dix siècles d'évolution paysagère. in: Cahier des Annales de Normandie n°33, 2003. Le Marais Vernier. Ten centuries of landscape evolution. pp. 1-256. https://doi.org/10.3406/annor.2003.2450

[3] A. Sgard. Qu'est-ce qu'un paysage identitaire ? in C.Burgard, F.Chenet (Eds), Paysages et identité régionale. De pays rhônalpins en paysages, 1990, Ed. La passe du vent, Valence - France, pp.23-34. halshs-00270702

[4] A. Frémont, *La Région, espace vécu.* 1976, Flammarion, 290 p.

[5] J. Debrie, B. Steck, L'enclavement, réévaluation théorique et application à l'Afrique de l'Ouest. L'Espace géographique, 2001/1 tome 30. pp. 26-36, https://doi.org/10.3917/eg.301.0026

[6] Institut National de la Statistique et des Études Économiques (INSEE), Données locales, Commune du Marais Vernier, 2022. https://www.insee.fr/fr/statistiques/

[7] S. Deprez, Commerce, consommation & territoire. Essai sur l'émergence d'une nouvelle convention consommatoire. Jeux géographiques, questions d'aménagement et enjeux d'urbanisme, Habilitation à Diriger des Recherches en aménagement de l'espace et urbanisme, 2022, Université Le Havre Normandie, 305 p.

[8] R. Lajarge, Territorialités en développement. Contribution aux sciences territoriales. Habilitation à diriger les recherches, Grenoble, Université Joseph Fourier, 2012.

[9] Préfecture de l'Eure, Projet de classement au titre des sites - Article L.341-2 du code de l'environnement - *Vallée de la Seine - Marais Vernier*, 61 p.

Coastal vernacular architecture: A symbiosis between man and nature example of the city of Tangier

JIDA Rime [1,a*], Taieb BOUMEAZA[1,b]

[1]Faculty of Literature and Human Sciences, Mohammadia, Morocco

[a]jida.rime@gmail.com, [b]taiebboumeaza2@gmail.com

Keywords: Coastal Architecture, Urbanization, Vernacular Architecture, Climatic Resilience, Tangier, Morocco

Abstract. It is a work of vernacular architecture in which you can see how people adapt to natural conditions as part of the mutual relationship between mankind and the environment. These time-tested practices, perfected over various generations, reveal the ability of local communities to use natural resources as a method of climate resilience. There the actual outcome and consequences are few wherein they are places where vernacular architecture and coastal regions meet, not only due to what materials can be used during construction within those kinds of environment, but, moreover, it needs to be more resistant towards natural disasters like storms, flooding, coastal erosion, and tsunamis. The old buildings tell a silent story of human communities that have — over time and space — adapted to the harshest conditions. The paper will discuss the process of vernacular architecture transformation to combat natural hazards in coastal regions, specifically focusing on Tangier City (Morocco). Situated in northwest Africa, bridging the Mediterranean Sea and the Atlantic Ocean, Tangier provides a diverse case study for examining how traditional architectural practices can coexist with environmental pressures. Using both speculative logic and an empathetic gaze, this piece will merge the brilliant ways that Tangierians have mitigated natural threats with what these practices tell us about today's urban-ecological conditions.

Introduction

Vernacular architecture is not just about constructing architecture; it is much more about the relationship that a man sustains with his environment. These traditional building practices have evolved over many centuries; they show how local people were able to effectively use the available natural resources as well as cater to climatic dangers. [1] In coastal regions, the significance of earthen architecture is even more heightened by the fact that the constructions are frequently threatened by storms, floods, erosion, and tsunamis. These vernacular buildings are testimony to people's ability to survive and innovate when faced with unfriendly physical contexts. In this article, we try to analyze the development of vernacular architecture about the nature of the coastal regions concentrating on the example of Tangier, Morocco. Tangier's geographical location on the northwest African coast and at the entry point of the Mediterranean and the Atlantic sea provides a viable area of study to examine how traditional architectural approaches affect the environment. **[1]** Thus, the purpose of this article is to demonstrate how inhabitants of Tangier have learned to organize themselves and adapt their techniques to counter natural dangers by using an analytical and empathetic approach along with pointing at the relevance of these practices in the context of current tendencies towards urbanization and climate change.

Context and Issues

Tangier situated on the North West of Morocco has much more to offer than a coastal territory; it is a city with deep historical and cultural significance and a strategic position. Indeed, the shores of Tangier which are comprised of breathtaking sandy beaches, rocky cliffs, and sloping steep shores also contain places with various severe environmental concerns. This region's natural risks are numerous, and all-encompassing – not only does it impact the physical environment, but it also impacts the structures and lives of people.

Hurricanes and storms during sea transport are one of the most frequent disasters in Tangier. Flooding can be triggered by strong winds and huge waves and it destroys homes, roads and other structures necessary for the community. Houses with thick walls and flat roofs are generally more performing against these factors than the common creosote houses which are not much adapted to the local configurations.

Another ever-growing problem affecting Tangier is the problem of coastal erosion. [2] The nature of waves and sea currents are such that they cause erosion of the coastline progressively thus posing danger to buildings and other structures that are built near the sea. Some of the folk architectural practices which started including construction on high podia and the use of durable materials like stone, then are the results of this phenomenon.

Climate change enhances this situation in Tangier by increasing the number of problems it has to face.

This exacerbates the effects of storm surges since with increased sea level, the normal frequency and intensity of storms have been noted to have risen. Such a rate of natural changes always puts pressure to transform construction techniques and borrowed technology where local communities must strive to meet the challenges of sustainability of houses and infrastructure. However, this region is also characterized by water hazards such as floods, and Tangier is also situated in one of the seismically active zones because it is in the vicinity of the subduction zone of the Eurasian and African lithospheric plates. [1,2] Some of the local structures include features to counter earthquake impact through the use of ductile components, and structures that can act as a form of shock absorber to the impact of the earthquake.

Importance of Vernacular Architecture

Tangier's popular architecture to an extent does much more than its counterparts to cope with the tasks that are created by the natural conditions and offers a clear theoretical program for contemporary architectural paradigms. These building designs bear ecological and economical resistance to ecological natural hazards by incorporating tested methodologies enhanced by local material resources. [3]

The present article has a primary intent to explain in a detailed way the Setback Measures used in the Popular Architecture of Tangier to counterbalance the Natural Disasters. Based on specific examples and numerous high-ranked scientific sources we will try to explain how these practices can be incorporated under the umbrella of modern risk management and urban planning regulations. Finally, we would like to prove that the practical knowledge, that people took with them from the specifics of vernacular architecture, can be as effective in the modern conditions of increased threats to the environment.

Adapting Vernacular Architecture

This discussion on the vernacular architecture of Tangier reveals an understanding of how people have had to overcome environmental constraints. This paper identifies how the people of Tangier use local construction materials, adaptive structures, and raised building designs as some of the ways that reduce natural disasters' impact on their houses. The architectural expressions used in constructing houses in Tangier are those materials that are locally available including; stone, terracotta, and wood. [3] Some of these materials are cheap and readily available in the market and the physical characteristics of these materials have also been made responsive to the climatic conditions of the area.

Stone: Stone is widely used for construction especially for the foundation and walls as they bear the structure load. Due to its density and ability to withstand erosion by water, it is most suitable for the construction of houses that are built along the shores of the sea and the lake. Thick stone

Materials Research Proceedings 47 (2025) 30-35 https://doi.org/10.21741/9781644903391-4

walls also make for good insulators and help in regulating the internal temperature to the external climate thus making a home less vulnerable to changes in climate.

Terracotta: As far as used mostly in roof tiles and bricks terracotta is significant as a natural insulator. These roofs are made from clay or red clay tiles which are popular in Tangier traditional houses and help in controlling temperature inside the houses during summer and winter seasons. It is this ability that makes good insulation necessary for a building to sustain a reasonable level of heat regulation without overdependence on heating or cooling devices.

Wood: As for wood, it is placed in the construction of houses and occupies a less significant position as compared with stone and terracotta but holds the main and decorative features of construction. That makes it suitable for further use in construction especially in the zone prone to earthquake shocks. Besides, wood as the material used in roof frames and or interior working result in firm structures that when warmed garnish the desired look.

Traditional architectural designs in Tangier reflect a deep understanding of environmental challenges and an ability to respond effectively to them. [4]

Flat or slightly pitched roofs: Sleek roofs, common in culture homes, have poor water holding capacity; however, this makes it easier to let rainwater drain hence minimizing chances of the water seeping into the house's structure. This feature is especially sensitive in certain areas where the occurrence of rainfall is unpredictable and in high concentration.

Thick Walls: Thick interior and exterior walls could also be made of stone or rammed earth, giving good protection against wind or sea storms. They are also an important part of the thermal regulation of the building allowing a better indoor climate.

Natural Ventilation: The layouts of the windows and doors have been put into consideration, and this is because the natural ventilation is properly directed. Besides, it assists in cooling inner space during hot days, as well as in diminishing the force brought to bear on constructions by gales, which in turn leads to a decrease in the risk of destructive forces.

Probably one of the most creative features of the Tangier traditional architecture is the establishment of houses on stilts to mitigate against flooding and coastal encroachment.

Raised construction: Floors are usually paved with stone or concrete while the level of occupation of the structure is above the sectional level of the streets. This technique assists in reducing the impacts of flood waters which in most cases flood the low-lying regions of the city during the storm surge or seasons of heavy precipitation. [3,4]

Reinforced foundations: By-laws for house foundations are such that this building will not give in to the erosive effects of seawater and earthquakes. Stone, steel, etc., are some of the materials that can be used to construct strong structures in which loads are distributed more evenly thus lasting longer against ground movements and erosive forces.

Concrete Examples of Adaptation in Tangier

As for the climatic hazards, the vernacular architecture of Tangier is quite exemplary concerning the ability of buildings to withstand natural hazards; it produces traditional architectural elements that protect against hazards. [4] Two examples of this adaptation are the Kasbah of Tangier and the traditional houses of the medina.

The Kasbah of Tangier.

The very well-known Kasbah of Tangier, which is perched on a hill above the water, is a very good representation of how the inhabitants of this region built their houses for protection against natural disasters. Old as it might be; dating back to the 12th century, this edifice was constructed as a fortress, and due to the intricate construction that was employed in the construction of the walls, it was able to bear the worst of any hostile environmental conditions as they sought to pull it down.

Thick Stone Walls and Deep Foundations: The thick stone walls of the Kasbah along with the deep basement help to withstand coastal knockdown and strong blowing winds. Stone hails from the area and is relatively easy to acquire while at the same time, should be used because it is strong and does not corrode easily by water from the sea. These great walls also have prime importance concerning thermal regulation or, in other words, the regulator or controller of the inner climate of the building regardless of the fluctuating climatic conditions.

Narrow and Winding Allies: The urban plan also has a determinant effect: the network of inaccessible lanes and serpentine streets of the Kasbah offers even more protection from sea storms. These alleys also assist in the reduction of the effects of strong winds on constructed structures since their impact is to some extent reduced. Besides, they help manage the flow of water by directing rainwater away from inhabited places which in turn reduces cases of Acqua Standing water.

Elevated Position: To prevent flooding, the structure of the kasbah is located on the top of a hill so there is no need to build extra walls around it. Thus, when it comes to storms and tsunamis, the Kasbah's location significantly decreases the risks mainly because it is built on higher ground away from areas where floods may reach. This is an excellent strategy, which demonstrates how the local geographic environment has been incorporated into architecture and construction to increase the structures' durability.

Traditional Houses of the Medina.

Another example of vernacular architecture suitable for protecting people and structures against natural threats is the traditional houses of the Tangier medina. Such structures, which are hundreds of years old, use architectural designs that afford a measure of shelter against unfavorable climatic conditions.

Thick-Rammed Earth Walls: Compact earth is used to construct the thick walls of the medina houses as illustrated below Other than that, the roofs of the medina houses are made of palm tree trunks and, compact earth in the construction of thick walls. This material ensures the provision of thermal and sound insulation and therefore residents benefit from changes in temperatures and noises from the outside world. Thick rammed earth walls also prove to be considerably very strong capable of withstanding powerful winds of storms.

Flat Roofs: Individuals can easily slide on the flat roofs which are a feature of traditional houses; this is due to the fast-draining process of the rainwater. They are to reduce the penetration of water and other structural hazards associated with water penetration into buildings. Additional spaces are also built and utilized as living spaces on flat roofs, and all persons can have the chance to have outdoors in complete security.

Interior Courtyards: Houses in Medina are traditionally arranged according to the inner light wells; they contribute a lot to ventilation and climate control. These courtyards create an area that will be shielded from the wind and rainfall, the place being a zone of leisure and togetherness. Besides, they help enhance natural convection thereby cutting down on the amount of artificial air conditioning.

Impact of Natural Hazards on Tangier

Although the present paper demonstrated how adaptive practices are enshrined in Tangier's built traditional form, natural risks remain pertinent concerns to this contemporary city.

Coastal Erosion: Mention should also be made of the continued problem of coastal erosion aided by the increased rush to the coast by people for settlement as well as climate change. Facilities that are built close to the water are especially affected and pressure on coastal resources is growing as a result this process is being advanced. There is a need for measures like use of sea walls and replenishment of beaches as a way of reducing the effects of erosion to infrastructure development.

Maritime Storms: As cyclones, storms, and hurricanes intensify and become more and more frequent, navigation and port infrastructure are under threat. Wind storms bring down structures, roads, and drains including those that have been constructed. These conventional construction techniques though robust have to be built often, newly innovated, and updated alongside other smart and better means to handle these incidences.

Rising Sea Levels: Climate change especially the increase in sea level becomes one of the major risk factors affecting the lowlands of Tangier. This position has been aggravated by climate change effects that have led to increased sea levels. Mitigation strategies for example the creation of natural barriers, and the construction of protective structures are all vital to, safeguard the coastal infrastructure and population.

Future Recommendations

Successful utilization of both traditional building practices and scientific knowledge would allow the Tangier to experience increased protection from natural phenomena and risks. Here are some key recommendations to achieve this goal: Here are some key recommendations to achieve this goal:

The Integration of Vernacular and Modern Techniques

1. Use of Sustainable and Adapted Materials: Efforts should be made to incorporate locally available materials like stone, terracotta, and rammed earth while at the same time also taking into consideration the benefits of using new materials that have higher durability in extreme climates. For instance, it is possible to adopt the reinforced concrete-based construction for the foundations and the other support works although retaining the preferred and more thermally optimizing vernacular materials. [5]
2. Adaptive Architectural Design: Choosing good sites by integrating computer-aided design and sustainable architectural principles, non-slippery floor coverings, specially designed public toilets and ramps, flat-roofed structures for quick draining of rainwater, and designing buildings around inner courtyards for natural ventilation. Incorporating contemporary techniques like installing of Green roofs or providing a rain water harvesting system can add support to such fundamental methods.
3. Urban Planning Policy Development: It is here that measures towards coastal protection which form part of the Urban planning policies should be implemented and put into practice after analyzing the natural forces and other potential environmental disasters. Sea walls or seawalls, dunes, and any other elements in effect act as barriers against erosion and floods, and coastal ecosystems like mangroves and salt marshes should also be involved.

Awareness Rising and Training of Local Communities

1. Resilient Practices Training: It is crucial to raise awareness and educate local communities about resilient building practices. Workshops, seminars, and short courses must be conducted to inculcate various conventional and contemporary methods of building construction that can withstand or resist the adverse effects of natural calamities.
2. Engage with the community: Local communities must be a part of the planning and execution process of any urban resilience project. The involvement of the community makes certain implemented solutions tailored to suit local needs and context, hence increasing their efficiency and acceptability. [5]
3. Promotion of Sustainable Practices: The promotion of sustainable practices in construction and urban planning, the use of local materials, reuse/recycling of waste from buildings, and low environmental impact construction technologies, are long-term resilience measures that can be implemented in the city.

Building Resilient Infrastructure

1. Resilient Infrastructure: from elevated roads and improved drainage, to reinforce public buildings that don't fall victim to the next storm or earthquake;
2. Monitoring & Maintenance: Establishing perpetual monitoring systems to identify developmental erosions, land subsidence, and infrastructural degradation signs become early precautionary measures. It is also very important to use the existing infrastructure sustainably and regularly maintain it for that functionality to continue.
3. Innovative Technologies: The use of innovative technologies (for instance early warning systems of storms & Tsunamis; smart building materials that adjust to the climate and renewable energy solutions) can allow for positive impacts in enhancing urban resilience.

Conclusion

The vernacular architecture of Tangier presents an inspiring model of resilience to natural hazards, illustrating how traditional building practices can effectively respond to environmental challenges. Local materials, adaptive designs, and robust construction techniques, as used in the Kasbah and medina houses, demonstrate a deep understanding of local climatic conditions and natural hazards.

Tangier's concrete examples demonstrate that vernacular architecture is not limited to heritage and cultural value, but offers pragmatic and sustainable solutions for coastal environments facing increasing risks. However, climate change, rapid urbanization, and coastal erosion pose considerable challenges that require integrated and multidimensional responses. To strengthen the resilience of Tangier and other coastal cities, it is essential to combine these vernacular techniques with modern and scientific approaches.

Tangier, with its rich architectural heritage and historical experience, can serve as a model for other regions facing similar risks. By integrating traditional wisdom with contemporary innovations, Tangier can continue to thrive while facing environmental challenges. This integration is not only a resilience strategy but also a celebration of cultural richness and human ingenuity in the face of the forces of nature.

References

[1] Benitez, S. (2019). *Traditional Coastal Architecture: Strategies for Resilience*. Journal of Coastal Research, 35(2), 345-357. https://doi.org/10.2112/JCOASTRES-D-17-00195.1

[2] El Hassani, A., & Bahaj, A. S. (2017). *Assessment of Coastal Erosion and Flooding Impacts in Morocco*. Coastal Engineering, 125, 123-134.

[3] Hamdi, N., & Garcia, M. (2021). *Integrating Vernacular Knowledge into Urban Resilience Planning in Morocco*. Environmental Planning and Management, 64(8), 1379-1395.

[4] Salim, R. (2022). *Climate Change and Coastal Urbanization in Morocco: A Case Study of Tangier*. Climate Policy, 22(1), 76-89.

[5] Stevens, M., & Hans, P. (2020). *Building Resilient Coastal Communities: Integrating Traditional and Modern Approaches*. Journal of Urban Planning, 45(3), 567-590.

Materials Research Proceedings 47 (2025) 36-47 https://doi.org/10.21741/9781644903391-5

The Ait Bouguemez Mountains in Moroccan central High Atlas: A laboratory for vernacular architecture - synthesis essay

Jamal CHAHROU[1,a]*, Mohammed MAYOUSSI[1,b], Farid JAA[1,c]

[1]Landscape Dynamics, Risks and Heritage Laboratory, Faculty of Letters and Humanities Sciences, Sultane Moulay Slimane University, Béni Mellal, Morocco

[a]jamalenvironement.2013@gmail.com, [b]Mohammedmayoussi@yahoo.fr, [c]Faridjaa@yahoo.fr

Keywords: Ait Bouguemez, Landscape Dynamics, Risks and Heritage Laboratory, Vernacular Architecture, Inventory, Collective Granary, Kasbah House

Abstract: This paper synthesizes our reflections, observations, conclusions, and results from our fieldwork in the Ait Bouguemez Valley and at the Landscape Dynamics, Risks and Heritage Laboratory (LDRHL) within the Faculty of Letters and Humanities Sciences in Béni Mellal, whose priority is the enhancement of built cultural heritage. Our team, composed of geography professors specializing in urban planning and heritage, conducted field trips to identify, study, understand, and safeguard vernacular architecture, which represents a unique heritage in Morocco. We have established a work program to inventory and present the architectural potential of the Ait Bouguemez Valley, selecting sites based on their heritage significance and degree of degradation. Additionally, this article proposes a quantitative method for the identification and assessment of granary collectives in the Ait Bouguemez Valley. This method proposes new criteria of evaluation relatively in conformity with the characteristics of the Moroccan mountain by referring to the various methods that were already carried out in the European countries. The objective is thus to reduce the subjectivity of the evaluator and to clarify deeply the criteria composing the assessment values (scientific value, additional values). Finally, the inventory we conducted in the valley enabled us to select six collective granaries.

Introduction

The Ait Bouguemez Valley remains a vitally important living environment. This is due in particular to its geographical location, climatic conditions, and probably to a combination of local traditions. In this valley, the intermingling of populations of diverse origins has been attested to for centuries. This kaleidoscope is the result of a long process of migrations, ethnic pushes, a turbulent history, and various social constraints. This mixture has given rise to authentic architecture, whose most original features can be seen in the kasbah houses and collective granaries. Despite the efforts made, this heritage needs a new lease of life. Today, it is imperative to ensure the commitment of the country's institutions to the preservation of this cultural heritage [1].

This study analyses the vernacular architecture of the Ait Bouguemez Valley, addressing some of the themes that characterize its diversity. Methodologically, it combines fieldwork based on fundamental studies of the valley's territory, culture, and vernacular architecture.

With this in mind, and through our Landscape Dynamics, Risks and Heritage Laboratory (LDRHL) - the body responsible for the conservation and enhancement of earthen architecture in the Béni Mellal-Khénifra region - we will set out some thoughts on the work carried out on vernacular heritage, as well as our vision for the future of these earthen architectures, following the various researches and numerous surveys carried out on the subject.

This article is divided into two parts: the first is devoted to conceptual definitions and methods, while the second presents the results.

I. Theoretical framework and research method

The starting point for this work is to explore the notion of vernacular architecture.

1. Vernacular architecture in Morocco

1.1. Definition and distribution of vernacular architecture

Vernacular architecture, long little recognized or despised in Morocco during the last century, has fortunately regained interest over the last two decades. It is, however, an architecture of exceptional qualities, testifying to a unique know-how and art of living. Contrary to certain preconceived ideas or ignorant insinuations, vernacular architecture has never been a solution imposed by economic difficulties on individuals and communities. It is first and foremost a highly sophisticated scientific response [2].

It was through intelligence and advanced scientific knowledge that mankind built with earth in arid, desert, or oasis environments, and used stone and hogback roofs, or at least sloped roofs, in snowy regions and mountainous areas with high rainfall. This observation is easily confirmed by observing the architectural map of Morocco: the habitat differs from Jbala to Oujda, from Khénifra to Azilal, from Essaouira to Assa, from Ifrane to Tafraout, from Nador to Taroudant, and from El Jadida to Dakhla...

An analysis of this map of Morocco reveals the concentration of vernacular architecture in oases and arid or desert climates. This architecture stretches from one end of the country to the other, following a continuous line from the extreme north-east to the south-west and the south in general, right up to the Saharan confines with Mauritania. This line lies to the east of our Atlas and Rif mountains, from Agadir to Ahfir. This vernacular architecture covers areas such as Figuig, Arfoud, Er-Rachidia, Tinghir, Alnif, Tazarine, Kelaat M'Gouna, Ouarzazate, Zagora, Tata, Assa, Guelmim and the rest of the Sahara. However, it should not be forgotten that in certain regions of Morocco, vernacular architecture can also be found in both arid and mountainous areas (Ait Bouguemez, Zaouit Ahncal, Ait Boulli, Megdaz...) [3].

1.2. Architectural potential to be developed

Morocco boasts a wide variety of traditional local architectural styles. However, earthen architecture remains one of the country's most distinctive features. It is characterized by features that give it exceptional value, such as adaptation to climatic conditions, fusion and integration with the surrounding landscape, the simplicity and antiquity of the architectural processes and techniques used the harmony of proportions and volumes, and the sobriety of the decorations with which the buildings are clad [4]. In the arid and semi-arid zones of southeastern Morocco, vernacular earthen architecture is widespread and takes three forms: [5].

1. Kasbahs: seigneurial dwellings reminiscent of perched fortresses, built in strategic locations;
2. Ksour: residential agglomerations concentrated around a public square and surrounded by a wall;
3. Igoudar: collective granaries or citadel granaries, fortified community warehouses [6]. They are considered among the most characteristic defensive settlement structures in mountainous and semi-desert regions, stretching from the southern slopes of the High Atlas to the southern foothills of the Anti-Atlas.

These stores are known locally by several Amazigh names: Ighrem in the central High Atlas, Tihouna in the eastern High Atlas, and Agadir in the western High Atlas and Anti-Atlas. The Agadir is a tribal edifice where each family owns a hut. It is a collection of individual possessions protected by a common defense to prevent looting. It is also a traditional institution linked to Amazigh culture, embodying cohesion and solidarity within groups.

1.3. Vernacular architecture: a special object of research

Vernacular architecture, one of the representations of spatial organization inherited from the past and a testimony to ancestral know-how and practices, is today one of the most important components of Moroccan architectural heritage. An expression of identity, it illustrates man's relationship with his environment. This architecture is part of an endangered landscape, affected by multiple factors, mainly linked to changes in lifestyle, which are characterized by an increasingly marked break with everything related to tradition. Indeed, the quest for modernity, through new models of habitat corresponding to contemporary life, confronts this habitat with threats that accelerate its degradation [7].

Moreover, the lack of popularity of this type of architecture, compared with others (such as earthen architecture, for example), also contributes to its decline, despite its great importance in Moroccan rural architecture. What's more, in the face of regulatory and technical deficiencies specific to rural heritage (legal texts, effective institutional framework, specific technical studies, standards, etc.), combined with the general public's lack of interest in heritage issues, traditional vernacular stone domestic architecture is undergoing advanced degradation and falling increasingly into disuse.

II. Materials and methods

1. Study area

The Ait Bouguemez valley is administratively part of the Tabant commune and is located in the central High Atlas, on the borders of the Azilal province, in a region that is still landlocked and typical of mountainous Berber Morocco. Located 170 km east of Marrakech in the heart of the central High Atlas, the Ait Bouguemez Valley landscape is characterized by two major topographical features. It is mainly dominated by a mountainous massif that occupies almost the entire area (over 80%), bordering the valley to the north by Mount Azurki (3,677 m), to the south by Ayt Abbas, to the west by Mount Tizal (3041 m), and to the east by Waougoulzat (3,763 m). The valley lies at the center of this narrower, less extensive range, and is around 35 km long, between 1,800 and 2,200 m above sea level.

This valley is the result of erosion caused by powerful Atlas glaciers, whose last period of activity dates back to 10,000 B.C. These intense geological processes have given the valley a Y shape, forming a deep, fertile plain. This plain is composed of deposits resulting from a Quaternary landslide. The mountain slopes, which are not very fertile due to their stony and pebbly formation, complete this landscape [8].

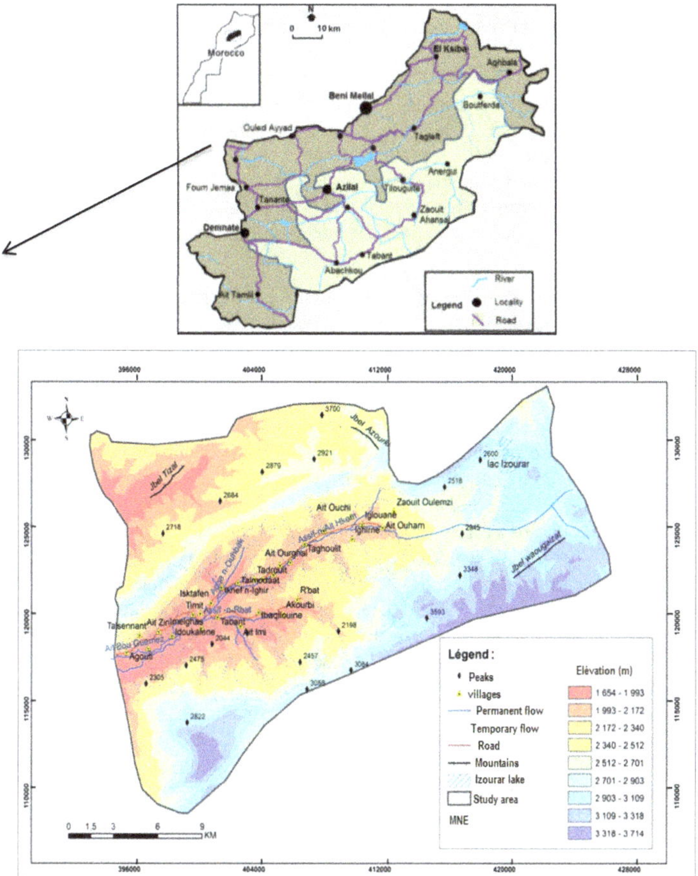

Fig 1: Location map of the study area in Morocco.

2. Methodology

This contribution is based on a case study of research carried out by the landscape dynamics and heritage laboratory in the Ait Bouguemez region.

The work was carried out following a series of bibliographical analyses on the theme of vernacular architecture, which enabled us to establish the conceptual and methodological framework. The data collection tool, such as the interview method, helped to overcome the problem of lack of data. In the field, investigations involved the valley's local population and stakeholders.

In the first phase of fieldwork, a database was drawn up. This consisted of collecting statistical data and qualitative benchmarks through interviews with the local population and local players on the theme of enhancing vernacular architecture.

The second phase of this work was designed to present the results of the inventory of heritage sites in the Ait Bouguemez Valley. These assessment results are based on the "IGUL" method [9],

Materials Research Proceedings 47 (2025) 36-47
https://doi.org/10.21741/9781644903391-5

the first institute to have formulated a methodology for assessing heritage sites. With the help of this work, we have created our evaluation sheet, better adapted to the architectural sites of the Ait Bouguemez Valley.

First, we'll present the general features of heritage sites and we'll describe their spatial distribution. We will then attempt to classify them as collective granaries or Kasbah houses. The evaluation proposed by the "IGUL" method begins with a detailed study of the scientific value, which includes multiple criteria (rarity, representativeness, integrity). The average of these criteria gives the overall scientific value of the heritage sites.

We then proceed to a precise treatment of the additional values, which result from the combination of all the other values (ecological, aesthetic, cultural, and economic).

3. Problematic

Since 2014, under the direction of Professor Mayoussi Mohammed, our laboratory has rapidly encountered the problem of identifying and assessing sites of particular interest. The first inventory of earthen architecture in the Aït Bouguemez valley was undertaken in 2015 by student Hicham Bouizgarne, who carried out an inventory of heritage in the Béni Mellal-Khénifra region, particularly in the Aït Bouguemez valley.

In 2015, our laboratory carried out inventories in the field of vernacular architecture based on new technologies (geographic information system). The subject was "the contribution of GIS tools to the enhancement of the architectural heritage of the Aït Bouguemez valley". This project was co-directed by professors M. Mayoussi and H. Redad.

From 2017 to 2022, as part of a thesis entitled "The architectural heritage of the Aït Bouguemez Valley (Central High Atlas -Morocco): Inventory, assessment and valorization", we conducted an identification survey of earthen architectural heritage in the Aït Bouguemez Valley.

The general aim of these inventories is to acquire in-depth knowledge of the architectural heritage in the Aït Bouguemez valley in historical, ethnographic, and architectural terms, and of its relationship with its immediate environment. These inventories will make it possible to define appropriate protection measures for the dynamic safeguarding of this ancestral heritage while taking into account the evolution of settlement in the valley. In all, the project has inventoried and studied 60 sites of particular interest. The results of this thesis will shortly be published. In general, the inventory work demonstrated the irreversible state of degradation of the earthen architecture. The granaries ceased to function a century ago, and the kasbah houses are no longer inhabited, with a few exceptions. This fragmentation has given rise to a different type of housing from that which existed before.

4. Results and discussion

Faced with the process of fragmentation leading to the disappearance of many kasbah houses and communal granaries, the Faculty of Letters and Human Sciences in Béni Mellal has set up a research laboratory whose mission is to promote and evaluate earthen architecture in the Béni Mellal-Khénifra region. However, despite the laboratory's efforts in the field of academic research, the process of breaking up vernacular architecture is intensifying by the day, and, as a result, the majority of earthen buildings are disappearing. Technical and financial problems are exacerbating the situation. It is currently difficult to intervene on a large number of sites and to give importance to all the buildings, as material and financial means have become limited for such achievements.

The vernacular heritage of the Aït Bouguemez valley, made up mainly of kasbah houses (tighrematines) and collective granaries (ighrmanes), bears witness to great skill in the design and construction technique of earth and stone architecture, perfectly adapted to the building materials available, to the type of climate prevailing in this mountainous area and even to the valley's landscape and environment.

The laboratory is experimenting with research work, notably that of H. Bouizgarne, M. Mayoussi, and M. Amili, who are drawing up a research project aimed at taking stock of vernacular architecture in the Aït Bouguemez region.

The research is also aimed at conserving and promoting heritage in the region, which boasts a rich vernacular heritage. My thesis, under the supervision of Professor M. Mayoussi and Professor F. Jaa, aims to evaluate and enhance the heritage sites in the Aït Bouguemez valley, which are already being worked on by the CERKAS association. The results of the assessment are based on considerations of the state of conservation of the unique vernacular architecture sites in the Aït Bouguemez valley, with the aim of intervening to enhance this architecture.

The research phase and field surveys, as well as the collaborative effort with professionals already working in the field, underlined the need for a relevant intervention strategy compatible with heritage buildings.

The assessment of the architectural sites inventoried in the Aït Bouguemez Valley was based on the inventory sheet proposed by the Institute of Geography University of Lausanne (IGUL)

The assessment began with the scientific qualification of the sites. This involved determining their morphology, state of conservation, local representativeness, whether or not they are rare in the study region, and finally their added value, such as aesthetic, ecological, cultural and economic value.

Generally speaking, the inventory work carried out since 2016 has focused on collective granaries. These buildings are indeed of particular interest as priceless heritage.

4.1. Scientific value of collective granaries

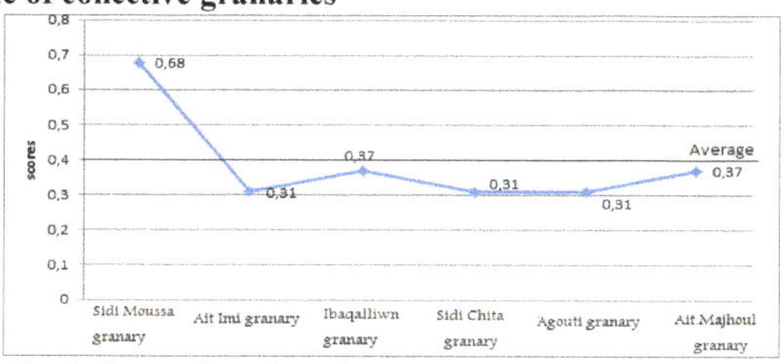

Fig 2: Scientific value of inventoried collective granaries

We note that only one site has an above-average scientific value, while five others have a below-average value. The site with the highest scientific value is the Sidi Moussa granary, in the Timit douar, with a score of 0.68. This value is closely linked to its high representativeness of the construction techniques and materials used in the region, and to its high integrity since the site is in a good state of conservation and has been restored by UNESCO. The site is extremely rare in the Aït Bouguemez valley. Conversely, some granaries have a low to medium scientific value (0.43), such as the Sidi Chita granary in the Idoukalne douar and the Aït Imi granary, with a score of 0.43. These collective granaries are linked to a local historical event of modest importance. In addition, a very low scientific value was recorded for the Agouti collective granary and the Aït Majhoul collective granary, with a score of 0.31. This weakness is mainly due to the poor state of conservation of the sites and/or the abundance of similar sites with no regional particularity.

We can deduce that the scientific value differs from one granary to another since there are very specific sites concerning construction techniques and materials used. These sites can be used to pass on scientific knowledge of construction techniques (decorative elements, wall and ceiling techniques, restoration, etc.). For example, the Sidi Moussa granary has a shape that is rare in national (circular). Sidi Moussa, one of Morocco's most famous saints, is renowned for his healing powers against madness.

Materials Research Proceedings 47 (2025) 36-47 https://doi.org/10.21741/9781644903391-5

Fig 3: Sidi Moussa granary, world heritage site, 2022

There are also some architecturally remarkable granaries. We're talking here about the Ibaqalliwn granary and the Sidi Chita granary; these sites are representative of the main features of the Ait Bouguemez valley (local building materials) and even of neighboring areas (Megdaz valley) [10]. The shapes are equally distinctive. The materials used to build these granaries are stone, adobe and cob [11].

Fig 4: Sidi Chita granary, 2022

Fig 5: Ibaqalliwn granary, 2022

The other granaries are also exemplary, providing us with information on the relationship between man and architectural heritage. They also tell us about periods of political and military insecurity in the valley (Agouti, Aït Imi and Aït Majhoul granary).

Fig 6: Ait Majhoul granary

Fig 7: Ait Imi granary

Fig 8: Agouti granary

With this in mind, we present the average of the criteria that make up the scientific value of architectural sites, to give a general idea of the importance of each criterion in the inventory and its contribution to the evaluation.

4.2. Detailed analysis of scientific value

4.2.1. The integrity of collective granaries (Ighremans)

The Sidi Moussa granary has high integrity, with a score of 1. Also, 1/6 of the granaries have a medium value, with a score of 0.5. On the other hand, 4/6 of the sites have a low value of around 0.25.

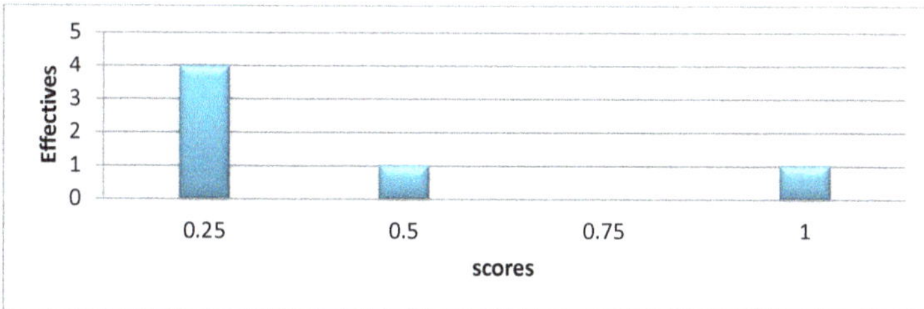

Fig 9: Scientific value: integrity criteria

The high integrity of the Sidi Moussa granary is linked to the site's good state of conservation, with little natural or, above all, human impact. Tourist infrastructures and activities are limited and exert no pressure on the granary. On the other hand, sites with low integrity (0.25) are precisely those granaries that have been destroyed by natural factors: this is the case for the granaries of Agouti, Ait Imi, Ait Majhoul and Sidi Chita, which have been destroyed by natural and human impacts.

4.2.2. Representativeness of collective granaries:

All the granaries are representative of the main features of the Ait Bouguemez valley (local building materials).

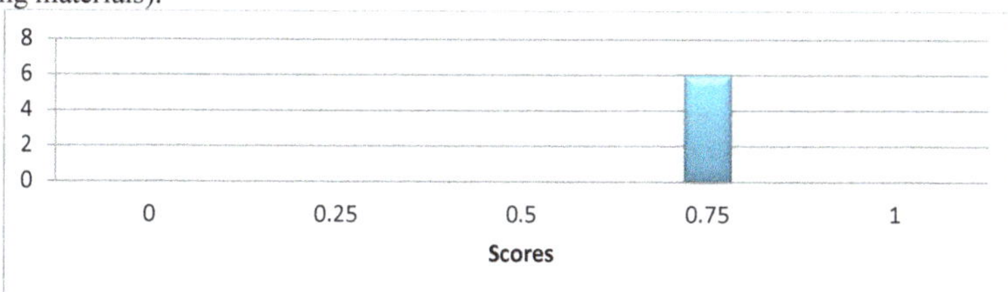

Fig 10: Scientific value: representativeness criteria

Most of the collective granaries are exemplary. The sites inventoried present a variety of architectural aspects, which characterize the study region (construction techniques and materials used). In this respect, 6 granaries are highly representative of the main features of the Ait Bouguemez valley (local building materials).

4.2.3. Collective granaries are relatively rare:

According to the inventory, collective granaries are different, and each one has its own characteristics in terms of shape or unique features. We have therefore observed only one granary, "Sidi Moussa", which has unique features at regional and national level.

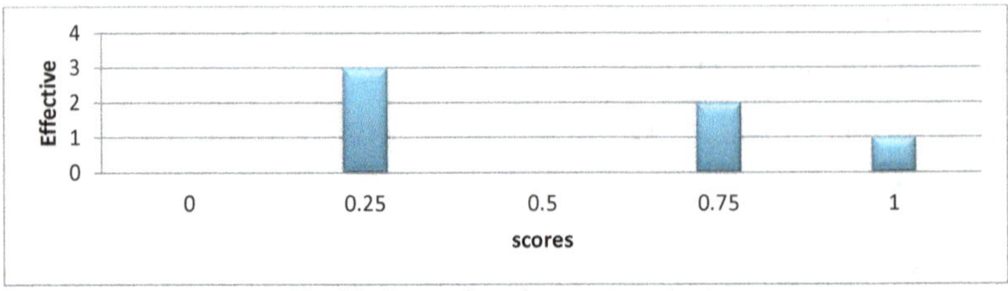

Fig 11: Scientific value: rarity criteria

5. Additional values for collective attic:

Additional values, such as the cultural, ecological, aesthetic, and economic values of an attic, highlight the links between the architectural environment and the natural and human environment in which the object is embedded. In this inventory, we assessed all the additional values based on information gathered in the field, testimonials from local people, and certain local documents. In this way, we obtained a numerical evaluation for the first three values (ecological, aesthetic, and cultural) and a simplified, qualitative evaluation for the economic value.

5.1. General analysis of the additional values of collective granaries:

The middling additional value of collective attics is low, reaching around 0.30. This can be explained by the state of abandonment of these buildings. Observation of the inventoried collective granaries revealed that the walls and roofs are the most dilapidated and ruined parts. Furthermore, the absence of literary and historical works, which could enrich the evaluation of two criteria: artistic and geohistorical, contributes to this weakness. In general, all the granaries inventoried are unpublished, and no documents mention them, except the Sidi Moussa granary in the Timit douar, discussed by Titouan Lamazou [12].

.

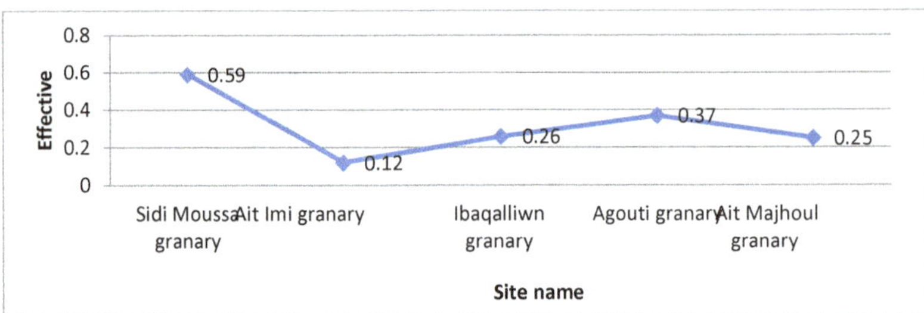

Fig 12: Additional value of inventoried collective granaries

The site with the highest additional value is well-known locally and regionally. It is the Sidi Moussa granary, which obtained a score of 0.59. Concerning the average additional values, we concluded that the aesthetic additional value registers a high score, with an average of 0.74. Cultural additional value ranks second with a score of 0.29, followed by ecological value at 0.10 and, finally, economic value at 0.083.

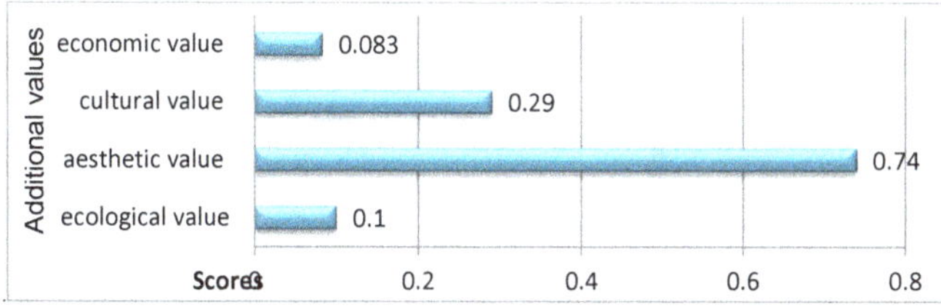

Fig 13: Average additional values of inventoried collective granaries

5.2. Results of additional values

5.2.1. Ecological value

This assessment analyses the ecological content of collective attics. The evaluation is based on the criterion of ecological impact, which depends directly on whether or not there is a bibliographical database. Sites for which we have found no information are given a zero value for this criterion.

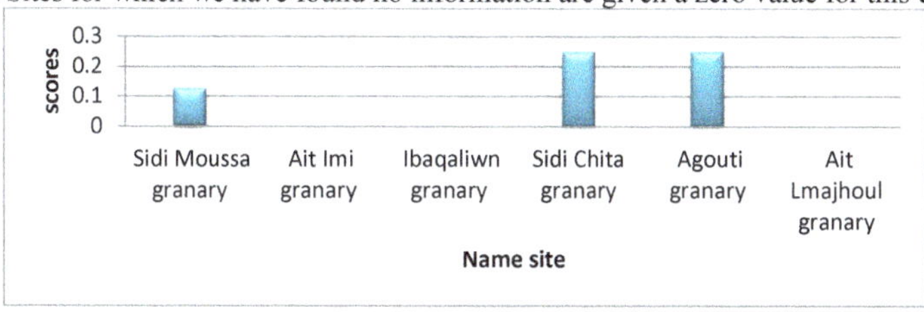

Fig 14: Ecological value

Figure 14 shows that only two sites scored low to medium. By contrast, the ecological value of the other granaries remains low, ranging from 0 to 0.125. Most granaries have no ecological impact at all. What's more, the majority of sites are unprotected due to the lack of rehabilitation and restoration work.

5.2.2. The aesthetic value of collective granaries

Given the relativity and subjectivity of architectural landscape beauty, which depends directly on the observer, the aesthetic value taken into account in this assessment is very difficult to estimate objectively. The two proposed criteria influencing the "beauty" of the landscape are the different viewpoints and the structure. The overall assessment corresponds to the treatment of these two sub-criteria. First and foremost, we're talking about the site's visibility and accessibility. Sites that are difficult to access and offer no viewpoints, due to visual obstacles (trees, ridges), always score zero. Secondly, structuring corresponds to the vertical and horizontal dominance of the site, which may or may not mark the local landscape (landscape and architectural elements).

Materials Research Proceedings 47 (2025) 36-47 https://doi.org/10.21741/9781644903391-5

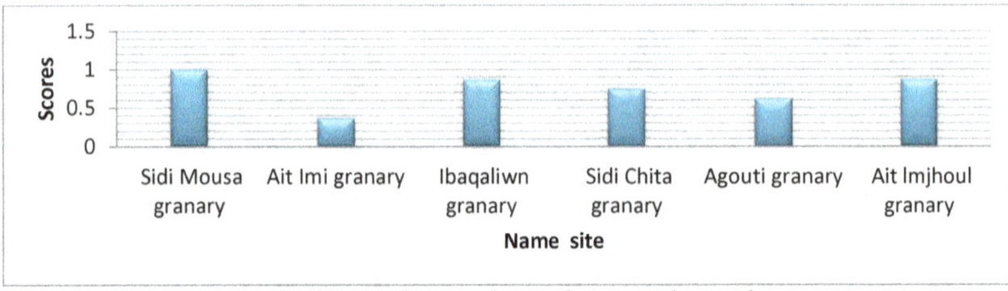

Fig15: Aesthetic value and scores obtained

According to figure n°15, the majority of collective granaries are aesthetically pleasing (0.74), with impressive shapes, such as those located on hills (Sidi Moussa and Sidi Chita granary). This type of shape also meets the criteria proposed by the method used.

5.2.3. The cultural value of collective granaries:

The cultural value of an attic is divided into three criteria: religious importance, historical and symbolic importance, and literary and artistic importance. The aim here is to highlight the cultural importance, in the broadest sense, that an architectural site has played or can play.

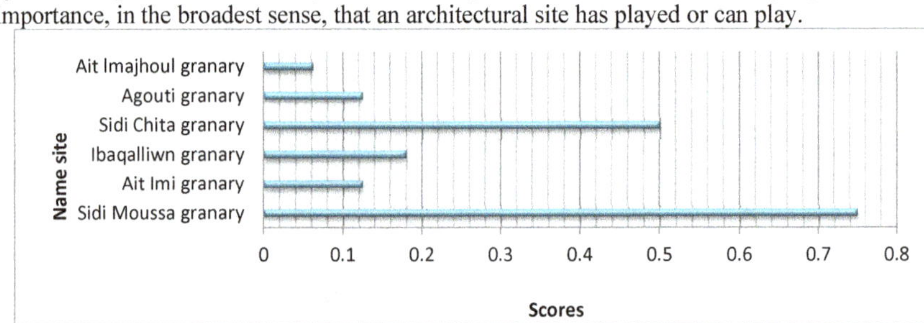

Fig16: Cultural value and scores obtained

The cultural value of collective granaries is above all marked by their historical importance. This cultural criterion reflects the relationship between man and nature. Its average value (0.29) indicates that granaries are generally of local importance. The granary with the highest cultural value is the Sidi Moussa granary, symbolic and charged with curative values, as it is considered one of the most renowned saints in Morocco, renowned for its healing benefits for madness. Conversely, the collective granary of average cultural value (0.5) is the Sidi Chita granary, which, according to oral tradition, originated in Figuig [13]. The local people attribute many miracles to this granary. The rest of the granaries have a low cultural value, mainly due to their lack of connection with a historical event, poor conservation of the sites, and lack of transmission of information about their history.

5.2.4. The economic value of collective granaries:

The economic landscape of the Ait Bouguemez valley is dominated by apple orchards and pastures. Only one granary has an average economic value (Sidi Moussa granary). This value results from the exploitation of this architectural asset (the entrance fees it charges tourists when they visit). In the inventory, we have mentioned the economic activity that concerns each granary, but the values are nil.

Conclusion

In conclusion, we can say today that the laboratory has reached cruising speed; the results obtained are indisputable testimony to its capabilities.

As far as I'm concerned, nine years of very varied experience, touching on all areas of vernacular architecture, the only experience spent in the Ait Bouguemez valley, the shortest but richest, is most interesting for a geographer in search of novelties drawn from the past and the ancient; it will profoundly influence my learning and evolution.

Noble and constructive are the stains accomplished by the Landscape Dynamics, Risks, and Heritage Laboratory to save and enhance what remains of a thousand-year-old heritage for future generations.

Enhancing architecture that bears witness to a past rich in both tangible and intangible lessons is an opportunity for all those involved.

The tasks are complex, the constraints numerous, but the personal pride and satisfaction are so great when the job is done.

Acknowledgments
This work would not exist without the help of a great number of great people. The authors would like to thank all those who contributed to this study.

References

[1] J. Chahrou, M. Mayoussi, La valorisation du patrimoine architectural dans la région béni Mellal khénifra cas de la vallée des Ait Bouguemez, mémoire du master université Sultane Moulay Slimane, Béni Mellal, (2018) ,237p

[2] A. Chebri, Les ksours et kasbas du Maroc, patrimoine séculaire exceptionnel, un savoir-faire, un savoir-vivre et une richesse singulière.2014 P10

[3]J. Chahrou, M. Mayoussi, F. Jaa, le patrimoine architectural de la vallée des Ait Bouguemez (Haut Atlas Central), inventaire, évaluation et valorisation, thèse du Doctorat, Université Sultane Moulay Slimane, Béni Mellal ,(2022), p. 360

[4] M. Ait Hamza, les architectures en pisé : identité et modernité, in Patrimoine culturel matériel dans la région Souss-Massa-Draa, N° 35, IRCAM, Rabat, (2013), p.61

[5] A. Bentaleb, Valorisation du patrimoine architectural en terre au Maroc présaharien: kasbah de Taourirt à Ouarzazate », in Patrimoine architectural au Maroc, IRCAM, (2013), p.56

[6] M. Boussalh, Patrimoine architectural en terre au Maroc : Proposition de création d'un équipement culturel intégré dans la kasbah Taourirt à Ouarzazate. Mémoire de DEPA. Alexandrie : Université Senghor (1999).

[7] L. Hracherrass L'habitat rural traditionnel vernaculaire en pierre: Retour sur deux expériences d'habiter la montagne au Maroc, (2018)

[8] K. Amrouch. Etude hydrogéologique de la vallée d'Aït Bougmez, Rapport de stage, faculté des sciences d'Agadir (2003).

[9] E. Reynard, A. Perret, L. Grangier., & L. Kozlik. Methodological approach for the assessment, protection, promotion, and management of geoheritage. In EGU General Assembly Conference Abstracts. (2007c). (Vol. 14, p. 3740).

[10] L. Emile, Mots et choses berbères, Notes de Linguistique et d'Ethnographie, Dialectes du Maroc. Paris (1920): Augustin Challamel.

[11] S. Naji, Art and berber Architectures of Morocoo, Edisud/EDDIF, Aix-en- Provence / Casablanca, (2001), 206 p

[12] T. Lamazou. Le Trésor de l'Atlas Denoël. Paris, (1955) Année de publication : 1985

[13] H. Bouzagarene, M. Mayoussi, La protection du patrimoine culturel et naturel, des Ayt Bouguemez mémoire de master université Sultane Moulay Slimane Béni Mellal, (2013). 300p

Materials Research Proceedings 47 (2025) 48-58 — https://doi.org/10.21741/9781644903391-6

Laâzib: A dry stone architectural expression Case of the Laâzibs on the Yagour Plateau in the High Atlas of Morocco

HRACHERRASS Loubna

National School of Architecture of Fez, Morocco

loubnarch@gmail.com

Keywords: Stone Construction, Laâzib, Mountain, Shelter, Habitat

Abstract. The Moroccan mountains are filled with stone constructions that showcase skills and construction techniques in harmony with the natural environment and human ways of life. Rarely studied by researchers, these architectural expressions in stone, now considered vernacular heritage, deserve to be explored and studied to uncover their uniqueness and specificities. Among these constructions is the laâzib, a form of rural dwelling characteristic of the high Atlas Mountains of Morocco. True imprints of humans on their territory, built by and for transhumance and serving both as a dwelling and a sheepfold, laâzib embody centuries-old expertise in dry stone construction Exploring this architectural expression in terms of construction methods, ways of life, and usage will be carried out through the study of the laâzibs located on the Yagour plateau, a pastoral area of over 70 km², perched between 1,900 and 2,700 meters above sea level in the High Atlas Mountains, south of Marrakech. A true crossroads for shepherds, Yagour annually hosts nearly 7,000 people from the surrounding villages and is home to several laâzibs. This article aims to explore one of the oldest examples of traditional rural dry-stone buildings in Morocco, deeply rooted in a natural environment characterized by its harshness, exceptional management system, and construction techniques infused with the genius of the place.

Introduction

The laâzibs presented in this article are located on the Yagour plateau, one of the highest areas in the High Atlas Mountains of Morocco. This pastoral territory covers more than 70 km² and is situated at an altitude between 1,900 and 2,700 meters, south of Marrakech.

This plateau, made up of highlands of Permo-Triassic red sandstone[1], overlooks the Zat[2] and Ourika[3] valleys. Crossed by water streams from bottom to top, Yagour offers one of the most remarkable natural landscapes in the High Atlas. Its vast high-altitude meadows make it one of the most important Agdals[4] (seasonal grazing lands) in the region.

---

[1] Type of rock formation that dates back to the Permian and Triassic geological periods, approximately 250 to 300 million years ago. This rock is characterized by its distinctive red color, which is due to the presence of iron oxide (rust) within the sandstone. In the context of the Yagour plateau, this red sandstone forms the base of the highlands, contributing to the area's unique landscape and rugged natural beauty. Its composition and age make it an important geological feature that not only defines the visual appearance of the terrain but also influences soil properties and vegetation patterns in the area.

[2] The **Zat Valley** is located in the High Atlas Mountains of Morocco, south of Marrakech. It is traversed by the Zat River, which originates in the surrounding mountains and meanders through diverse landscapes, including traditional Berber villages, orchards, and terraced farmlands.

3The Ourika Valley is located in the High Atlas Mountains of Morocco, about 30 kilometers south of Marrakech. The valley is carved by the Ourika River, which originates from the snow-capped peaks of the Atlas Mountains and flows through the valley, creating a series of beautiful waterfalls.

[4] An **Agdal** is a traditional land management system in Morocco, primarily found in the High Atlas Mountains and other pastoral regions. It refers to communal lands or pastures that are collectively managed by local communities, often with strict access and use regulations. These lands are typically set aside for seasonal grazing and are an essential part of the pastoral and agricultural practices of the Berber communities.

Materials Research Proceedings 47 (2025) 48-58 https://doi.org/10.21741/9781644903391-6

With the arrival of good weather at the beginning of summer, hundreds of inhabitants from 47 surrounding villages, representing the five tribal fractions of the Mesiuwa of the mountain (Mesiuwa nudrar), move up to the plateau with their livestock, including cattle, sheep, and goats. A true crossroads for shepherds, Yagour annually hosts nearly 7,000 people from nearby villages and is home to several laâzibs.

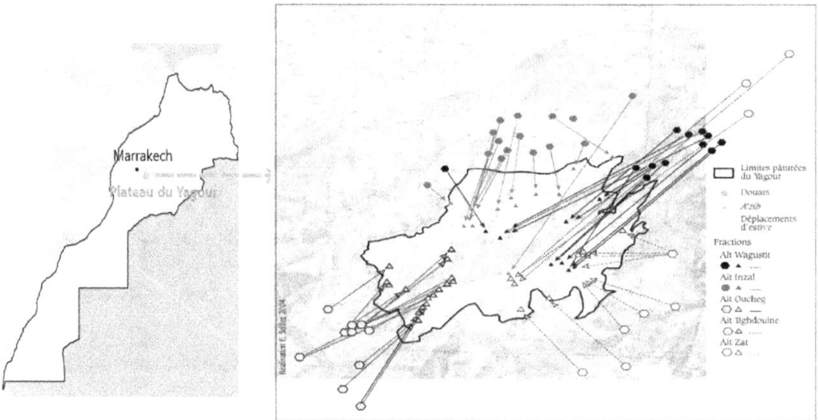

Fig. 1 Distribution of L'Agdal of Yagour among the 5 Mesiuwa tribal fractions.((Source : 1/50,000 topographic data and field surveys. E. SELLIER, Agdal Program,2024)

Laâzib: when the men mark their territories

While exploring the expanses of this Agdal, stone constructions with unusual shapes immediately catch the eye. Scattered across the meadows, leaning against large rock blocks, embedded on slopes, or grouped into clusters, these unique structures shape the landscape of the plateau. These are the Laâzibs, a term derived from the Arabic word "âzib," meaning bachelor, and the verb "azaba" (عزب), which means to distance oneself. These meanings explain the use of the term Laâzib to describe these constructions, which were initially inhabited by single shepherds, isolated from their home villages.

Over time, these Laâzibs in the High Atlas have evolved to accommodate entire families during the grazing season, incorporating various modifications to suit their needs. Now central points in the appropriation of pastoral lands, Laâzibs are true hearths, symbolizing the imprint of Man on the territory. They reflect an ancient history, marking a form of seasonal settlement for pastoral activity rooted in nomadic traditions.

Fig. 2 Locations of some Laâzibs at the yagour,(Source : The sheepfolds of Yagour and its slopes, made by NAIT LAHCEN, M., 2020

The Laâzibs origines : a toponymic lecture of sites

Like the other Laâzibs in the High Atlas, the Laâzibs of the Yagour Agdal, are seasonal constructions serving as shelters and refuges for shepherds. In other regions of the country, this term refers to different types of structures[5]. The word is also used in some North African countries, notably in Algeria[6] and Tunisia. Our research to determine the origins of these constructions took us back to the protohistoric period, an era marked by the discovery and use of metallurgy, including iron, bronze, and copper. Indeed, the site under study is rich in rock engravings dating back to the Bronze Age (between 2500 and 4000 BC). These engravings were likely made by the shepherds of the time, who settled in the area and were probably the first builders of these structures.

Archaeologists attribute these engravings, mostly dating to the first two millennia BC, to Berber pastoralists (RODRIGUE, 1999). The various depictions of weapons and tools, as well as the drawings of cattle found there, support this hypothesis. Moreover, according to some accounts from local inhabitants, these engravings were inherited from Christian Portuguese[7] who lived in this part of the mountain before the advent of Islam, during the period when the Agdal was established. According to local beliefs, these engravings also have a spiritual value, protecting the area.

[5] According to the sociologist P.Pascon a laâzib corresponds to a manor-like residence, similar to a feudal manor: a structure that serves both as a farmhouse and a dwelling, which "stands apart" and distinguishes itself from the tenant farmers' village, consisting of noualas (reed huts) or slightly less precarious earthen constructions. Thus, in the Chaouïa region, a laâzib is equivalent to the dar (house) of a large landowner.

[6] According to the Berber dictionary, in Kabylia (a Berber region in northern Algeria), laâzib refers to an isolated dwelling or a hamlet distinct from the taddart (village).

[7] This was the name given to the former inhabitants of this part of the mountain before the arrival of Muslims in these lands.

Materials Research Proceedings 47 (2025) 48-58 https://doi.org/10.21741/9781644903391-6

During our field visit, accompanied by a geologist, he confirmed the dating of the engravings. Research on this part of the High Atlas, of which he was a member, had proven through geological studies, primarily on the patinas[8] found on these engravings, that they date back to the Bronze Age.

This scientific confirmation thus corroborates the previously cited archaeological studies, linking the presence of pastures/transhumance and engravings in this part of the Moroccan High Atlas. The shepherds who used these sites were not only the creators of these engravings but also the first builders responsible for the origins of the Laâzibs.

Fig. 3 Photo of a rock engraving on the Yagour Plateau, (Source: own picture, 2022)

Fig. 4 Photo of the patina on rocks in the immediate environment. (Source: own picture, 2022)

The Architectural Organization of Laâzibs: When Simplicity Defines Uniqueness

Laâzibs, these atypical stone constructions, are primarily built from dry stone. Despite their low height (constructed on a single level) and simple, rustic appearance, they do not go unnoticed by visitors.

As we moved through the meadows, stone structures appeared before us on either side, either standing alone or organized into clusters. Comprising one or two rooms, these laâzibs occupied a significant portion of the Agdal. Spatially, the laâzibs in this region are generally found in two configurations: individual laâzibs, which consist of a shelter for the shepherd and an enclosure for the livestock, and shared laâzibs, where two or three shelters share the same enclosure.

The relationship between the built space (habitat) and the enclosure is not always the same. Sometimes, the livestock area takes up more space at the expense of the shepherd's shelter, while at other times, the balance is more even. However, the enclosures are consistently the larger areas in all laâzibs. The size of the built spaces varies, with a minimum area of 5m x 7m. The covered spaces, which serve as accommodation for the shepherds, are simple, consisting of a single room without interior partitions or flooring. The ground is naturally covered by the sandstone slabs abundant in the region, which form the base upon which the laâzibs are built.

[8] In geology, patina represents the physical and chemical alterations of a rock's surface due to the action of climatic elements. The color of the same type of rock can vary greatly depending on the patina it develops over geological ages.

Materials Research Proceedings 47 (2025) 48-58 https://doi.org/10.21741/9781644903391-6

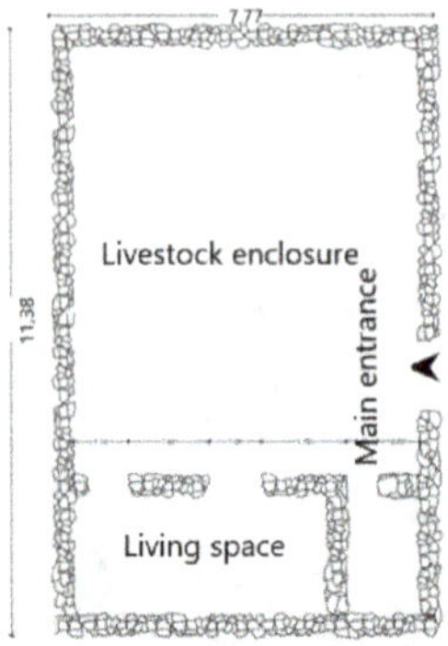

Fig. 5Schematic plan of a collective laâzib with two living spaces for shepherds and one sheepfold. Source (own production)

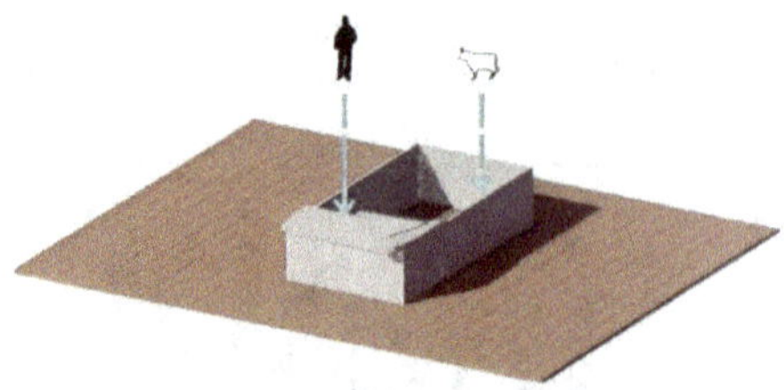

Fig. 6 3D restitution of a collective laâzib, showing the relationship: Living spaces / Sheepfold. Source (own production)

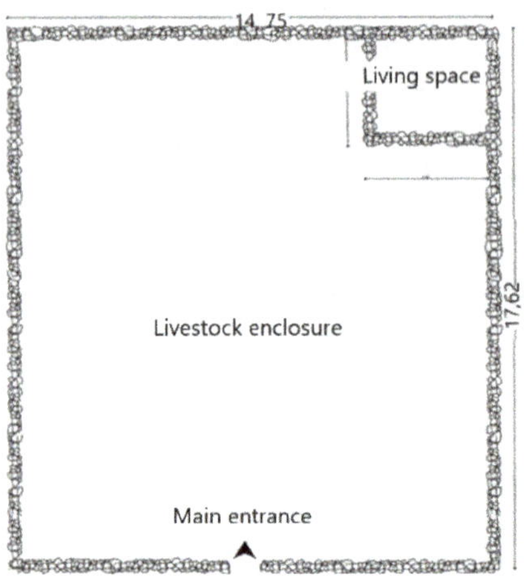

Fig. 7 Schamtic plan of an individual laâzib

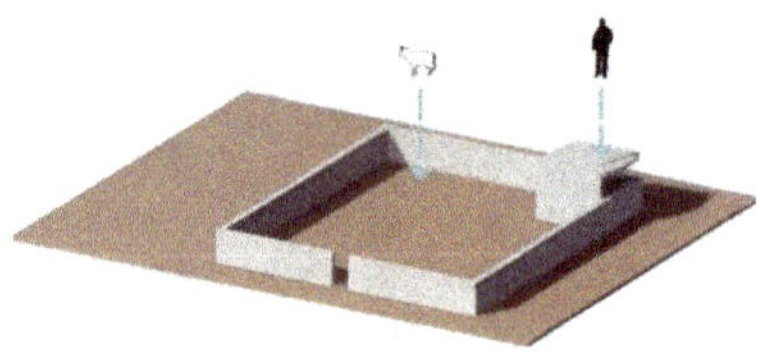

Fig. 8 3D restitution of an individual laâzib, showing the relationship: Shelter/Shepherd. Source (own production

Volumetrically, the laâzibs of Yagour, when seen as standalone structures, often take the form of parallelepipeds with somewhat irregular edges due to the use of uncut natural stone, which makes achieving perfect angles challenging

However, while exploring the Agdal, we observed some laâzibs with more regular shapes and straighter angles, distinguishing them from the rougher-looking ones. Our research suggests that these laâzibs are relatively newer compared to the older, rougher structures. Throughout our site visit, we also found other laâzibs that merge harmoniously with the natural environment; some are built against large boulders, while others are tucked beneath mountain slopes.

Fig. 9 Photo of a laâzib with parallelepiped shapes and well-defined contours.(Source:own photo)

Fig. 10 Example of a laâzib set against large stones.(Source :own photo)

Fig. 11 Example of a laâzib nestled on a side of a mountain. (Source :own picture)

Regarding the facades, they are generally simple and devoid of ornamentation. The openings are basic (simple holes in the walls) that serve essential functions such as access and ventilation, with lintels made of natural wood trunks or stones placed horizontally. These openings do not have doors; to close them, stones are simply used. Additionally, the ratio of openings to solid walls, and the heights of the openings, are not standardized. In most cases, the height of the openings does not exceed 1.5 meters, which forces the inhabitants to bend down to pass through them.

Materials Research Proceedings 47 (2025) 48-58 https://doi.org/10.21741/9781644903391-6

Fig. 12 Photo of a window of a laâzib.(Source: own Photo)

Fig. 13 Photo of an entrance door without a door.(Source : own photo)

Fig. 14 Schematic illustration of a laâzib facade displaying the door height relative to human scale. (Source : own illustration)

These laâzibs do not have accessible terraces; their roofs are generally flat and covered with thin layers of stone. The floors are supported by beams made from local tree trunks and consist of a layer of reeds topped with a layer of earth, covered with a plastic film that separates it from another layer of earth, and finally topped with flat stones. The introduction of plastic films is certainly a recent development, observed in many laâzibs on this plateau, aimed primarily at improving the waterproofing of the roofs. This raises the question: is it a distortion of a purely traditional vernacular construction method, or an adaptation using new materials and techniques to enhance performance? The debate remains open. However, aside from this observation, the laâzibs of Yagour remain among the most faithful illustrations of traditional rural vernacular architecture in Morocco, preserving their original characteristics to a significant extent.

Fig. 15 Photo of the roof of a laâzib. (Source own Photo)

Fig. 16 Photo illustrating the use of plastic sheets on a roof.(Source : own photo)

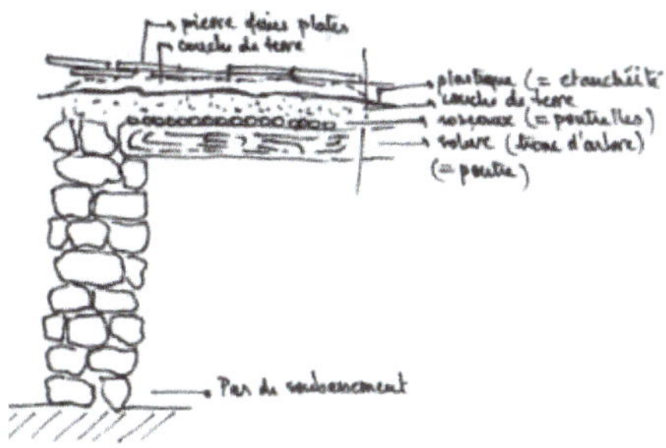

Fig. 17 Schematic cross-section illustrating the roof details of a laâzib.(Source: own illustration)

Construction Techniques

The rocks used in the construction of the laâzibs of Yagour are detrital rocks. These are mainly reddish sandstone stones, known for their hardness, high quartz content, and variable porosity. Due to their physical and chemical properties, these stones have been used in construction for thousands of years. Abundantly available on the Yagour plateau, they are the main construction material for the laâzibs. Other natural materials used include wood, primarily cypress, which is commonly found in the area. In some cases, the builders of these structures use a natural binder composed of a mixture of earth and water as a joint. The walls are built directly on the ground without foundations, using a thoughtful assembly of stones. Typically, large stones are placed at the base of the walls to support the structural load. Then, other stones are stacked layer by layer, respecting their natural orientation. In other words, the builders carefully lay the stones parallel to their stratification, ensuring that the most recent layer is on top. This arrangement helps minimize water infiltration through the stone layers, which better preserves the walls.

Fig. 18 Roof's supporting elements, featuring cypress beams. (Source :own photo)

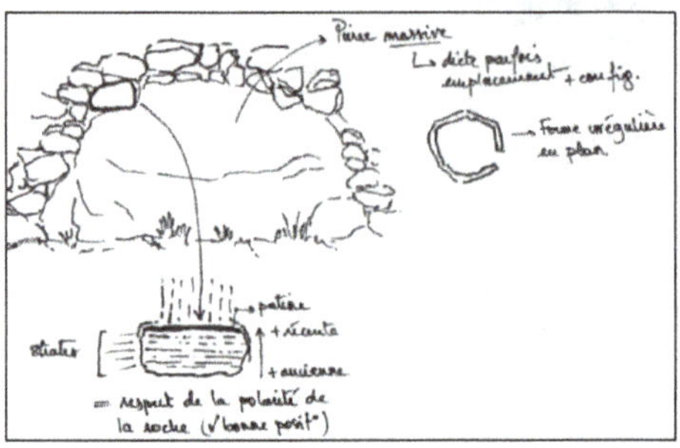

Fig. 19 Illustrative diagram showing the stone placement method in a laâzib wall. (Source: own illustration)

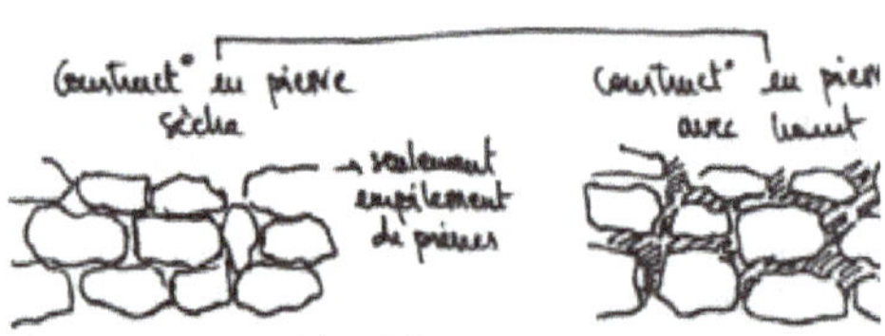

Fig. 20 Sketch of a dry stone wall without mortar (left) and a stone wall with mortar (right). (Source: own illustration)

Conclusion

The example of the laâzibs of Yagour that we have presented illustrates a unique form of traditional vernacular stone architecture that has endured for centuries. Despite its longstanding existence, this type of dwelling remains largely unknown, likely due to its location in the remote and difficult-to-access mountain peaks. It represents one of the oldest chapters of spatial occupation in this area—a seasonal and rational use of land with well-preserved natural resources. Dating back to the Neolithic period, this rustic habitat, which continues to function today, exemplifies the persistence of a simple yet effective means of anchoring to the land. Its minimalist architecture, primarily based on local materials (stone), harmonizes perfectly with its mountainous environment.

Built in the same way and faithful to ancient construction traditions—with few exceptions—this habitat demonstrates the possibility of extending and preserving original building models as long as they continue to meet the needs of their users, specifically the fundamental need for shelter during the grazing season. This may be the key to their preservation: in this remote part of the High Atlas, few social and economic changes have impacted the area, which has somewhat shielded these laâzibs from potential transformations. Additionally, the distinctive nature of the territory (Agdal), on which they are built, is governed by a highly conservative community management system, one of the most traditional in the country's history.

While literature has highlighted the uniqueness of this Agdal and contributed to its recognition as heritage—through the works of researchers who have emphasized the potential and particularities of this territory as a social and sylvopastoral organization, and in environmental, geological, and archaeological terms—the laâzibs, despite being a significant form of habitat in this area, have not received the recognition they deserve. Beyond being mere piles of stone, these simple temporary shelters for shepherds represent one of the earliest manifestations of human settlement in the area, enduring to this day.

Though simple in appearance, these constructions carry valuable lessons, particularly in terms of rational use, reflecting an admirable craftsmanship that deserves to be preserved and valued. Recognizing this heritage as an important rural architectural element to safeguard is crucial. Patrimonialization, in this case, becomes one of the most plausible means for the better conservation of this habitat.

References

[1] A. Ait Hmida, A. Bouizgaren, M Y. Lalaoui Rachidi, L. Sikaoui, Living conditions of rural communities and management of natural resources in the Haut-Atlas in Morocco. New Medit.vol 6, n°1. (2007)

[2] A . Laurent, M. Alifriqui, P. Dominguez, D. Genin, A Pastoral Monument Facing the Challenge of Heritage Preservation: The Agdal of Yagour in the Moroccan High Atlas. (2013)

[3] A. Fougerolles, The Central High Atlas: The Mountain of Mountains. (1991)

[4] C. Crépeau, M. Tamim, Pastoral Communities and Housing Systems in the High Atlas of Beni-Mellal (Morocco). *Annuaire de l'Afrique du Nord,* Volume XXV. (1986)

[5] G. Maurer, Man and the Atlas Mountains in the Maghreb. *Annales de Géographie*, No. 587, pp. 47–72. Retrieved from http://www.jstor.org/stable/23454958.(1996)

[6] J. Dresch, L'habitat rural indigène en Afrique du Nord .L'Information Géographique. (1936).

[7]L.HRACHERRASS, traditional vernacular stone rural habitat in the atlas mountains of morocco: specificities and uses of a declining heritage. Doctoral Thesis.(2023)

[8] M. Aderghal, R. Simenel, The Construction of Autochthony in Morocco: From Indigenous Tribes to Amazigh Peasants. *Espace Populations Sociétés*.(2012)

[9] M. Mahdi, P. Dominguez, The Agdals of the Moroccan Atlas: A Cultural Heritage in Danger. *Bulletin Économique et Social du Maroc*, pp. 327-347.(2009).

[10] M. Coste-El Omari, The Rock Engravings of Oukaïmeden. Retrieved from: https://patrimoinesdemarrakech.wordpress.com/2021/12/11/les-gravures-rupestres-de-loukaimeden/(2021).

[11] M. Peyron, Rural Habitat and Mountain Life in the Midelt High Atlas (Morocco). *Revue de Géographie Alpine*, Volume 64, No. 3, pp. 327-363. (1976)- P. Frey, Learning from Vernacular: An Architecture Exhibition. Press Dossier. (2010)

[12] P. Pascon, Types of Habitat and Land Use Planning Issues in Morocco, R.G.M, Vol 10 (1968).

[13] P. Pascon, Rural Studies, Ideas and Surveys on the Moroccan Countryside. Moroccan Society of Publishers United. (1980).

[14] R. Segonzac, In the Heart of the Atlas: Mission to Morocco 1904-1905. Paris: Larose. (1910)

[15] A. Simoneau, Research on Rock Engravings in the Moroccan High Atlas. *Bulletin de la Société Préhistorique Française*, pp. 642-653. (1968).

[16] T. Zouhair, Y. Missenard, O. Saddiqi, D. Frizon de Lamotte,High Atlas of Marrakech, Oukameiden Circuit. *Notes and Memoirs of the Geological Service*. Ministry of Energy, Mines, Water, and Environment of Morocco. Volume 2, pp. 77-88. (2011)

Materials Research Proceedings 47 (2025) 59-68 | https://doi.org/10.21741/9781644903391-7

Developing eco-construction at the risk of vernacular building cultures Case of Moroccan mountain villages in post-earthquake context

Kenza BELAHNECH[1,a *], Pierre FERNANDEZ[1], Khalid EL HARROUNI[2]

[1] LRA (ENSAT), ED TESC, Université Jean Jaurès, Toulouse, France

[2] Ecole Nationale d'Architecture de Rabat, Rabat Instituts, Morocco

[a] Kenza.belahnech@toulouse.archi.fr

Keywords: Natural Construction, Vernacular Lifestyles, State Intervention, Building Professions, Cultural Shift

Abstract. Growing environmental concerns have renewed interest in natural materials such as stone and earth within the building sector and public institutions. However, there are apprehensions regarding the decline of traditional technical knowledge and the inclination of rural populations to construct dwellings using local resources in remote areas. This paper presents a methodology for studying building cultures and investigates the effects of state intervention and construction sector stakeholders on vernacular building practices of mountain villages. The study examines eight ecological worksites in the Assif El Mal district of Chichaoua, Morocco, as part of the post-earthquake reconstruction program. Research methods included focus groups with beneficiaries, interviews with local master builders, analysis of design documents, and site observations. The findings indicate that these interventions introduce administrative complexity into the vernacular construction process, while also challenging the credibility of local builders. The design team manages the oversight of construction inspired by local techniques, although they face challenges, in implementing improvements for seismic resistance and building durability. The paper offers practical recommendations to enhance coordination between state and professional management and vernacular networks of mediation and organization, thereby reinforcing local knowledge and technical autonomy in natural building.

Introduction

Vernacular architecture is characterized as a modest construction style that reflects the distinct features of a specific location, utilizing locally sourced materials and techniques passed over generations. It is closely tied to local traditions and socioeconomic conditions, embodying the lifestyle and values of the communities it serves [1]. Notably, earthen and stone construction techniques are significant in this context. However, there is a concerning trend towards the decline of these materials in rural housing, particularly in developing countries like Morocco, where villages are increasingly opting for reinforced concrete structures. This shift results in the erosion of vernacular knowledge and a decline in the quality of traditional building practices [2,3].

At the same time, since the 1980s, there has been a global movement toward using geo- and bio-sourced materials in construction due to their environmental benefits, such as low energy consumption and enhanced user comfort [4]. In line with this trend, Morocco adopted anti-seismic building regulations for earth and stone construction in 2013, aiming to promote sustainable construction techniques and support the preservation of patrimonial sites.

This research emerges from a paradox: while rural communities in Morocco are gradually abandoning ecological building materials and methods, the sustainable building industry is advocating for their revival. This prompts an important research question: How do efforts by institutions and the professional construction sector to standardize and regulate natural materials influence the traditional practices of these communities?

This study is particularly relevant following the 6.9 magnitude earthquake that hit Morocco in September 2023, damaging around 50,000 homes in mountain villages across the High Atlas. In this post-disaster context, there is a need for collaboration between two distinct socio-technical realms: the traditional organic community and the regulated professional sector, to rebuild resilient settlements that adhere to legal standards.

Our fieldwork focuses on the Assif El Mal district in Chichaoua province, where beneficiaries of the state-assisted reconstruction program have opted to use local raw earth and stone for rebuilding. Through focus groups, interviews, design documentation, and site observations, this paper examines the interactions among beneficiaries, local builders, and professional stakeholders. Understanding these dynamics will enhance cooperation for strengthening local technical autonomy and building knowledge. This research is part of a PhD thesis that aims to analyze changes in rural building cultures in Morocco and contribute to their preservation.

Theoretical framework: The process of production of building cultures

A building culture develops gradually through a slow process of experimentation, shaped by specific environmental factors, such as context, site, climate, and resources. These practices, transmitted across generations, embody the identity and cohesion of the communities that create them, reflecting both their tangible and intangible cultural values [1]. In her analysis of vernacular building cultures that gave rise to Moroccan Berber architecture, Naji identifies two types of production systems. The first, the "closed self-sufficient system," relies on local cultural, physical, and technical resources, with master builders (*maalmin* in Arabic, plural of *maalem*) playing a central role. This system is rooted in community networks of mutual aid and is typically used for ordinary rural dwellings. The second type, the "outwardly adaptable system", utilizes external resources and models, with a greater emphasis on monetization, often applied in building kasbahs, mosques, and other significant constructions commissioned by authorities [5]. However, shifts such as the commodification of cultural practices, urbanization, and globalization have blurred the lines between these systems, complicating their categorization today.

Based on Ferreira's thesis, building cultures can be studied under the following components: Available resources – Building processes – Systems of mediation and organization – Systems of transmission and diffusion. Systems of mediation and organization encompass various mechanisms related to management, communication, social interaction, and the agreements established between parties. These systems outline the organization of labor, influencing the conditions and levels of cooperation among stakeholders involved in the construction process. In contrast, systems of transmission reflect a society's capacity to share and transmit its knowledge and expertise, thereby solidifying its potential across spatial and temporal dimensions [6].

By examining these components, this paper interrogates the evolving building practices in light of the introduction of manufactured materials and the emergence of regulated professions and processes, particularly in the context of post-earthquake reconstruction.

Case study: Four villages of Assif El Mal district

Assif El Mal is an administrative district in the province of Chichaoua, situated within the Marrakech-Safi region. Nestled on the northern face of the Western High Atlas, it is roughly a 1 hour and 40-minute drive from Marrakech and about 40 km from Al Haouz earthquake's epicentre. The district comprises 24 villages, primarily located at an elevation of around 900 meters above sea level. This study specifically focuses on four villages of the district: Anebdour, Imi N'Ouassif, Imighzer, and Anamer (Fig.1), with populations consisting of 170, 136, 68, and 30 households respectively [7].

Settlement morphology and local architecture. The settlement pattern in these villages is densely clustered, shaped by the terrain and proximity to the Assif El Mal river, with the road running parallel to the river (Fig.1- 4). The villages integrate seamlessly into the surrounding landscapes. Construction techniques vary depending on material availability and site accessibility. In Anebdour, the stone predominantly used is a dark shale stone, extracted directly from the bedrock (Fig.1- 5). In some instances, it is combined with bright and rounded river stones (Fig.1-3). It was noted that coatings were rarely applied to stone walls. In the nearby Imi N'Ouassif village, rammed earth is a prevalent method of construction (Fig1-Image). It is typical for these structures to utilize stones in the foundation and the wall bases. Stones are used to provide a good overhang. The latter is sometimes masoned with earth or with a cement-based mortar (Fig.1-1).

Figure 1: Four villages of Assif El Mal (location and local architecture)

Hybridizations

In rammed-earth dwellings, the use of cement was observed in rendering floors and the lower part of interior walls. Plaster coatings were employed on their higher part. More recently, cement has been used extensively in reinforced concrete post and beam structures (Fig.1-2) and slab construction for dwellings in addition to prefabricated elements like concrete bricks and beams. It is also found as part of the mortar in stone masonry, in roofing screeds and, in parapets.

Ecological sites of the reconstruction program. In the villages examined, data provided by commissioned architect Khalil Morad El Ghilali (Atelier BE) reveals that 37 of approximately 300 earthquake-affected dwellings are set to be reconstructed using a stone/earth load-bearing technique as part of the reconstruction initiative. While this may seem like a small number, it is exceptional within the broader program, which predominantly relies on materials such as cement, rebar, and concrete blocks as well as the post-and-beam building method.

Of the 37 dwellings set for reconstruction, 8 sites were visited. Their conditions within the villages, along with corresponding photos, are shown in the figures below. In what follows, these sites will be referred to as eco-reconstruction sites, and their project owners as eco-beneficiaries.

Reconstruction using stone with a lime-earth mortar was common to all of them, except in Imi N'Ouassif where rammed-earth was applied. They were typically constructed either within the footprint of the demolished house or adjacent to it, often in confined areas, along cliff edges, or on steep terrain. Most visited sites were at the early stages of digging trenches and building foundations. Most of the planned reconstructions consisted of a single room or two, intended to be used as a living/bedroom and kitchen, sometimes with adjoining bathrooms.

The satellite views, dated October 2023, display the original house footprints (marked in yellow) and the planned reconstruction area (marked in red). However, the footprints of the areas to be rebuilt are not accurately represented.

Figure 2: Eco-reconstruction sites in Imi N'Ouassif.

Figure 3: An eco-construction site in Imighzer.

Materials Research Proceedings 47 (2025) 59-68 https://doi.org/10.21741/9781644903391-7

Figure 4: Eco-construction sites in Anebdour.

Figure 5: Eco-construction sites in Anamer.

Elements of context: The state-assisted reconstruction program

After the September 2023 earthquake, multidisciplinary teams compiled a list of eco-beneficiaries eligible for financial state support and technical assistance to rebuild their homes. A national call for architects and engineers led to the formation of teams to oversee the reconstruction in 2,930 affected villages, with the case study involving consultancies from Casablanca and Rabat. Initial site surveys included topographical measurements and construction placement assessments, while a public laboratory evaluated seismic risks and provided recommendations. Eco-beneficiaries received architectural and technical plans via local authorities and applied for building permits, specifying their intention to use local natural materials. Once legal documents were signed, they could use the first subsidy instalment to buy materials, hire a *maalem*, and potentially recruit additional help. The design team intervened at three key construction milestones: soil approval, foundation and wall base approval, and floor/roof structure completion. At each stage, the eco-beneficiaries had to notify the ***muqaddam*** (a civil servant acting as an intermediary between the population and local government), while waiting for validation by a design team member.

Field research methodology
The following fieldwork was carried out in September 2024.
Investigating the shift in building cultures. In addition to a visual examination of vernacular architecture (see section above), a focus group and interviews were conducted in the four villages of our study area.

- Focus group with eco-beneficiaries: This group consisted of seven out of fourteen beneficiaries from Anebdour, including six males and one female, with ages ranging from 53 to 87. All participants were born and raised in Anebdour.
- Interviews with *maalmin*: With the help of villagers, a list of sixteen *maalmin* who have contributed to building stone and earth dwellings over the decades was compiled, and six were interviewed. They were all males, with ages ranging from 34 to 62. They had all learned with a more experienced community member in the village without any formal school or training.

For each of the two methods, questions were centered around:

1. Materials, tools, and equipment used traditionally/ in the current context,
2. Building techniques used traditionally/ in the current context,
3. The processes of mediation, organization and transmission of the local building culture before and within the reconstruction program.

Studying the reconstruction worksites. Based on visits of the eco-reconstruction sites and a study of the technical plans provided to the eco-beneficiaries, the specific materials, techniques and processes of the reconstruction program were examined. Architectural design elements, such as windows and doors, as well as the overall layout of the houses—including the reconstructed areas—were not thoroughly evaluated.

Results
Local inhabitants and *maalmin* have reported that cement began to be used in dwellings at least 40 years ago, albeit in small quantities (typically for floors and as a coating for rammed-earth indoor walls). They also noted that plastic sheeting for roofing was introduced around the 1990s. These hybridizations in materials and techniques align with two cultural shifts: the cessation of the community-building tradition known as *tiwizi* (which reportedly ended around the 2000s) and the implementation of building regulations concerning reinforced concrete structures (noted around 2010). Collectively, these recent changes constitute what we refer to as an "altered vernacular building culture," which will serve as a baseline for identifying the differentiating elements introduced by the state-supervised and professionally assisted eco-reconstruction sites. The collected data are compiled in Table 1.

Table 1: Comparative description of vernacular and state-assisted eco-construction

Component	Altered vernacular building culture	Elements of differentiation of the eco-reconstruction sites
Techniques	Stone building technique: Load-bearing 50cm wide walls of stone masonry using onsite earth as a mortar Rammed-earth technique: 50cm-wide walls made of onsite earth manually compacted in wooden frames.	Respect of the predominant local technique (load-bearing stone masonry/rammed-earth depending on locally available material, wall thickness) For both: Greater foundations' depth preconized (as deep as 1.10m in zones of high seismic risk) With rammed-earth: Higher wall base.

	For both: Foundations with dry stones and wall bases with stone masonry. Recent variation for stone building: cement used in mortar for foundations and/or walls Roofing with: wood beams, woven canes, cardboard, plastic sheets, earth	Burnt lime and earth mortar replace cement-based ones. For both: Addition of anti-seismic features. Horizontally: glass-fiber mesh (under the wall), galvanized wire armature (above the wall). Vertically: iron bars inserted at the top of the walls, every 50cm. Improved impermeability with a 5 layers roof system (including a cement screed)
Materials	Local shale stone, onsite sieved earth, poplar wood (deciduous tree), woven canes, plastic sheeting (since the 90s'), and round river stones. A local grey soil reputed to be "as waterproof as bitumen" was reported to be historically used in roofing. Potential for other materials: Old deposits of plaster and lime were reported.	Same materials… Minus: Plastic sheeting, reputed grey soil, poplar wood, cement used in mortar Plus: Burnt lime, quarry sand, anti-seismic grid, elastomer membrane, resinous wood, reinforced iron bars, galvanized wire armature, cement, tiles (for roofing)
Tools and equipment	String, trowel, hammer, axe (for stone building), plumb line, water level Bigger tools: shovel, pickaxe, crowbar For rammed-earth: A wooden rammer and locally made wooden formwork	Same for stone buildings. Possibly a jackhammer. For rammed earth: Possibly, a mechanical damper and a formwork with metallic fixating elements.
Stakeholders	Project owner (family members), *maalem* chosen by the owner, *muqqadem* (knows everyone and mediates with the local authorities) Helpers provided (or paid) by the owner.	Same but with more nonlocal stakeholders (surveyor, laboratory, public actors, engineer, and architect from design offices in remote cities).
Systems of mediation and organization	Onsite discussions (needs, orientation, interior layout, openings position, etc) between *maalem* and project owner. The *maalem* knows what to do. Basic tools are brought by him. Bigger equipment is lent by the project owner. Permitting process: Tacite approval of local authorities is usually given as a notification for restoration of old buildings. Objections in case of proximity to the road, conflict between heirs over the land.	The *maalem* is given the plans. Some tools exogenous to the area are lent by members of the design team. Permitting process: Design following program specification Application for planning permission Specifications of site constructability Worksite verification by project team at 3 critical checkpoints (see above).

Systems of transmission	Oral transmission of know-how, through repetition and proven experience, from generation to generation. Inter-knowledge network of local experienced *maalmin* and younger practitioners who are more proficient with new technologies and media. No written documents drawings, or calculations.	Scientific references, calculation templates, drawing standards and global engineering culture are conveyed by specialized programs, industry firms, and public institutions. Technical signed documents: Seismic hazard map Architectural drawings Engineering plans

For more clarity, stakeholders, tools, and equipment were presented separately.

In terms of techniques, the main input from the design team was to add features to improve two elements of the vernacular building model: its seismic performance and roofing impermeability. While the main local materials were retained, burnt limestone was introduced as a significant change. The recommendation to use resinous wood instead of local poplar wood warrants further investigation, as does the potential for utilizing the local grey soil, known for its watertightness.

In terms of mediation and organization, the construction process became more burdensome due to the involvement of additional stakeholders, often located far from the worksites, and the intermediation of local authorities at various steps. These administrative layers, however, helped ensure professional oversight of the works.

Regarding systems of transmission, older *maalmin* struggled with interpreting the provided plans and lacked the autonomy to source specific materials or repair unfamiliar equipment. However, younger community members stepped in to support them, reactivating intergenerational collaboration around building practices. Additionally, engaged members of the design team offered further support, fostering solidarity around the choice of natural materials.

Discussion and recommendations

The involvement of multiple stakeholders seems to diminish the decision-making power of project owners and maalmin, who previously enjoyed greater autonomy in the construction process. As a result, maalmin are increasingly relegated to an executive role, distancing them from their historical position as the main technical reference in their communities. This shift has elevated the authority of local state bodies, placing them at the center of regulated interactions that now dictate the pace and structure of the building process. If the different administrative steps are validated, this transfer of authority should result in the guarantee of the work no longer falling on the *maalem*, but instead being assumed by the design team or the local authorities.

The remoteness of the sites, compounded by the large number of projects assigned to the professional stakeholders, makes it challenging for them to effectively oversee the works. Additionally, the difference in technical culture between *maalmin*, particularly older ones, and city-based professionals likely contributes to a disconnect between the two groups. This gap complicates the formation of a new building culture that acknowledges and integrates the technical legitimacy of both parties.

Based on these findings, the following recommendations can be made:

- **Site coordinator role**: Establish a dedicated position within the local community to oversee multiple eco-construction sites. This person would ensure proper plan interpretation, efficient communication, material supply, and progress synchronization.

- **Enhancing practical knowledge**: Strengthen the skills of young *maalmin* through training initiatives that encourage knowledge sharing with engineers and architects, including training in reading plans,
- **Additional state support for ecological worksites**: Provide financial and logistical support for frequent site visits by the design team to ensure on-to the-ground adjustments and maintain construction quality.
- C**ulturally relevant media**: Utilize relevant media, like videos in the local language, to share technical guidelines and inform key construction steps.
- Situated and informed design: Develop programs for engineers and architects to work effectively in rural and mountainous contexts, focusing on local building culture and resources.
- Sector **development**: Promote academic and scientific research in stone, earth, and wood construction, as consultancies specializing in these materials are currently rare in Morocco.

Some of these recommendations are already being implemented by professionals and civil society actors, as seen at the newly established Atlas Training Centre in Eco-construction, located in Tahanaout.

Conclusion

The reconstruction efforts following the September 2023 earthquake in the High Atlas exemplify the increasing involvement of the state, along with professional organizations, in rural building practices—a trend that has developed over recent decades. It is crucial to recognize that the program, which promotes self-building, relies on beneficiaries' choice of local *maalmin* mobilizing their shared knowledge of the sites and materials. In the ecological reconstruction sites studied, the introduction of new materials and construction methods has addressed specific weaknesses in traditional building features.

On one hand, the intermediation of local authorities and the addition of new stakeholders have altered traditional systems of mediation, organization, and knowledge transmission, making it more challenging for *maalmin* to maintain their roles as technical references. On the other hand, the participation of eco-construction specialists has revitalized local knowledge and skills related to traditional materials and techniques, contributing to a more positive perception among rural communities regarding the value of their local resources and practices.

Rather than serving as a comprehensive study, this fieldwork forms part of an ongoing ethnographic investigation aiming at informing how professionals in sustainable construction can effectively support and enhance the technical resilience of rural communities.

References

[1] ICOMOS, Charter on the built vernacular heritage, Mexico (1999). https://www.icomos.org/en/179-articles-en-francais/ressources/charters-and-standards/164-charter-of-the-built-vernacular-heritage

[2] A. Marsh, Y. Kulshreshtha, The state of earthen housing worldwide: how development affects attitudes and adoption, Building Research & Information. 50 (2022), 485–501. https://doi.org/10.1080/09613218.2021.1953369

[3] S. Naji, Pour une éthique de la préservation. Architectures du bien commun [For an Ethics of Preservation. Architectures of the Common Good], MetisPresses Editions, vueDensembleEssais Collection, Italy, 2019, pp.14-41.

[4] K. Bouroubat, La construction durable : Étude juridique comparative / Maroc-France [Sustainable construction: A comparative legal study / Morocco-France], Law Thesis, University of Paris-Saclay, 2016. https://tel.archives-ouvertes.fr/tel-01617586

[5] S. Naji, Art et architectures berbères du Maroc [Art and Berber Architecture of Morocco], first ed., Eddif&Edisud, Casablanca et Aix-en-Provence, 2001.

[6] T.L. Ferreira, Architectures vernaculaires et processus de production contemporains : formation, expérimentation et construction dans une communauté rurale au Brésil [Vernacular architectures and contemporary production processes: training, experimentation, and construction in a rural community in Brazil], Architecture thesis, University of Grenoble (France), University of Sao Paulo (Brazil), 2014 , pp.73-84. https://theses.hal.science/tel-01310505

[7] Higher Planning Commission (HCP), Profil sociodémographique de la zone sinistrée suite au tremblement de terre survenu au Maroc le 8 septembre 2023 [Socio-demographic profile of the disaster-affected area following the earthquake that occurred in Morocco on September 8, 2023], 2023, https://www.hcp.ma/file/238776/

Materials Research Proceedings 47 (2025) 69-80 https://doi.org/10.21741/9781644903391-8

Intermediate spaces in Moroccan vernacular architecture: Climatic and cultural challenges

FARS Sara[1,a*], BEN BALLA Chaymae[2,b], MOULOUDI Hicham[1,c], ALJEM Sanae[1,d]

[1]National School of Architecture of Rabat, Morocco

[2]INAU, Rabat, Morocco

[a]sara.fars@enarabat.ac.ma, [b]benballachaymae@gmail.com, [c]h.mouloudi@enarabat.ac.ma, [d]s.aljem@enarabat.ac.ma

Keywords: Vernacular Architecture, Intermediate Spaces, Climatic Comfort, Cultural Identity, Local Materials

Abstract. Intermediate spaces in Moroccan vernacular architecture are important in addressing cultural, functional, and climatic challenges, acting as transitional zones between public and private spheres. These spaces, which include terraces, courtyards, patios, and thresholds, not only facilitate social and economic interactions within local communities but also play a crucial role in regulating the thermal comfort of dwellings. This study, focusing on specific cases in Sebt el Gardan in the Taroudant region, highlights the diversity and functionality of these intermediate spaces. It demonstrates how they contribute to both thermal regulation and the facilitation of social interactions while emphasizing the importance of local materials such as rammed earth and mud in their construction. Traditional techniques showcase how vernacular architecture meets contemporary needs while preserving its cultural identity. The central hypothesis of this research is that the climate and communal ways of life largely influence the typology of intermediate spaces in Moroccan vernacular architecture. Through a documentary approach and case studies, the study aims to deepen the understanding of spatial and social dynamics within communities, while underscoring the need to preserve this often underestimated cultural heritage. Thus, intermediate spaces are not merely architectural elements but dynamic vectors of Moroccan cultural heritage, capable of adapting to socio-economic and historical changes, reinforcing their relevance in the discourse on sustainable development and cultural heritage conservation.

Introduction

Moroccan vernacular architecture exemplifies a deep adaptation to environmental and social needs, seamlessly integrating sustainable and cultural solutions. Among its key features, intermediary spaces serve as crucial elements between public and private spheres, offering both climatic regulation and socio-cultural functions. These spaces not only moderate temperature and ventilation but also facilitate social interactions and uphold traditions. In contemporary contexts, however, these intermediary spaces face challenges due to modern construction practices, urbanization, and evolving comfort needs. The central question is whether these spaces can continue to fulfill their climatic and cultural roles within changing environments. This study explores how traditional materials (like rammed earth, wood, and stone) and architectural features (such as patios and courtyards) ensure thermal comfort while supporting social cohesion. It investigates their dual role: as passive climate regulators and as cultural anchors that promote social ties, aiming to assess their relevance and adaptability today.

Two main hypotheses will guide this reflection. First, intermediary spaces play a crucial role in climate regulation, allowing for optimal heat management in arid environments. Second, these spaces serve as places of transition between the public and private realms, reflecting the cultural identity of regions and strengthening social ties within communities.

The study is based on a qualitative and comparative approach, combining a thorough theoretical analysis of key concepts with in-situ case studies. It includes the review of existing research, documentary analysis, and field observations in a douar in the Taroudant region, at Sebt El Gardan, to identify variations in intermediary spaces according to climatic, cultural, and social contexts. Interviews with local inhabitants will also help understand the use and perception of these spaces.

In summary, this article analyzes the climatic and cultural roles of intermediary spaces in Moroccan vernacular architecture, highlighting how, through the use of local materials and adapted architectural features, they ensure thermal comfort while meeting the socio-cultural needs of the inhabitants. The case studies in Sebt El Gardan, in the Taroudant region, will illustrate the functional and symbolic advantages of these spaces in preserving local identities and promoting sustainable architecture.

1. Intermediate Spaces: Architectural Strategies in Response to Climate Challenges

The notion of intermediate spaces

The concept of intermediate spaces in architecture, although rooted in older traditions, truly began to take shape in the mid-20th century, particularly influenced by the Team 10 group of architects and urban planners. Formed in 1953 as a reaction against the rigid modernist doctrines of the International Congress of Modern Architecture (CIAM), Team 10 advocated for a more contextual and human-centered approach to architecture. [1]

Among its most influential members, Aldo van Eyck and the couple Alison and Peter Smithson played key roles in shaping the concept of "intermediate spaces." Aldo van Eyck, a pioneer of the idea of "in-between space," viewed these areas as essential for creating meaningful inhabited places. He considered them transitional zones that encourage social interaction and a sense of belonging, proposing that architecture should create thresholds conducive to encounters and interactions, transcending the simple dichotomy between interior and exterio [2]. Alison and Peter Smithson further developed this notion through their "streets-in-the-air" concept, where intermediate spaces were envisioned as social interaction zones within dense urban environments, strengthening the community fabric. They saw these in-between areas as a way to structure everyday life within residential buildings. [3]

Beyond Team 10, other theorists such as Kevin Lynch and Christian Norberg-Schulz have enriched the discourse on these spaces. In *The Image of the City*, Kevin Lynch highlights the importance of "pause spaces" in urban organization, noting that these areas not only provide rest for users but also help strengthen orientation within the city. [4] Climate management in intermediate spaces – Cases 1 & 2Norberg-Schulz, with his theory of *genius loci* (the spirit of the place), adopts a phenomenological approach to space, suggesting that intermediate spaces are crucial for a complete architectural experience, serving as transitional areas where individuals perceive and feel the character of a given environment.[5]

More recently, the research of Valérie Lebois on intermediate spaces in collective housing, especially within the context of social housing in Paris, introduces new issues related to the design and management of these spaces. She proposes the concept of co-production, emphasizing the importance of collaboration between various stakeholders (users, architects, managers) to enrich and improve the appropriation of intermediate spaces. Lebois distinguishes three typologies: Basic spaces, which fulfill elementary functions without encouraging active appropriation; Complementary spaces, such as gardens and common rooms, that promote social interaction; and Specific spaces, like community gardens or gyms, which require active user participation. [6]

The theoretical contributions of figures such as Pierre Rabhi, with his idea of "happy sobriety," bring a new ecological dimension to this discussion. Rabhi sees intermediate spaces as sustainable

Materials Research Proceedings 47 (2025) 69-80 https://doi.org/10.21741/9781644903391-8

solutions in response to the scarcity of natural resources, advocating for their promotion as simple but effective alternatives in a world facing climate challenges. [7]

These various contributions demonstrate that intermediate spaces play a crucial role in architectural design, not only for thermal comfort (thanks to passive strategies) but also for preserving social and cultural practices. They represent a significant contemporary challenge: how to reconcile tradition and innovation to create spaces that meet both modern needs and pressing ecological concerns. Their adaptability and ability to foster social and cultural interactions make them essential elements for future architecture and urban planning.

Ecological Adaptation

Intermediate spaces play a critical role in climate-responsive architecture, especially in hot regions. Designed according to age-old traditions, these buffer zones moderate thermal variations and contribute to the stability of indoor temperatures, promoting passive thermal comfort. Features such as courtyards, patios, or loggias ensure a gradual transition between the exterior and interior.

Hassan Fathy's theories emphasize the effectiveness of traditional architectural elements, such as *malqaf* (wind catchers) and *mashrabiyas* (lattice screens), in the passive management of ventilation and cooling. These devices are designed to address local climatic constraints while respecting social and cultural values. [8] Amos Rapoport and Paul Oliver have also highlighted the importance of intermediate spaces in thermal regulation. Rapoport emphasizes the interaction between climate and vernacular construction techniques, [9] while Oliver demonstrates that these spaces are essential not only for climate moderation but also for fostering social interaction. He argues that vernacular architecture provides a coherent and adapted response to community lifestyles, where intermediate spaces balance functionality and thermal comfort. [10]

From a more philosophical standpoint, Christian Norberg-Schulz argues that these intermediate spaces express a deep relationship between humans and their environment. They create zones of connection between the dwelling and the natural landscape, ensuring a fluid continuity between nature and living space. [11]

Christopher Alexander, with his theory of patterns, offers a systematic reading of the design of intermediate spaces based on recurring architectural patterns. Among these patterns, Pattern 112, concerning the transition between public and private spaces, is particularly relevant for understanding intermediate spaces in vernacular architecture. These transitions, such as Moroccan patios, create buffer zones where light, sound, and texture subtly guide the passage between different levels of intimacy. Pattern 127 describes the progression from more public to more private spaces, while Pattern 140 illustrates the concept of private terraces linked to street life. These ideas are directly applicable to the analysis of intermediate spaces, particularly in Moroccan vernacular dwellings, where community life and privacy coexist harmoniously. [12]

Bernard Rudofsky, in *Architecture Without Architects*, highlights the adaptability of vernacular architecture, celebrating examples such as Moroccan architecture with its patios and natural ventilation systems. These elements optimize the use of local resources while addressing climatic and social needs, responding to both climatic requirements and combining functionality with aesthetics to make the most of the environment. [13]

Finally, Lisa Heschong's research on thermal regulation confirms that intermediate spaces are essential for maximizing natural ventilation and controlling sunlight exposure. By promoting passive techniques such as shading and air circulation, these spaces contribute to effective thermal management without relying on energy-intensive systems. [14]

Vernacular architecture, particularly in regions with extreme climates, represents an exemplary model of adaptation, linking human ingenuity to environmental constraints. By combining these perspectives, it appears that vernacular architecture embodies a form of collective and contextual intelligence, blending functionality, aesthetics, and sustainability. The work of Rudofsky and Heschong highlights that these approaches, although rooted in ancient traditions, offer modern and applicable solutions for sustainable energy management and passive thermal comfort. Therefore, far from being a mere historical curiosity, vernacular architecture provides crucial lessons for contemporary architecture facing current climate challenges. These theoretical contributions offer a rich framework for exploring the complexity and value of intermediate spaces in Moroccan vernacular architecture, which constitutes essential elements of architectural resilience and sustainability, addressing contemporary climate challenges while preserving local practices.

Architectural Responses to Climatic Needs

Semi-arid climate areas are subject to extreme temperature variations and harsh environmental conditions, often requiring architectural solutions that can effectively adapt to these challenges. In regions with Mediterranean influences, where the climate presents significant fluctuations between hot summers and cold winters.

The architectural choices in Sebt El Gardan, within the Taroudant region, are significantly influenced by the local climate. The area experiences a semi-arid to Mediterranean climate. In spring (March to May), temperatures range from 18°C to 28°C, accompanied by ample sunshine and northeasterly winds shaped by pressure gradients between the Atlantic Ocean and the Atlas Mountains. Rainfall is limited, often occurring as sporadic showers, particularly in March. During summer (June to August), temperatures frequently rise to 40°C, especially in July and August, with up to 12 hours of daily sunshine. The "chergui," dry winds from the east or southeast, exacerbate the heat, leading to periods of drought. In autumn (September to November), temperatures moderate to between 20°C and 35°C, and northwesterly winds bring cooler air, with increased rainfall in November. The winter months (December to February) are mild, with temperatures between 10°C and 20°C, while nights can be cooler due to the proximity of the mountains. This climate, characterized by strong sunlight and variable seasonal winds, greatly shapes agricultural practices, lifestyles, and architectural organization. Studying Sebt El Gardan offers valuable insights into the relationship between climate, culture, and architecture in Moroccan vernacular design

Vernacular architecture often incorporates strategies to optimize natural resources such as wind, light, and shade. These strategies aim to ensure thermal comfort throughout the year by taking advantage of local environmental conditions.

Intermediate spaces play a crucial role in this adaptation strategy. By managing airflow and reducing direct sun exposure, these spaces act as buffer zones that help to stabilize interior temperatures while also ensuring the privacy of residents. Features such as patios, inner courtyards, *skifas*, and terraces are vital in facilitating natural ventilation and in organizing the transitions between public and private spaces. The effectiveness of these strategies can be seen in the case studies of Sebt El Gardan, located in the Taroudant region of southern Morocco, where the principles of climate-responsive design are demonstrated.

Materials Research Proceedings 47 (2025) 69-80 https://doi.org/10.21741/9781644903391-8

Figure 1: The skifa– Author

The *skifa* is a significant architectural element in Moroccan vernacular architecture, functioning as a transitional space between the public exterior and the private interior of a dwelling. Originating from traditional design principles aimed at addressing cultural, social, and environmental needs, it typically serves as a vestibule or antechamber near the main entrance. Its layout often involves a narrow passage, sometimes with a bend, to limit direct visibility into the house, thus ensuring privacy, a core value in Moroccan and Islamic culture. The design prevents outsiders from having an unobstructed view into the private living areas, offering an immediate sense of separation from the street.

Beyond privacy, the *skifa* plays an essential role in climate adaptation. By functioning as a buffer zone, it helps regulate indoor temperatures. The enclosed nature of this space can reduce the influx of hot air during the summer and cold air during the winter, contributing to a more stable indoor environment. Its walls, often constructed from traditional materials such as rammed earth or stone, add to the thermal mass, absorbing heat during the day and slowly releasing it during cooler nights. In addition to these practical functions, the *skifa* holds cultural significance as a space where social interactions and daily activities occur.

The vernacular architecture of the region is characterized by the intelligent use of intermediate spaces, which play a crucial role in thermal regulation and social organization. The first case study, illustrates an evolution of these principles, integrating modern materials while preserving traditional elements in utility spaces. Ancestral techniques, such as the use of rammed earth and adobe, demonstrate resilience in the face of contemporary challenges. The second one, maintained in its traditional form, showcases vernacular construction techniques that, despite their simplicity, effectively address local climatic constraints.

The mix of materials in the first house, which combines modernity and tradition, demonstrates a hybrid response to climatic needs. While the modern residential sections incorporate contemporary materials, the enclosure, maintained in raw earth and wood, illustrates the resilience of ancestral techniques. Thanks to their thermal inertia, these traditional materials absorb heat during the day and gradually release it at night, ensuring interior comfort. The intermediate spaces continue to play a fundamental role, addressing both climatic needs (ventilation, shade) and domestic activities.

In the second house, preserved in its vernacular integrity, traditional techniques prove to be particularly suited to local climatic constraints. Buffer spaces, such as thresholds and skifas, illustrate a fine-tuned control of thermal exchanges. They optimize the circulation of cool breezes while blocking hot air currents, thus promoting effective natural ventilation. The multifunctional terrace, for instance, is essential not only for daily life but also for its contribution to thermal comfort, particularly by facilitating the evacuation of hot air in summer.

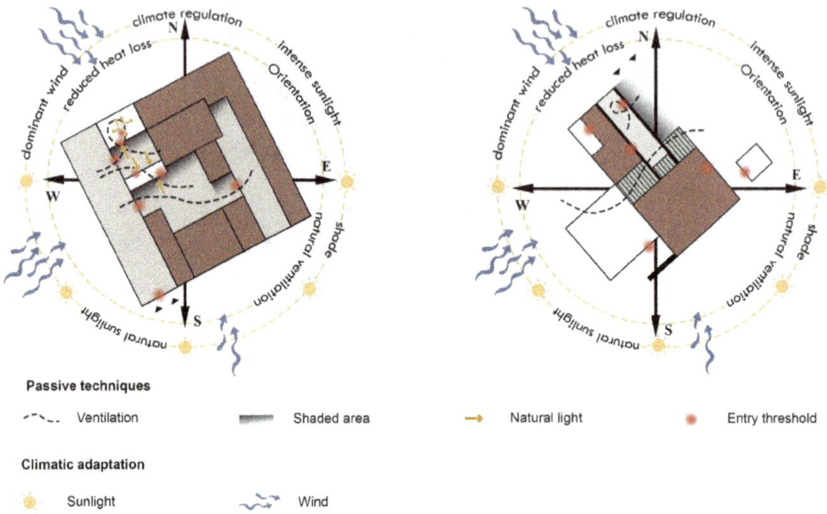

Figure 2: Climate management in intermediate spaces – Cases 1 & 2 - Author

The intelligent use of local materials, combined with passive thermal regulation techniques, ensures optimal comfort while minimizing the environmental impact of construction. This traditional know-how demonstrates a continuous adaptation to environmental conditions, aiming to provide a living environment that is both functional and environmentally respectful. These case studies demonstrate how vernacular architecture continues to meet the thermal needs of inhabitants by incorporating design principles suited to environmental constraints. The ability of these intermediate spaces to provide climatic comfort while adapting to the evolving urban context makes them relevant architectural solutions for current challenges related to resource management and climate change. They embody a sustainable and resilient response to contemporary environmental issues.

Intermediate spaces in Moroccan vernacular architecture go beyond climatic functions and also embody places of cultural transition. They play a central role in social organization by fostering interactions and preserving traditional practices. This cultural dimension, intrinsically linked to architecture, will be explored in subsequent sections.

2. Intermediate Spaces: A Bridge Between Culture and Vernacular

Social interaction and cultural norms – Some reflections

The concept of intermediate spaces, in cultural and social terms, has been explored by numerous theorists, each bringing a unique perspective to understanding their role in structuring human interactions. Henri Lefebvre, in *The Production of Space*, establishes that space is inherently social, produced by specific cultural practices. Intermediate spaces, as they appear in his work, facilitate a smooth articulation between public and private realms, offering an ideal setting for nuanced social interactions while preserving a certain degree of intimacy. [15] Lefebvre's approach resonates with Edward T. Hall's work, who, with his concept of proxemics, highlights the importance of social distances in space organization. [16]

Materials Research Proceedings 47 (2025) 69-80 | https://doi.org/10.21741/9781644903391-8

While Lefebvre sees space as a social product, Hall adds that physical distances within this space are culturally determined, and intermediate spaces play a central role in regulating human relationships according to variable cultural proximities. [17] Together, these theorists reveal how these transitional zones meet both universal and culture-specific needs. Furthermore, Bernard Rudofsky, in his exploration of vernacular architecture, argues that the organization of intermediate spaces in traditional architecture, particularly in Morocco, reflects these social and cultural concerns. [18] In continuity with Hall and Lefebvre, he explains that spaces such as thresholds and courtyards are not merely passageways but are designed to balance collective use and individual privacy. Serge Chermayeff and Christopher Alexander follow this line of thought by introducing a hierarchy of spaces from public to private, showing that intermediate spaces, such

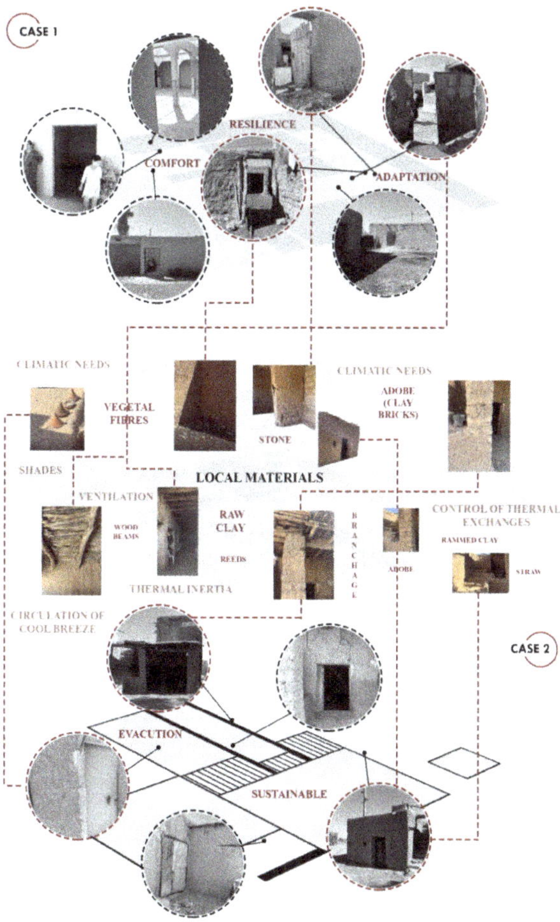

Fig. 3 : Intermediate space and local materials - cases 1 & 2

as courtyards or thresholds, are essential for managing tensions between community life and private life.[19]

Building on these ideas, Jan Gehl emphasizes the importance of the human dimension of space as a whole, noting that intermediate spaces, such as the narrow alleys and patios of Moroccan medinas, are privileged places for social interactions. [20] While Gehl places more emphasis on emerging social dynamics in public and semi-public spaces, he nevertheless aligns with other theorists by insisting on the quality of these spaces as catalysts for meaningful human interactions.

Amos Rapoport explores the cultural determinants that influence architectural form, highlighting intermediate spaces as reflections of local social and cultural practices. By doing so, he enriches previous approaches by demonstrating how courtyards and patios embody social values such as sharing and the preservation of privacy. [21] Gaston Bachelard's phenomenological approach complements this analysis by focusing on the emotional perception of space. [22] His concept of topophilia, or attachment to places, helps understand that intermediate spaces are not only functional but also carry a strong psychological and cultural symbolism, connecting Rapoport's reflections on the cultural dimension of these spaces with those of Gehl and Hall on their social significance.

The work of Françoise Navez-Bouchanine makes a significant contribution to the analysis of Moroccan housing and the social dynamics that shape it. One of her major analyses concerns the appropriation of intermediate spaces, such as courtyards, patios, and alleys in traditional Moroccan neighborhoods. These spaces, situated between public and private realms, play a fundamental role in organizing social interactions and preserving local cultural identities. Bouchanine shows that residents continuously reinvest in these spaces to meet their social and economic needs. Her analysis enriches previous reflections by highlighting the resilience and flexibility of these spaces in response to modern urban realities. [23]

These intermediate spaces ensure cultural continuity by demonstrating their capacity to adapt to new urban realities while preserving social practices deeply rooted in tradition. Indeed, these reflections show that intermediate spaces are not merely passage zones but essential elements of both architecture and social life. They represent a balanced intersection between public and private space, between proximity and distance, while ensuring the continuity of cultural practices in a constantly evolving urban context. As spaces of social, climatic, and cultural regulation, they remain a relevant model for sustainable, human-centered architecture.

Intermediate Spaces as Vectors of Interaction

The vernacular architecture of Morocco, particularly in the Taroudant region, reflects deeply rooted social and cultural dynamics, with a spatial organization that embodies local values and practices. Intermediate spaces, such as thresholds, patios, courtyards, and terraces, are designed to structure social interactions and preserve privacy according to cultural norms. These areas are crucial for organizing the gradual transition between public and private spaces while modulating social interactions based on the principles of spatial hierarchies.

Materials Research Proceedings 47 (2025) 69-80 https://doi.org/10.21741/9781644903391-8

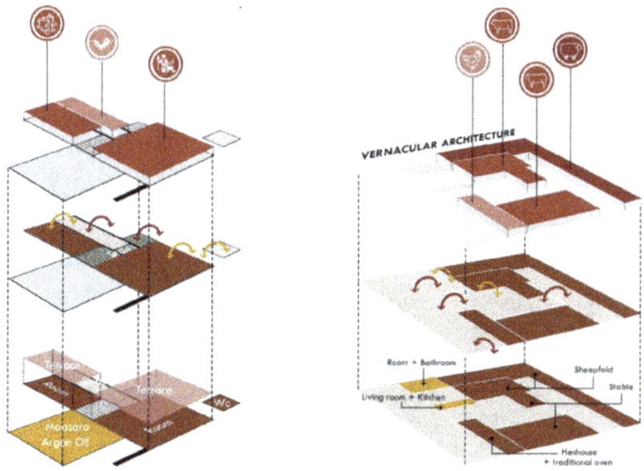

Fig 4: Intermediate space , typology and function – Cases 1 & 2- Author

These spaces are not merely functional architectural elements but also social mediation sites. Sociologist and urban planner A. Deboulet describes these areas as essential zones in vernacular habitats, facilitating interactions between interior and exterior while preserving residents' privacy and cultural norms. These spaces extend beyond a simple passage function: they are transitional areas that modulate proximity relationships. [24] Deboulet's work aligns with broader reflections on proxemics and social space dynamics. Drawing on the analyses of Edward T. Hall, she emphasizes the importance of these intermediate spaces as catalysts for social interactions. In Moroccan culture, concepts such as *hchouma* (honor) and *hijab* (modesty) dictate the use of these spaces, which simultaneously reinforce social bonds and protect residents' privacy.

In the two houses studied, intermediate spaces fulfill significant social and functional roles. The design of these areas embodies a deep cultural emphasis on social organization and privacy maintenance. Thresholds, for instance, serve as boundary markers between public and private realms, managing the flow of visitors and activities. Men's gatherings typically occur near the main entry or in living rooms, separated from areas where women carry out daily tasks, thus respecting cultural norms around gender and privacy.

These spaces not only facilitate neighborly interactions but also serve as social filters, regulating access and maintaining appropriate proximity between residents and guests. Christopher Alexander's patterns, specifically Patterns 112, 127, and 140, illustrate how spatial transitions can promote harmonious relationships within the habitat. The skifas and entrance thresholds are pivotal in articulating social relationships, enabling formal reception while preserving domestic privacy.

Women often engage in household and social activities within these intermediate spaces. Inner courtyards or patios become hubs for preparing food, crafting, and socializing with close friends and family. The multifunctional terrace serves as an extension of the living space, doubling as a communal area where women can dry agricultural products, such as argan nuts, thereby supporting traditional livelihoods. Terraces and courtyards extend beyond the immediate family circle, facilitating community interactions where neighbors gather for conversations or mutual assistance. Their integration into the social structure underscores the communal aspect of Moroccan culture, fostering a sense of belonging and collective responsibility. These spaces are also settings for family meals, celebrations, and shared moments, particularly during festivities like Eid al-Adha, where they play a central role in cultural practices and social values.

Materials Research Proceedings 47 (2025) 69-80 https://doi.org/10.21741/9781644903391-8

The design and use of intermediate spaces in Moroccan homes thus reflect a cultural continuity that balances the need for social interaction with the preservation of family privacy. They uphold traditional practices while adapting to modern requirements, maintaining a strong connection to the cultural heritage of the region.

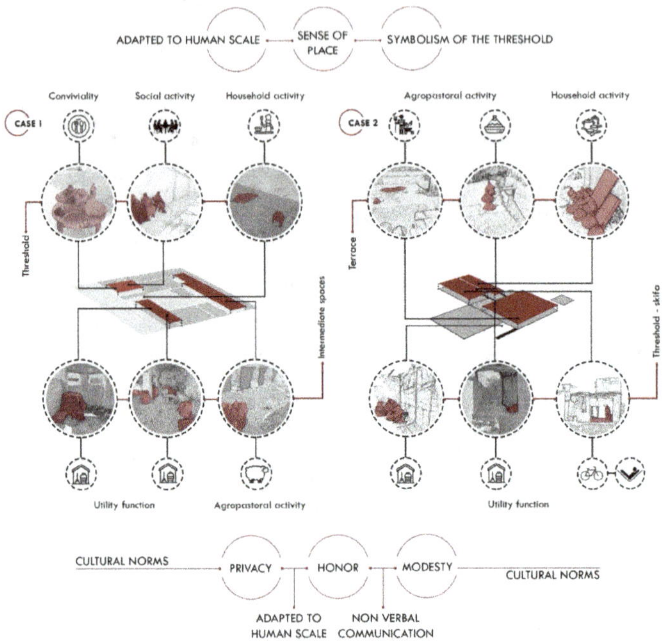

Fig 5: Intermediate space , Uses – Case 1 & 2 – Author

While sociocultural aspects are central, it is also essential to recognize the influence of climatic conditions on these transitional spaces. Architectural solutions developed to accommodate social dynamics often coincide with adapted climatic strategies. This creates a close relationship between cultural, social, and environmental issues. Such synergy allows for the design of spaces that promote residents' well-being and the preservation of vernacular practices while responding to local climatic constraints.

Conclusion - Towards New Tensions of Modernization

The findings of this study underscore the multifaceted roles of intermediate spaces in Moroccan vernacular architecture, where these zones facilitate social interactions, uphold privacy, and reflect respect for family and community structures. However, the forces of modernization and urbanization exert pressure on these spaces, challenging them to adapt while retaining their cultural significance. This tension between tradition and functionality is a recurring theme, echoing Françoise Choay's insights on cultural preservation. The key challenge lies in reconciling the need to preserve cultural practices with the necessity of adapting these spaces to meet contemporary social and environmental demands. An in-depth analysis reveals that these intermediate spaces serve not only as crucial elements in the spatial organization of traditional Moroccan dwellings but also as sustainable solutions for thermal comfort through passive strategies such as natural ventilation and thermal inertia. By utilizing local materials and promoting collective use, they ensure a sustainable approach aligned with both climatic needs and social practices. The case of Sebt El Gardan illustrates how these spaces are far more than mere architectural features, they are active participants in the preservation of local traditions while addressing modern climatic challenges.

This study confirms that the resilience of these spaces lies in their ability to adapt to both societal transformations and the impacts of climate change. As modernization advances, these traditional elements offer relevant responses by integrating passive strategies with contemporary uses. Yet, the modernization of materials and construction techniques necessitates a balanced approach, ensuring that the unique characteristics of these spaces are preserved while exploring innovative solutions to current ecological issues. Thus, a contemporary reinterpretation is essential to address new climatic constraints while maintaining the cultural integrity of these spaces. Further research is needed to refine architectural strategies, enabling a sustainable future that honors the heritage of vernacular practices in the face of global change.

References

[1] Team 10. *Team 10 Primer*, édité par Alison Smithson. Cambridge: MIT Press, 1968

[2] Van Eyck, A. *The Child, the City, and the Artist*. Amsterdam: Stedelijk Museum, 1962

[3] Smithson, A., & Smithson, P. *The Shift: Architecture After the Functionalists*. London: Academy Editions, 1974

[4] Lynch, K. *The Image of the City*. Cambridge, MA: MIT Press, 1960

[5] Norberg-Schulz, C. Genius Loci: Towards a Phenomenology of Architecture. New York: Rizzoli, 1980

[6] Lebois, V. Espaces Intermédiaires et Habitat Social: Typologies et Usages dans les Quartiers Parisiens. Paris: Éditions Parenthèses, 2016

[7] Rabhi, P. *Vers la Sobriété Heureuse*. Paris: Actes Sud, 2010[8] Fathy, Hassan. *Architecture for the Poor: An Experiment in Rural Egypt*. Chicago : University of Chicago Press, 1973

[8] Oliver, Paul (éd.). *Encyclopedia of Vernacular Architecture of the World*. Cambridge : Cambridge University Press. 2006

[9] Norberg-Schulz, Christian. *Genius Loci: Towards a Phenomenology of Architecture*. New York: Rizzoli, 1980

[10] Alexander, C. *A Pattern Language: Towns, Buildings, Construction*. Oxford University Press, 1977

[11] Rudofsky, Bernard. Architecture Without Architects: A Short Introduction to Non-Pedigreed Architecture. University of New Mexico Press, 1964

[12] Heschong, Lisa. *Thermal Delight in Architecture*. MIT Press, 1979

[13] Lefebvre, Henri. *La Production de l'Espace*. Paris : Anthropos, 1974

[14] Edward T. Hall détermina expérimentalement l'existence de quatre distances chez l'humain, le passage de l'une à l'autre étant marqué par des modifications sensorielles : ce sont les distances intime, personnelle, sociale et publique. Chacune des distances comporte deux modes, le proche et le lointain.*

[15] Hall, Edward Twitchell. *The Hidden Dimension*. New York : Doubleday, 1966

[16] Rudofsky, B. Architecture Without Architects: A Short Introduction to Non-pedigreed Architecture. New York: Museum of Modern Art, 1964

[17] Chermayeff, S., & Alexander, C. Community and Privacy: Toward a New Architecture of Humanism. Anchor Books, 1963

[18] Gehl, J. *Life Between Buildings: Using Public Space*. Van Nostrand Reinhold, 1971

[19] Rapoport, A. *House Form and Culture*. Prentice-Hall, 1969

[20] Bachelard, G. *La Poétique de l'Espace*. Presses Universitaires de France, 1957

[21] Navez-Bouchanine, Francoise. *L'urbanisme marocain à l'épreuve du temps*. Casablanca : Editions Le Fennec, 1997

[22] Deboulet, Agnès. *Villes et sociétés : l'ordinaire du politique*. Karthala, 2006

[23] Choay, Françoise. *L'Allégorie du patrimoine*. Paris : Éditions du Seuil, 2009

[24] V. Biau: *Architecture et transition climatique : Enjeux et pratiques contemporains,* Paris: Éditions Hermann (2012).

[25] J. Chelhod: *Les structures du sacré chez les Arabes,* Paris: Maisonneuve et Larose (1986).

[26] J. Dethier: *Habitat de terre: Tradition et modernité,* Paris: Centre Pompidou (1981).

Materials Research Proceedings 47 (2025) 81-90 https://doi.org/10.21741/9781644903391-9

Endangered building cultures in Morocco: The case of Tazotas in Doukkala region

Samia NAKKOUCH [1,a], Mohamed FILALI[1,b]

[1] Rabat National School of Architecture (ENA Rabat), Morocco

[a]s.nakkouch@enarabat.ac.ma, [b]m.filali@enarabat.ac.ma

Keywords: Tazotas, Dry Stone Building Culture, Endangered Local Know-How, Transformations, Anthropogenic Factors, Morocco

Abstract. Moroccan vernacular architecture reflects the diversity of cultural, social, and environmental specificities of local communities and regions. Authentic local know-how utilizing various techniques such as rammed earth, adobe, and stone building is contextually developed by local traditions, materials, and climate conditions. The survival of these traditional building cultures, in a contemporary context of modernization of construction techniques, depends on several anthropogenic factors. The replacement or extinction of these traditions would imply a significant loss of cultural heritage, architectural diversity, and local knowledge concerning sustainability and environmental adaptation. The Doukkala region known for its vast agricultural plains, situated in the east-southern countryside of Casablanca-Settat region, has witnessed the development of a unique dry stone construction technique known as Tazotas. The remaining structures featuring this specific architecture are currently facing constant transformations and destruction due to inhabitants' practices and adaptations, which endanger the remaining traces of this unique local vernacular tradition. This study aims to explore the characteristics of this exceptional building culture identifying practices and transformations likely to conserve or imperil its remaining landmarks. The results are based on an empirical study that involved extensive fieldwork, surveys, material samples testing, and interviews with residents and Tazotas owners.

Introduction

In Morocco, studying vernacular built heritage has gained interest during the last two decades, aiming to rediscover the richness of regional expressions in terms of building cultures and local know-how [1]. Traditional building techniques are shaped by geographic configurations, climate conditions, available materials and social traditions and practices. These parameters influence the inscription of a community in a space, which evolves over time to a distinct character of a place that embodies the cultural heritage and identity of a territory [2].

The stakes of studying vernacular built heritage also involve an environmental dimension. It provides insights into practices and climate adaptations rooted in local construction cultures. In an era of uncertain climate changes and increasing risks, investigating local and sustainable solutions through vernacular heritage offers valuable insights into resilience, resource efficiency, and environmental balance. The minimized ecological impact while maximizing the use of available materials and local knowledge is highly sought-after in contemporary architecture studies, fostering sustainability and climate-responsive design in modern construction [1, 2].

The evolution or survival of vernacular architecture is influenced by complex factors, including globalization, urbanization, and economic and cultural changes. The impacts of industrialization and modernity on traditional building cultures have been widely criticized, highlighting how standardization of construction often undermines local knowledge and craftsmanship, leading to a disconnection between people and their environments. With vernacular landscapes increasingly facing hybridization and the invasion of standardized construction techniques and materials, the erosion of local identity has resulted in a homogenized built environment, where architecture no

Materials Research Proceedings 47 (2025) 81-90 https://doi.org/10.21741/9781644903391-9

longer reflects unique cultural and social contexts Therefore, learning from vernacular architecture as a contextual response to local constraints involves understanding deeply rooted configurations and adaptations to specific geographic, cultural, and climatic conditions, emphasizing sustainable practices [3, 4].

The Moroccan landscape rich in vernacular heritage, encompasses various techniques and materials such as rammed earth, mud brick, stone masonry, and straw. These traditional materials and methods have been gradually replaced due to the impacts of industrialization, the rise of mass-produced materials, along changes in social and cultural norms. This shift has significantly altered the relationship between social practices and building culture, prioritizing uniformity and efficiency over cultural specificity. Therefore, acknowledging these pressures of modernity and urbanization is essential to comprehend the transformations affecting how local communities engage with their heritage and building traditions, and the anthropogenic factors contributing to the decline of built heritage [5].

Considering these factors provides a holistic view of contexts that shape how communities interact with their architectural heritage. The conservation and revival of local practices and landmarks has become a global concern, aiming to reconnect building culture with social and cultural practices that are deeply intertwined. Addressing these influences is also essential for fostering a sustainable relationship between modern development and traditional practices [6, 7].

The case study of Tazotas illustrates one of the traditional building cultures that is in decline due to contemporary challenges. As a unique dry-stone architecture in Morocco, it is still challenging times and shifts of social and cultural norms, undergoing various transformations. The remaining vestiges of this authentic heritage are endangered by the rupture in the transmission of the local know-how through generations, which complicates the process of conservation.

The article highlights the Tazota's authenticity of Tazota as a building culture and adaptation to local materials and climate conditions [8]. The methods of investigation are both quantitative and qualitative, including observations, surveys, and technical drawings, as well as laboratory tests on samples collected at the investigation site at the materials lab in the National School of Architecture in Rabat. Interviews have also been conducted with the local population, especially those owning Tazotas, to investigate transformations and adaptations made by residents that are rarely qualitative.

Tazota: a unique dry stone building culture

The name "Tazota" refers to an authentic dry stone building culture implemented in the countryside of Casablanca-Settat region, exclusively in the plains of Doukkala, 20km from the city of El Jadida, toward both Azemmour and Marrakech (Figure 1).

Tazotas are prominent features in the rural landscape of Doukkala. They are identified across a vast area in the region that encompasses several communes, extending from Oulad Rahmoune in the north to Oulad Aissa in the south, and reaching Oulad Hamdane in the east[1].

[1] The implementation scope is defined through the classification of satellite image of the region with Qgis engine. Using a pixel based method, it has detected over 700 tazotas in the Doukkala plains that where corrected, adjusted and verified in the field. The previous studies on tazotas present approximately the same scope, but less extended [10]. The results of classification show two separate concentration areas (over 10 tazotas/1km) in the north and the south of the scope. Compared to previous delimitations, there is a discontinuity in the concentration area. To note a disappearance of tazotas especially around the national road RN1. This may be interpreted as a consequence of the transformations along the RN1 road which is pressured by the urbanization process.

Materials Research Proceedings 47 (2025) 81-90 https://doi.org/10.21741/9781644903391-9

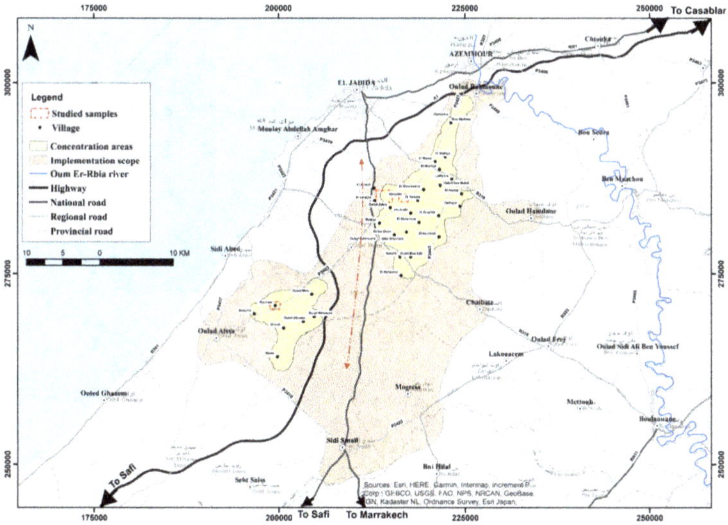

Figure 1: Implementation areas of Tazota in Doukkala plains, Morocco (Author)

The presence of Tazotas in the rural landscape of Doukkala is notable due to their distinctive shape and dimensions, resembling stone mounds. Previous studies [8, 9, 10] have proposed various hypotheses regarding the origin of the name "Tazota" and its construction date, which remains uncertain. The most accurate suggests that Tazota or Tazudea has its origins in Berber culture and dates back at least a century, meaning an inverted bowl in the Berber language. Another hypothesis relates the emergence of Tazotas to the colonial period when individual ownership of farmland was imposed by the regime. This theory is based on geographical studies conducted during the same period that didn't mention the existence of Tazotas in the region (studies like Villes et tribus du Maroc, Les Doukkala of Michaux Bellaire, and Etudes Rurales, Idées et enquêtes sur la campagne marocaine of Paul Pascon), suggesting that they appeared until the beginning of the twentieth century, and their name is a deformation of the French idiomatic expression "Tas au tas". Other theories relate the emergence of Tazotas to the Portuguese era, as "casitas" meaning small houses, highlighting similarities in territories that knew Portuguese occupation. The hypothesis that matches with the native's interviews suggests that the first Tazotas were constructed in the nineteenth century, as an initial form of settlement for a nomadic population in a fertile region where agriculture prospers [10]. Some inhabitants suggested that it was also a distinctive sign at that time between the owners of the lands who lived in Tazotas and tents, and the outsiders like migrants and farmworkers who lived temporary in Noualas[2], before the development of extended habitations.

[2] Noualas are temporary habitations made of reeds and thatch, for one person or a family that have migrated or have an unstable situation, or seasonal work on a farm.

Materials Research Proceedings 47 (2025) 81-90 https://doi.org/10.21741/9781644903391-9

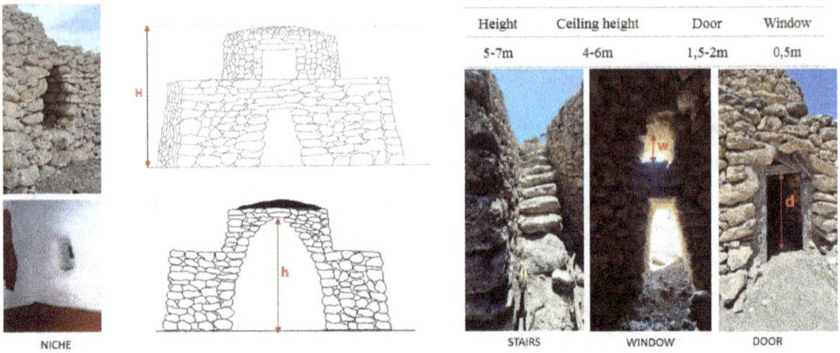

Figure 2: Shape and components of typical Tazota (Author)

With a spherical shape (sometimes trapezoidal), Tazotas take various configurations in terms of volume and composition, including frustums of cones (common), or layered frustums of cones, complete with doors, windows, stairs for access, and niches when needed (Figure 2). These variations reflect the unique touch and expertise of each builder.

Commonly used as shelters and storage areas for agricultural and pastoral activities. The functions of Tazotas have evolved to meet the changing needs of the local population. The first Tazotas were built as dwellings and shelters for landowners [9, 10]. A typical enclosure would contain at least two Tazotas for habitation and a Toufri for storage[3]. The introduction of other construction techniques and changing family structures led to the emergence of additional forms of housing, to which the Tazotas became annexed. As a result, different implementations are identified.

Whether a Tazota is annexed to a farmhouse or is built in the agricultural compound separately, its function has varied over time according to the farmer's needs (Figure 3).

Figure 3: Tazota implemented in the farmland (1) Tazota annexed to the residential compound (2)

The common shape represents a Tazota annexed to the farmer's habitation. In this case, it is used as a storage room for provisions and domestic furniture (Figure 4).

A unique site known as "the seven Tazotas" encompasses most of the variations of Tazotas.

It encompasses seven Tazotas in a rare configuration that are well conserved and protected by an enclosure. The entrance to the site provides access to a large courtyard. Two of them are terraced

[3] A Toufri refers to an underground gallery approximately three meters deep that connects one or more spaces by a corridor at the same subterranean level. Access is provided by a slope or stairs. Its rectangular shape covers the stairs using the same construction technique as the Tazota. It is used for the preservation of grain and straw. According to Sergio Gnesda, a researcher in vernacular architecture, the term Toufri in Berber means "place of preservation"[8].

Materials Research Proceedings 47 (2025) 81-90 https://doi.org/10.21741/9781644903391-9

separately from the five others, which are used as shelters for livestock. In tems of shape, they feature layered frustum of cones.

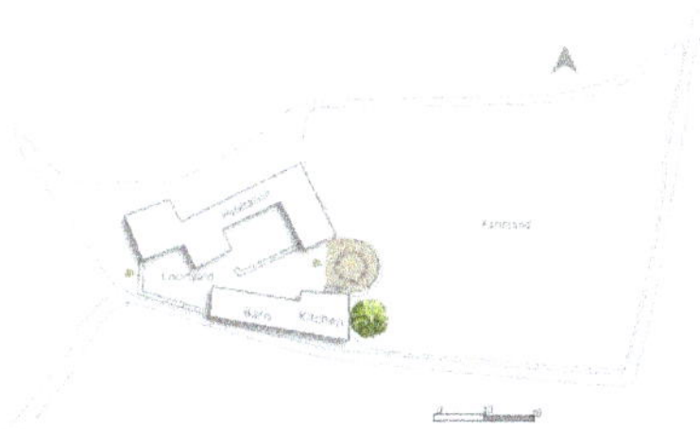

Location	:	Douar Lalla Merzaka, Commune El Haouzia; Province El Jadida
Orientation	:	West
Surface	:	42,77m² (total) 23,4 m² (usable)
Height	:	4,5m
Comprenents	:	Door, Window, Stairs
Use	:	Storage of family provisions

Figure 4: Example of the common type of Tazotas, douar El Hrabza (Author)

The five others feature the common shape with a frustum of cone. One of them is flat with a chicane entrance. The five Tazotas have common stairs instead of one for each (Figure 5). They were used as hay and crop storage areas[4].

Location	:	Douar Rouahla Commune Ouled Hcine Province El Jadida
Orientation	:	Southwest / southeast
Dimensions	:	Variable - Tazotas frustum of a cone - Tazotas with layered frustums of cones
Components (variations)	:	Doors, Windows, Double window Stairs / Double stairs / common stairs Niche / additional spaces
Use	:	Touristic site annexed to a guesthouse

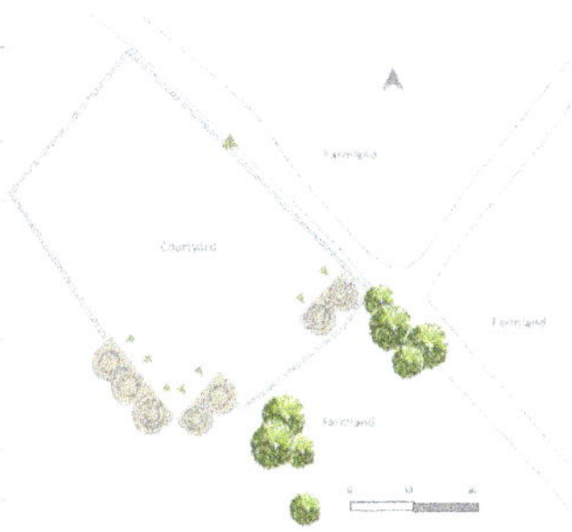

Simple tazota, with niche, without cone

Tazotas with layered frustums of cones

Commun stairs

Figure 5: The seven Tazotas, representative of Tazotas variations, douar Rouahla (Author)

[4] According to the seven Tazotas owners, the site of the seven Tazotas once functioned as a farm with all its components, based on what they remember being told by their ancestors. the five tazotas used to be once living rooms, with one of them having a chicane entrance that were used as a kitchen Over time, it only partially served as a barn and silo for crops, and the workers occasionally took shelter there.

Materials Research Proceedings 47 (2025) 81-90 https://doi.org/10.21741/9781644903391-9

An original constructive technique

Tazota features a dry-laid structure made of natural stones, without the use of mortar or binder. Stability is ensured through a subtle arrangement of stones supported by smaller wedging stones.

The construction process of Tazotas starts with clearing stones from fields, and then sorting and stacking them based on size. Larger stones serve as the base for the wall, as well as for lintels and corner blocks while wedging stones enhance the stability of the walls for the next layer.

The desired size and shape of Tazota are defined at the beginning of the process by delineating a circle or square on the ground. Although Tazotas do not require foundations, a superficial excavation of 10 to 30 centimeters is conducted before laying the first course of the walls.

The construction of the walls begins with laying the first course of the wall starting from the diameter of the Tazota. A stairwell with a regular slope is built to access the terrace of the first cone and to facilitate the construction of the next one.

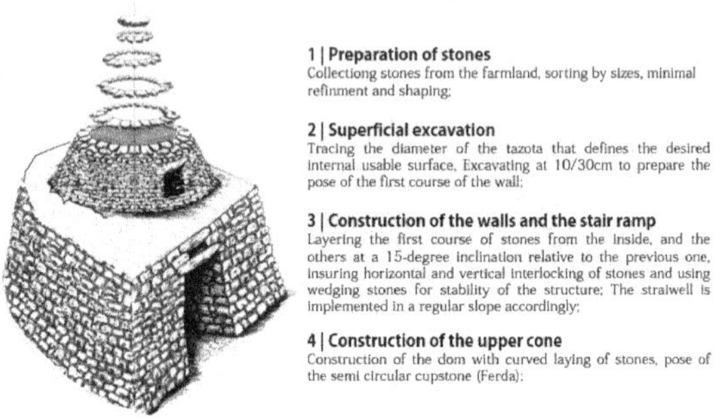

Figure 6: Construction process of a Tazota (Author)

The wall either extends vertically until it reaches an edge at which point it curves, or the curve may begin at the base stone, with an inclination of fifteen degrees in outward to lay the next course of the wall. This design is intended as a protective measure in case the Tazotas are damaged, as it prevents collapsing inward. Additionally, the vertical and horizontal arrangement of the stones helps prevent water infiltration into the Tazota. The stones decrease in size as they curve to form the dome (Figure 6).

The upper cone features a batter of concentric rings becoming narrower and tighter toward a final capstone with a diameter of 40 to 50cm in diameter, named "Ferda". The Tazota is sometimes covered with a thin layer of straw and rammed earth or white clay (Biada), which is applied as roofing or internal plastering[5].

Constructing a Tazota requires at least one maâlem who is often an experienced local farmer mastering the technique of Tazota, supported by two additional workers. The construction duration can range from 8 to 10 days [10].

An adaptation to local materials and climate

The use of stone material is widespread in the Doukkala plains, considering the abundance of a calcareous substrate that surfaces the ground. It is utilized not only in Tazotas but also for enclosures, barns, and farmhouses, which characterize the region's rural architecture and landscape [11].

The Doukkala plains are located in the geological unit known as the Moroccan Meseta, consisting of a series of low plateaus that slope toward the coastline, situated between the central massif and the ocean.

[5] Ferda and Biada are the common jargon used by locals when describing Tazota building process.

The altitudes in this area range around 300 meters, while the coastal region extends 10 to 40 kilometers inland. The area contains calcareous and gypsum formations along with a luminous cover, with thicknesses reaching up to 30 meters [12].

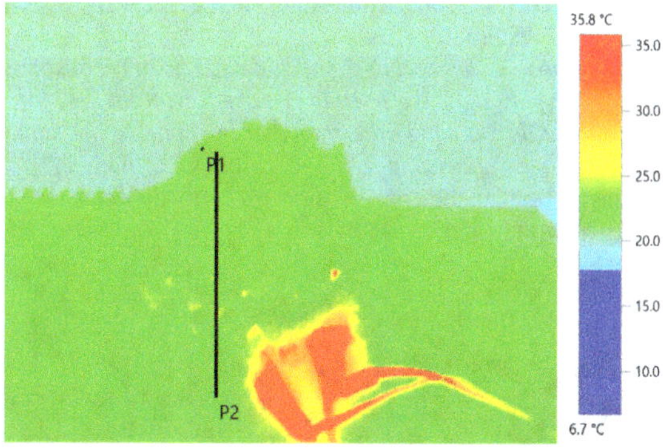

Figure 7: Thermographic image of Tazota (Author)

The Tazota building technique is adapted not only to local materials but also to local climate conditions, through its particular shape combined to the thermal and physical characteristics of the stone material.

Their semi-spherical dome design, typical of hot and arid climates, plays a crucial role in maintaining thermal comfort inside. The rounded shape and thick stone walls of the Tazotas provide a significant thermal advantage in the humid continental climate of Doukkala[6].

Their design optimizes heat dissipation while ensuring adequate thermal inertia, guaranteeing comfortable living and storage conditions despite external temperature fluctuations[7].

This enhances thermal comfort for occupants, particularly in hot and arid or semi-arid regions where substantial temperature variations occur between day and night.

Tazotas between decline and revival

In the last two decades, the deterioration of Tazotas heritage accelerated significantly. Changes in local conditions related to social, economic, and cultural values have drastically affected the relationship with this unique heritage.

Interviews conducted in the area of investigation highlight that the majority of remaining Tazotas are abandoned, left with no particular function, either in a conserved state or partially in

[6] The Doukkala plain falls within a hot semi-continental climate, characteristic of northern Morocco AND north of the Atlas Mountains. Notable differences in average daily temperature amplitude are highlighted by temperature curves, which can reach 40°C. The "chergui," an easterly wind originating from Saharan depressions, often blows across the region. Although temperatures are comparable to those of Morocco's arid zones, the air's humidity factor makes direct comparisons difficult. The ground is warmer than the ambient air, with a difference of 2/3 degrees in winter and up to 10 degrees in summer. The interior of the plain shifts between arid and semi-arid conditions, with a marked seasonal climate contrast [11,12].

[7] The results of the thermographic analysis conducted in the laboratory of ENA Rabat, utilizing samples of Tazotas stone collected in the field and a prototype of a scaled-down model to evaluate thermal performance and visualize temperature differences on both external and internal surfaces. Thermography is particularly well-suited to Tazota constructions, as the absence of plaster exposes the stone directly to temperature variations. This technique is especially useful for dry stone structures, where thermal bridges at unfilled joints can influence the thermal stability of the Tazota. The figure displays an infrared thermography image representing temperature distribution on a surface using color gradients. The temperature scale in the image ranges from 6.7°C to 35.8°C. Region 1 shows elevated temperatures, around 35°C, due to heat leakage through the entrance door. A significant thermal gradient is observed from the bottom to the top of the Tazota, likely related to material composition or structural elements (shape non-uniformity) affecting heat dissipation. Stone has a relatively high thermal capacity, allowing it to absorb substantial thermal energy without a rapid increase in temperature. In Tazota constructions, the stone serves a dual purpose: it absorbs part of the solar energy, reducing the direct impact of heat inside, and it slowly redistributes this heat, creating a natural form of thermal regulation.

ruins[8]. This decline is primarily due to the loss of local know-how that held inhabitants to intervene in damage and restore it.

"There is no maallem left in the region capable of reconstructing Tazotas. The last maallem we know of died years ago. We tried to restore it with the help of some construction workers but in vain. They couldn't arrange the stones of the cone in the same way. The Tazota is not repaired now, and water is infiltrating it. We're keeping it although"[9].

The disruption in the transmission of construction know-how is linked to various factors, including economic and social ones.

Tazotas integrated in housing in douar El Haddada (Author)

Abondonned Tazota in ruin in douar El Herbaza (Author)

Tazota as a touristic site in douar Rouahla (Author)

Figure 8: Tazotas at different states (Author)

With the rural exodus of the young generations and the communities shifting towards urbanized lifestyles, Tazotas became at times impractical for some inhabitants, after they had met for long specific rural needs. Being no longer maintained, repurposed, or inhabited, their deterioration is accelerated [13]. Land division issues between heirs have led in many cases to the abandonment of Tazota and the recycling of its stones.

Some of the other appropriation observed are the restitution of the residential function, especially in the configurations where Tazota is annexed to the main residential enclosure. Some families have transformed the internal space of Tazotas into living or guest rooms, appreciating the coolness of the atmosphere inside, plastering the interiors with Biada (white clay on site), and adding furniture. These forms of appropriation have also highlighted some non-qualitative practices, such as the external plastering of the Tazota with cement mortar, devaluing the character of the dry stone architecture. The habitation function is either permanent or seasonal, in the case of the families having second residencies in the region.

For certain interviewees, as the cultural identity shifts with the adoption of other social and cultural norms, Tazotas are no longer seen as a main cultural landmark. For some owners, they are "irrelevant" to their contemporary needs. The recycling of Tazota stones into farm enclosures or new stone buildings was also observed. Some owners also took an interest in selling them as a source of income due to their difficult conditions. Tazotas are quantified in terms of stone weight and sold per ton, for the amount of 4000-5000 dirhams.

[8] The investigation scope covered and collected informations about fourthy tazotas in the region.

[9] Interview with H.A., inhabitant of douar Bou Mediane, commune of Haouzia, June 2024.

Materials Research Proceedings 47 (2025) 81-90 https://doi.org/10.21741/9781644903391-9

"We inherited this farm from my grandfather who also bought it at that time with its Tazota already constructed. I used to store my harvests in the Tazota where I also relaxed while working in the fields. Having aged, with my children and grandchildren living far away, and due to the accident I had recently, I'm not able to work in farming anymore. I sold the Tazota to another inhabitant in exchange for 4000 dirhams. He will come next week to take the stones. He wants to use the stones for his habitation, he doesn't want to build with concrete"[10].

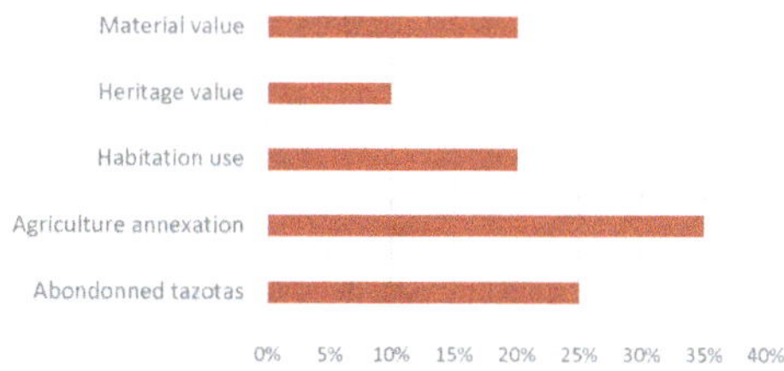

Figure 9 : Identified uses/ appropriation of Tazotas in the area of investigation (author)

Another form of appropriation that has emerged consists on the promotion of the cultural value of Tazota heritage through private projects and businesses (rest stations, guest houses etc.). The widely known example in the region is the case of the seven Tazotas located in douar Rouahla, whose owners chose to invest in the cultural value of the local heritage (Figure 8).

The interpretation of such transformation tendencies shows that 45% of the Tazotas investigated are abandoned, sold or recycled, which is a relevant indicator of how endangered the remaining traces of Tazota heritage are (Figure 9).

The other forms of appropriation, such as investment in the cultural value of heritage or residential comfort, are tight to a particular social configuration that has a role in the emergence of these practices. The interviews highlight a certain social status, such as notable families who chose to invest in their family heritage, or those with stable living conditions, with a certain level of education. The proximity of other networks such as local councils or cultural administrations has influenced their choice of investment or raised their awareness of heritage value. Non-conservative practices are generally imposed by difficult living conditions, a lack of expertise, and the inability to keep the Tazota functional, which leads to recycling, selling for material value or abandonment.

The remaining Tazotas are functional and annexed to pastoral and agricultural activities as long as they are standing, and farming activities are maintained.

Conclusion

The case of Tazotas demonstrates an original building culture has long been rooted in the rural life and landscape of the Doukkala region. As a contextual response to local needs and the environment, it is imperiled today due to many anthropogenic factors that have transformed the relationship between local communities and their built heritage. Understanding the changes in cultural practices that influence the relation of communities to their built heritage and environment and how they interact with it, helps identify the factors and processes that contribute to these transformations.

This transformed relation to the built heritage has influenced its deterioration in several ways. As local building materials and techniques are being replaced with industrial materials, it disrupts the traditional relationship between people and their environment, eroding the practice and knowledge of vernacular construction. The relationship between social practice and building culture has thus drastically changed, prioritizing uniformity and efficiency over cultural

[10] Interview with owner of Tazota in douar Guedadra, commune El Haouzia, June 2024;

specificity, and leading to the loss of local know-how. These changes in cultural and social values and norms as rural communities shift towards urbanized lifestyles, imperil the survival of vernacular heritage. If no longer maintained, repurposed, or inhabited, it accelerates their decay. The cultural detachment from built heritage has diminished the relevance of the Tazota as a main component of the daily life of local communities.

On the other hand, renewed forms of appropriation and uses of Tazotas inform about the active role of local communities in reviving them, as they readapt these structures to meet contemporary needs despite the difficulties of conservation. This process fosters a collective responsibility towards maintaining traditional practices and enhancing awareness of their historical value, ensuring that this heritage continues to be a vital part of their cultural landscape.

Conservation policies and practices rarely acknowledge the anthropogenic factors transforming the relationship between communities and their built heritage, which are essential to evolve from static and tangible interventions considering vernacular heritage as a mere landscape component to dynamic approaches that associate intangible attributes and incorporate more comprehensive dimensions of heritage values, especially those of social and cultural significance.

The revival of vernacular heritage as living expressions of community identity and collective memory depends on community involvement, which is gaining momentum in conservation strategies. Considering communities as local ambassadors of their heritage is likely to ensure that its cultural assets remain relevant and adapted in a contemporary context, to better support the resilience and sustainability of the vernacular traditions.

References

[1] Fédération MAJAL, Les traces des pratiques et de savoir-faire écoresponsables : architecture et urbanisme traditionnels au Maroc, Rabat, 126p, 2016.

[2] F. Kenneth, Prospects for a Critical Regionalism, Perspecta 20, 1983. https://doi.org/10.2307/1567071

[3] P. Frey (dir), Learning from vernacular: pour une architecture vernaculaire, édition Actes sud, 2010.

[4] H. Fathy, Construire avec le peuple, édition Actes Sud, 1999.

[5] ICOMOS, Charter on the built vernacular heritage, 1999.

[6] B. Rudofsky, Architecture without architects, 1964.

[7] I. Illich, Le genre vernaculaire. Seuil, 1983

[8] CERAV, Témoins d'architecture en pierre sèche au Maroc: les Tazotas et les toufris de l'arrière-pays d'El Jadida, Études et recherches d'architecture vernaculaire N°16,1996.

[9] Direction d'architecture, Les architectures régionales : région du centre, édition Okad, 244p.

[10] Direction d'architecture, Sur les traces de la Tazota, Cahiers d'architecture et d'urbanité, 244p. 2006.

[11] J. Goulven, La région des Doukkala dans: Annales de Géographie, t. 29, n°158. pp. 127-138, 1920. https://doi.org/10.3406/geo.1920.9179

[12] A. Michard, Elements de la géologie marocaine, Edition du service géologique du Maroc, Rabat, 1976.

[13] P. Oliver, Built to meet needs: cultural issues in vernacular architecture, Architectural press, 2006. https://doi.org/10.4324/9780080476308

Materials Research Proceedings 47 (2025) 91-97 https://doi.org/10.21741/9781644903391-10

Rural vernacular dwellings: Balancing heritage and contemporary influences. A study of the Tiylit commune in Tamazouzte, a rural commune within Al Haouz province

Houda ZERRAF[1,a *], Inna Yaya AISSATOU HAMADOU[1,b]

[1]National School of Architecture of Rabat, Morocco

[a]h.zerraf@enarabat.ac.ma, [b]innayaya@yahoo.fr

Keywords: Vernacular Architecture, Time, Change, Forms, Functions, Rural Dwellings, Morocco, Al Haouz Province

Abstract. This research on vernacular architecture highlights the crucial importance of "time" in the continuous adaptation to changes in socio-cultural, economic, technological, and environmental spheres. Examining the development of these structures provides a valuable understanding of their profound links to historical contexts. In Morocco, significant changes have occurred in both urban and rural dwellings, especially in recent decades. The rural settlement of Tiylit, situated in the Al Haouz province near the High Atlas Mountains, serves as an illustration of how rural vernacular dwellings have evolved. Field investigations uncover various adaptation strategies. Some houses remain intact and continue to fulfill their original purposes, while others have fallen into disrepair and been converted into livestock enclosures. Many dwellings have undergone hybrid construction, integrating new materials into older structures. These transformations are driven by social, economic, and technical factors, as well as a shift in residents' perceptions of comfort.

The Significance of "Time" in Vernacular Architecture Research

In the examination of vernacular architecture, the element of "time" plays a crucial role, as this architectural style continuously evolves and adapts. Dell Upton and John Michael Vlach (2008) proposed five key questions for investigating vernacular architecture, with "*How does it c*hange?" being particularly significant. The other inquiries address its definition, construction methods, functionality, and perception. [1]

Usually, studies of vernacular architecture concentrate on static forms and functions, often portraying these structures as immutable or ideally constant. This approach overlooks the fact that these forms are tied to specific historical contexts. Discussions about their evolution usually focus on decline and decay, emphasizing the pejorative image of change rather than exploring the underlying processes and causes of these transformations.

Scholarly works have shown considerable interest in understanding the mechanisms and reasons behind changes in vernacular architecture (Dell Upton, John Michael Vlach, 1986; Asquith & Vellinga, 2006; Carter & Cromley, 2005) [2]. These authors stress the significance of the "*time-place*" or "*historic-geographic*" context, which sets the framework for analyzing vernacular architecture. Carter and Cromley (2005) exemplify this concept by examining the California bungalow house type from 1880 to 1990. Initially, this form was relatively rare, allowing for the study of only a few examples. However, by 1920, the bungalow had emerged as the most popular new house type in the state—and indeed, globally—suggesting that a study commencing at this later date could encompass hundreds, if not thousands, of buildings. [3]

The historical aspect of vernacular dwellings is particularly significant, as it unveils changes in form, function, and usage over time. This evolution prompts a deeper investigation into how vernacular architecture adapts to new uses and practices while preserving links to traditional

Materials Research Proceedings 47 (2025) 91-97 https://doi.org/10.21741/9781644903391-10

elements and incorporating spatial innovations. Such exploration enhances our comprehension of the dynamic interplay between heritage and contemporary needs in adapting vernacular architecture.

Moreover, exploring how vernacular architecture responds to ecological, technological, and cultural changes offers valuable insights into the nature of these traditions and the processes that have led to their disappearance, adaptation, revival, or persistence over time. This is to question the common discourse that often contrasts tradition with modernity—a misleading dichotomy, as the tradition itself is a continuous and dynamic process.

Throughout history, Moroccan housing, both in cities and rural areas, has continuously adapted to various influences, including physical, institutional, and socio-economic factors, as well as advancements in technical fields of construction. These changes have been influenced by different historical periods and ruling regimes. However, in the early 20th century, particularly with the establishment of French and Spanish protectorates, the transformation of relationships between nature, society, and socio-economic dynamics accelerated significantly.

The 21st century has brought significant changes in these dwellings, most notably through the introduction of new building materials. The shift in household energy sources—from olive or argan oil to candles, and eventually to gas and electricity—coupled with access to potable water, has fundamentally altered both the functions and perceptions of comfort within these homes.

Rural architectural development is highly context-dependent, requiring a focus on specific case studies. We have chosen rural homes in the Haouz province for examination, an area that has generated significant discussion about traditional building materials and their earthquake resilience following the seismic event of September 8, 2023.

An Overview of the Field Study's Area

The dwellings within the rural settlement of Tiylit, situated in the province of Al Haouz, serve as a compelling illustration of the ongoing evolution of rural housing. Located in the municipality of Tamazouzte (Fig. 1), Tiylit comprises several sub-settlements, three of which have been subjected to detailed analysis (Fig. 2). Positioned at an altitude ranging from 300 to 600 meters at the foothills of the High Atlas Mountains, this region provides a rich context for examining how inhabitants adapt vernacular architecture in response to socio-economic and cultural changes.

Materials Research Proceedings 47 (2025) 91-97 https://doi.org/10.21741/9781644903391-10

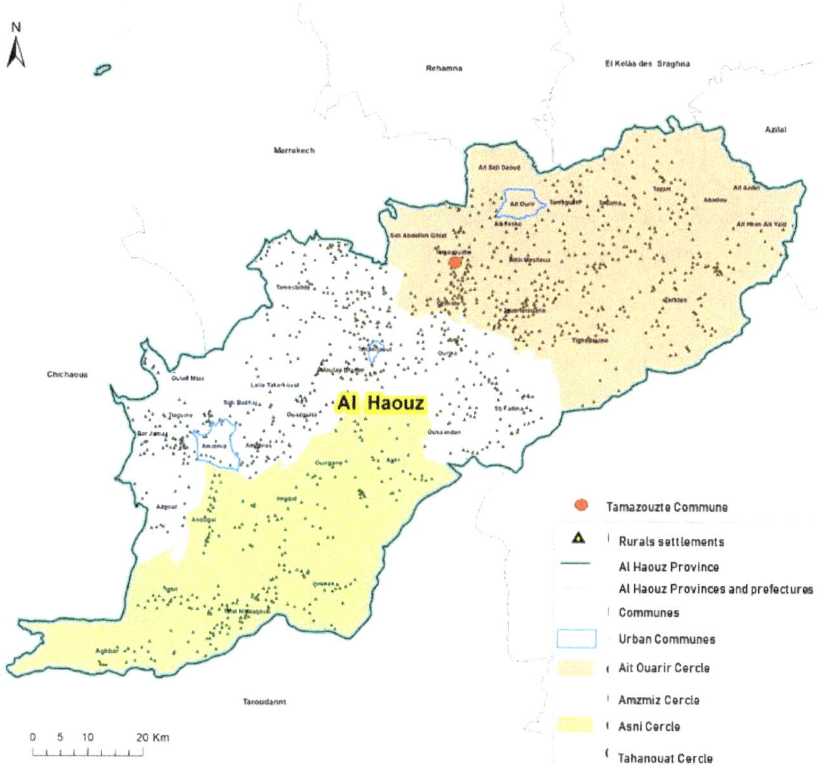

(Fig. 1): The location of the commune of Tamazouzte in the province of Al Haouz

The configurations of the three sub-settlements have experienced substantial transformations since the onset of the 21st century. Notably, a trend of densification has emerged, marked by the construction of new concrete houses often built alongside the original dwellings. Simultaneously, the reorganization of vernacular dwellings has occurred, accompanied by the construction of new structures in the spaces formerly occupied by these traditional homes.

Materials Research Proceedings 47 (2025) 91-97
https://doi.org/10.21741/9781644903391-10

(Fig 2): Evolution of the three visited sub-douars between 2003 (on the left) and 2024 (on the right

Main Formal Characteristics of Vernacular dwellings in the Study Area

Derouich A. and Melehi Y. (2008) distinguish between the building styles of Dern (high mountains) and Dir (foothills and plains). Dern's architecture is dense and predominantly made of stone, characterized by two-floor habitations where animals are housed on the ground floor and people live above. In contrast, Dir's architecture is less compact and utilizes earth-based materials such as adobe and rammed earth. Dir houses often incorporate a central courtyard [4]. While this categorization is useful for methodological purposes, it also highlights certain similarities between the two regions concerning construction materials, spatial organization, interior arrangements, and function. This is particularly evident in the area under study, where the traditional characteristics of Dir houses coexist with structures reminiscent of those found in Dern architecture. The homes examined in this research showcase a diverse array of building materials, including stone, earth, or combination of both. These buildings range from spacious houses with courtyards to those featuring upper-level rooms known as *Tamsrit*, typically reserved for owners or guests. Also, the new dwellings are made of reinforced concrete and generally consist of two levels (Fig. 3).

(Fig. 3): Photos of dwellings in the visited sub-settlements showcase both traditional structures built with local materials and the typical new concrete houses

Materials Research Proceedings 47 (2025) 91-97 https://doi.org/10.21741/9781644903391-10

The Evolution Scenarios of Rural Vernacular Architecture

Derouich A. and Melehi Y. (2008) emphasize that houses in Dir are typically built-in stages, expanding gradually as additional space is needed. Despite this incremental approach, homeowners usually envision the final form from the outset. The construction process can extend over 15 to 20 years or even longer [5]. In this case study, local narratives suggest that these dwellings evolve slowly, with progress influenced by both spatial requirements and resource availability. As a result, the building process may continue indefinitely, with some houses being as old as their oldest inhabitants.

Studies of the area's traditional architecture reveal a variety of adaptive techniques used to meet evolving needs and environmental factors. Some dwellings have been well-maintained, continuing to serve as complete residences or retaining specific traditional functions, such as food preparation and storage. Conversely, other structures have deteriorated and have been either abandoned or repurposed as livestock enclosures (Fig. 4). In certain cases, modern construction techniques, especially reinforced concrete post-and-beam systems, have been incorporated into the original structures, primarily impacting the exterior walls (Fig. 4).

(Fig. 4): a. A disused part of an old house transformed into a livestock enclosure (in the left). b. Integrati on of the reinforced concrete system with walls built from traditional materials (in the right)

In the majority of dwellings, all scenarios can intertwine within the original house (Fig. 5). As a result, one can find well-maintained areas that reflect careful upkeep, alongside sections that have fallen into neglect, and others that have been completely dismantled and replaced with new structures. This patchwork of preservation and transformation illustrates the evolving nature of the architecture in the region. Furthermore, even the various adaptations of vernacular forms seem to require a significant period of time to develop into new architectural expressions.

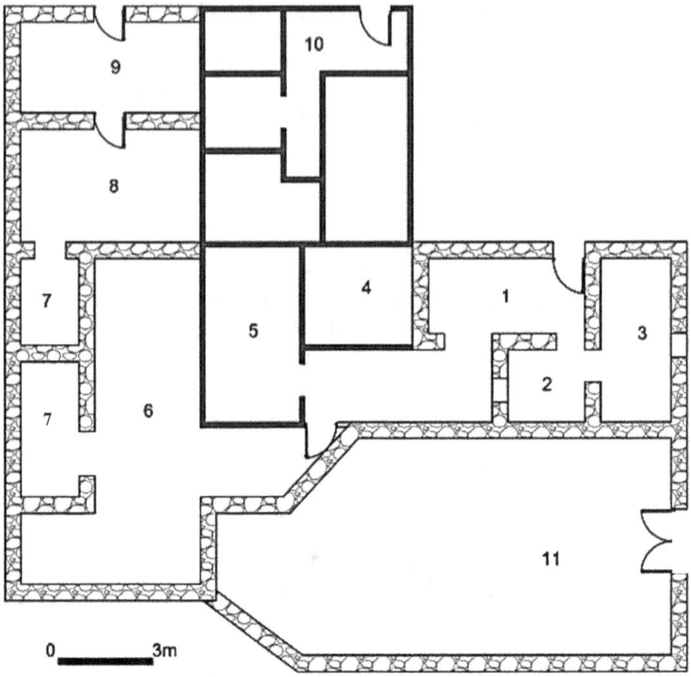

Entrance, 2. traditional kitchen, 3. storage space, 4. Room, 5. Reception room, 6, 7 et 11. Livestock enclosure, 8 et 9 storage space 10. new concrete house for the son's household

(Fig 5): Various adaptations of different spaces of the original house

Traditional dwellings are undergoing a significant transformation in the configuration and functions of their interior spaces. A prominent feature of these homes is the enclosed courtyard, which serves as a vital architectural element. These residences skillfully integrate contemporary and traditional kitchen spaces. Additionally, expansive reception areas, typically positioned near the entrance, create an inviting and impressive ambiance. The function of bedrooms is also evolving, with a shift towards more personalized spaces that cater to the specific needs of occupants. An emerging trend is the addition of extra levels to accommodate increasing spatial demands.

Factors Influencing the Evolution of Vernacular Dwellings in the Context of the Study

The area under study has undergone significant transformations in both architectural styles and living arrangements. These changes stem from the intricate interplay of social, technological, and economic influences.

One notable social shift is the trend of "de cohabitation," wherein younger individuals increasingly choose to construct separate, modern residences instead of adhering to traditional multi-generational homes. This shift is closely linked to notions of social advancement; the ability to build a contemporary home is often seen as a symbol of success. Within this context, these modern dwellings serve as potent indicators of social status and achievement, reflecting broader societal values and aspirations. Technological progress has also played a crucial role in these transformations. The increased availability of contemporary building materials and enhanced construction methods have significantly expedited the building process. These modern materials are often viewed as more hygienic alternatives to traditional earth-based construction, which is frequently stigmatized as a marker of low social standing. Furthermore, there is a widespread belief that homes built with earth are unhygienic and present risks, particularly due to their association with scorpion infestations during the summer months.

Economic factors are equally influential in shaping these architectural transitions. The ability to adopt modern housing styles largely hinges on individual financial resources. Those with limited economic means often find themselves compelled to remain in older, traditional homes, while individuals with greater financial stability prioritize the construction of modern houses as both a necessity and an aspiration. This desire reflects a broader shift in how housing is perceived; modern homes are increasingly viewed as symbols of improved quality of life and enhanced comfort. This evolving perspective serves as a key driver of ongoing changes in architectural practices and lifestyle choices of the inhabitants.

Conclusion

Time plays a crucial role in studying vernacular forms of architecture and is equally significant in the evolution of these structures, particularly in habitations that continuously adapt and transform. Field research conducted in the rural settlement of Tamazouzte, in El Haouz province, reveals that each vernacular dwelling evolves with distinct characteristics shaped primarily by the financial and social conditions of its inhabitants.

This study of vernacular dwelling evolution raises an important question: Can these new architectural forms be considered an emerging vernacular? These contemporary adaptations may indeed represent "*the best solutions*" to the challenges of modern living, pointing to the necessity of a multidisciplinary approach to defining exactly what is 'contemporary vernacular architecture'.

However, as time progresses, these accelerated transformations bring traditional architecture to a critical juncture. The ongoing adaptations risk undermining the preservation of forms with historical and architectural value.

References

[1] Upton, Dell, and John Michael Vlach, eds. Common places: readings in American vernacular architecture. University of Georgia Press, 1986. Page 3

[2] Asquith, Lindsay, and Marcel Vellinga, eds. Vernacular architecture in the 21st century: Theory, education and practice. Taylor & Francis, 2006. https://doi.org/10.4324/9780203003862

[3] Carter, Thomas, and Elizabeth C. Cromley. Invitation to vernacular architecture: a guide to the study of ordinary buildings and landscapes. Vol. 6. Univ. of Tennessee Press, 2005, page 11

[4] Derouich Aziz., Melehi Youssef., Les architectures régionales : Région de Tensift, Rabat, Okad, 2008.

[5] Derouich Aziz., Melehi Youssef., Les architectures régionales : Région de Tensift, Rabat, Okad, 2008.p. 53.

The vernacular facade as a contemporary urban issue case of the Bâlima hotel in Rabat

BERDOUZ Karima[1,a*], MESSOUS Ouafa[1,b] and TLEMÇANI MEKKAOUI Nezha[1,c]

[1]National School of architecture, Rabat, Morocco

[a]k.berdouz@enarabat.ac.ma, [b]o.messous@enarabat.ac.ma, [c]n.tlemcani@enarabat.ac.ma

Keywords: Architecture, Vernacular, Urban Facade, Bâlima, Morocco

Abstract. Moroccan vernacular architecture is much more than a simple object, it is a process of continuous evolution, where spatial, human, and temporal dimensions intersect. Anchored in its geographical and cultural context, it reflects the lifestyles, available materials, and ancestral techniques of a community. Vernacular architecture is not decoration, it is the result of an ingenious adaptation of man to his environment. Between 1912 and 1940, Morocco experienced a new period in designing public space [1,2,3] the urban facade became the main vector of a new image of the city, exposing the know-how of local crafts while opening up more to the public space as imagined by the administration of the French protectorate present in Morocco at this period. This movement of "staging" of architecture has sparked rich debates, between preservation of vernacular identity and adaptation to the demands of modernity. Architecture evolved, gradually abandoning traditional Moroccan styles to adopt more modern European influences. However, this transition also gave rise to an architectural movement specific to Morocco: the "Arabisance movement" [4], which attempted to create a "hybrid" architectural style, seeking to adapt Moroccan vernacular forms and details to European requirements. The study of the Bâlima hotel in Rabat, through the prism of its emblematic facade, invites us to reflect on the means of reconciling progress and preservation, of bringing the past and the present into dialogue to shape cities in the image of rich Moroccan traditions. The facade is not a simple object, but the reflection of an energy in constant metamorphosis, of an identity in perpetual reinvention. Our reflection will revolve around Moroccan vernacular architecture. Architecture perhaps never has truly completed form. It is becoming a perpetual movement, where matter and spirit constantly intertwine. The vernacular is a process in progress, a complex alchemy where spatial, human, and temporal dimensions combine. Vernacular architecture, in its very essence, embodies the most authentic expression of the constructive traditions and local know-how of a given region. Anchored in its geographical and cultural context, it reflects the lifestyles, available materials, and ancestral techniques specific to a community. Far from being a simple decoration, this vernacular architecture is the result of an ingenious adaptation of man to his environment, guided by practical, social, and aesthetic imperatives. Moroccan vernacular architecture was profoundly reinterpreted and reinvented to respond to the urban challenges of the pivotal period of the French Protectorate, between 1912 and 1940. This transition marked an important turning point in the way of thinking and shaping public space, confronting Moroccan vernacular architecture as new influences and aspirations. While the traditional medinas kept an introverted and intimate approach to housing [5], the urban facade of the new city became the main vector of a new image of the city, exposing local artisanal know-how while opening more onto public space. This movement of "staging" of architecture has sparked rich debates, between preservation of vernacular identity and adaptation to the demands of modernity. This period was marked by the arrival of French architects and town planners responsible for reorganizing urban space according to the principles of modern planning. Cities like Rabat, Casablanca, and Fez have seen the birth of new districts designed according to

Materials Research Proceedings 47 (2025) 98-102 https://doi.org/10.21741/9781644903391-11

orthogonal plans, facade alignments [6], and public spaces inspired by Haussmannian town planning.

Architecture in Morocco between 1912 and 1940

Architecture in Morocco in the early 20th century evolved, gradually abandoning traditional Moroccan styles to adopt more modern European influences. This transition has given rise to tensions and debates on the preservation of Moroccan architectural identity in the face of the imposition of foreign urban models. The facade then became the main showcase of the cities, exposing the know-how of Moroccan artisans while opening up more to new influences.

This urban heritage remains a source of inspiration for Moroccan urban planning today. It shows how vernacular architecture can nourish a sustainable urban vision, but also gave birth to an architectural movement specific to Morocco, which roots the city in its identity while opening it to the future[1].

The Arabism movement thus represents an attempt to create a "hybrid" architectural style, seeking to adapt Moroccan vernacular forms and details to Western requirements and construction techniques. This movement deeply marked the Moroccan architectural landscape and continues to influence reflections on the country's architectural identity [7].

Beyond the physical transformations, this period also marked a turning point in the conception of urban development in Morocco. The planning of cities has become a central issue, reflecting the political, economic and societal issues of the colonial era.

Today, the legacy of this pivotal period continues to influence Moroccan architecture and urban planning, fueling reflections on the preservation of cultural identity and the harmonious integration of heritage in the contemporary city.

Hôtel Bâlima, why the hotel?

The facade is not a simple object, but the reflection of an energy in constant metamorphosis, of an identity in perpetual reinvention. The horizontal space of the city, the vertical tree structure of the actors who shape it, and the multiple temporalities which intertwine there, together compose this living framework of the vernacular.

The Bâlima hotel is not only a building, but the place of a meeting of colonial authorities and local artisans, traditional and modern architectural styles and eras which succeed one another and overlap. Its facade is not a simple decoration, but the reflection of a subtle negotiation between antagonistic forces [8], of a dialogue between past and present, of a quest for balance between rootedness and openness.

Through the analysis of this building, we will explore how the vernacular facade, far from being a simple decorative element, becomes a real urban issue. We will see how it contributes to the construction of the city's identity, the preservation of cultural heritage, and reflection on sustainable urban development.

This study will also allow us to address broader questions: How to reconcile architectural tradition and contemporary needs? What role can vernacular architecture play in 21st century urban development? How to preserve cultural authenticity while meeting the challenges of modernization?

[1] Lyautey described colonial architecture in Morocco as one of the rare spectacular manifestations of a state style in the 20th century. Its architect Laforgue also described it as frank, logical, and tasteful architecture. He believed that traditional Moroccan art deserves to be admired, preserved, and restored, but not imitated, qualifying the borrowing of Arab architecture as a major action.

Hotel Bâlima, between yesterday and today

The BALIMA company, whose name is the acronym of the names of its three founding partners BARDY, LIOREL and MATHIAS, was founded in the early 1920s. It is a company whose main activity was real estate transactions mainly in the center of city of Rabat, and the emblematic building that has been talked about throughout its existence was the BALIMA Hotel.

The French administration, responsible for development, has put in place specifications, where buyers are obliged to comply with the drawings made by the fine arts and historic monuments services, these drawings " imposed on investors, then encouraged and invited, during the construction of buildings located in this area. The main investor in this area was the BALIMA company.

The BALIMA Hotel, built on land initially intended to house a residential building, benefits from a strategic location, and major constraints in terms of urban regulations and architectural ordinances, which have provoked many reactions. It is located at the bottom of the "small square[2]" which has been created on the side of the Lyautey courtyard opposite the Palace of Justice (the current Moroccan parliament).

With a proposal for the BALIMA Hotel set back and a height exceeding the authorized 29 m, it was considered more like "a belfry[3]" than a residential building with its disproportionate dimensions of a height of 30.55 m later reduced to 29 m. The BALIMA Hotel was considered by the municipal administration as a way to "brighten up[4]" the architectural order present in this part of the city, it was even considered by the Fine Arts department as an element to "furnish space[5]", alongside the buildings located upstream of the station square.

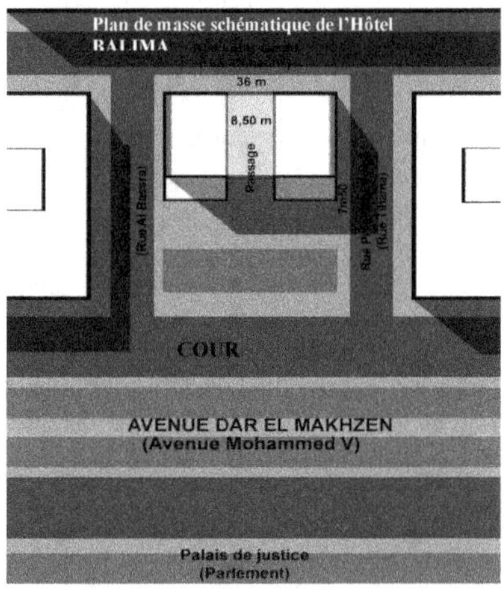

Figure 1: schematic ground plan of the location of the Balima hotel on the "Cour Lyautey" square, Karima Berdouz 2018.

"Passage under Saba", a passage through time

By analyzing the Bâlima hotel, we see that, although it is a modern building, the French fine arts and historical monuments services sought to integrate and reinterpret key elements of architecture. traditional of the medina of Rabat. The "passage under Saba[6]", for example, gives this building a particular character, recalling the covered transition spaces so characteristic of the old town of Rabat. This reinterpretation of a typically vernacular element creates a tangible link between past and present, between tradition and modernity.

Likewise, the square which separates the Bâlima hotel from the current parliament is inspired by the

[2] The designation of the "Petit Square" was recorded in the archives of the building authorization for the Balima hotel in 1926.

[3] Word used in authorization documents in the archives of the municipality of Rabat.

[4] A term found manually written on the building authorization archives of the Bâlima Hotel to notify the refusal of authorization and the reason for the refusal.

[5] A term found manually written on the building authorization archives of the Bâlima Hotel to notify the refusal of authorization and the reason for the refusal.

[6] Arabic word designating a street passage under a home according to ancestral local practices in Morocco which designates that the neighborhood has a right of passage on third-party land to access its housing, the owner of this land has the right to cover this passage to build his accommodation upstairs while leaving the passage open.

concept of an open "patio"[7], so prevalent in traditional Moroccan domestic architecture. Thus, the architect was able to skillfully combine ancestral local references with a more contemporary architectural language, establishing a fruitful dialogue between the old and the new, between the interiority of the medina and the exteriority of the new city. This subtle fusion of vernacular and modern codes makes the Bâlima hotel a living manifesto of Moroccan architectural identity, always in search of a balance between tradition and innovation.

By carefully studying the Bâlima hotel, we cannot fail to note the deep links which unite the traditional functioning of the medina of Rabat to the urban design of its new city, imagined by Marshal Lyautey. We thus find in the plan of Rabat characteristic elements of vernacular architecture, an organic urban framework, spaces of transition between public and private, as well as a harmonious integration of buildings and vegetation.

The study of the Bâlima hotel, through the prism of its emblematic facade, invites us to reflect on the means of reconciling progress and preservation, of bringing the past and the present into dialogue to shape cities in the image of the rich Moroccan traditions.

Thus, the facade of the Bâlima hotel is not a simple surface, but a living palimpsest, where traces of history, the gestures of the craftsmen, the aspirations of the sponsors are superimposed. It is in this incessant dynamic that the very essence of the vernacular resides, well beyond any static or definitive conception of architecture.

Urban planning, a political issue

Beyond the physical transformations, this architectural and urban transition marked an epistemological turning point in the conception of the development of cities in Morocco. The planning of cities has become a central issue, closely linked to the political, economic and social questions of the colonial era [9]. It was no longer just a question of building, but of shaping urban space as a reflection of the balance of power, aspirations and tensions of society.

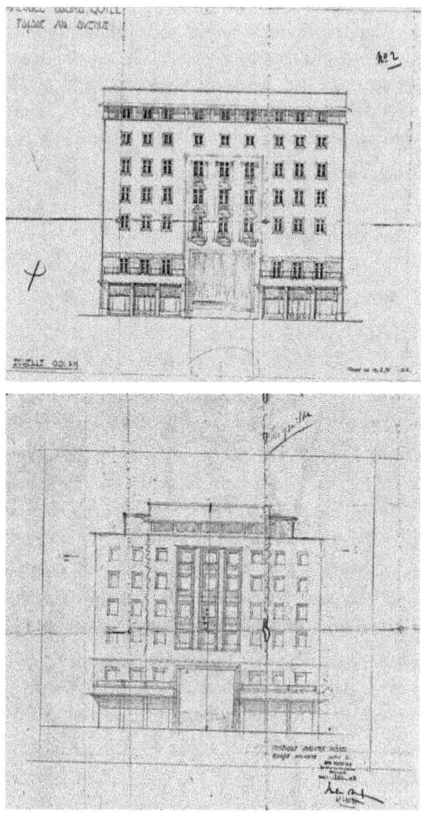

Figure 2 : Different proposals for the facade of the Bâlima hotel proposed to municipal services in 1931. Sources: the archives of the Commune Hassan, Rabat 2018.

Today, this architectural and urban heritage continues to profoundly influence the Moroccan landscape, nourishing profound reflections on the very nature of urbanity. What place should we give to the past in building the future? How can we preserve the soul of a city while adapting it to the demands of the present? The harmonious integration of heritage in the contemporary city

[7] Spanish words designating the interior courtyard of the Arab Muslim house are very present in the medinas in Morocco and Andalusia in Spain.

becomes an identity challenge as much as a technical one, inviting us to rethink the foundations of our understanding and our apprehension of urban space.

Beyond built forms, an entire system of knowledge, practices and representations are thus called into question. Morocco's urban heritage reflects the complexity of human experience, the clashes of civilizations [10], the desire for roots and openness. He invites us to consider the city not only as an arrangement of stones and mortar but as the theater of a quest for knowledge, of an incessant search for balance between tradition and modernity, between individual and community.

Thus, studying this architectural heritage is not only an exercise in history or conservation but an act of philosophical reflection on our way of designing, producing, and experiencing urban space. Because in these facades, these streets, and these squares, lies part of our collective understanding of the city, a reflection of our quest for meaning and harmony within the built environment.

References

[1] LAFORGUE Adrien, l'architecture au Maroc, le miracle marocain La région de Rabat, Editions de la vérité marocaine, 1932.

[2] La pacification du Maroc, 1907 – 1934, J.LADREIT DELACHARRIERE, publications du Comité de l'Afrique française Paris 1936

[3] RIVET (D.), Lyautey et l'institution du protectorat français au Maroc 1912-1925, Paris, 1996 Ed. L'Harmattan, Coll. Histoire et perspectives méditerranéennes, 267 p.

[4] Karima BERDOUZ, Le patrimoine colonial au Maroc. Le façadisme comme approche de sauvegarde, 244, 2022, ISBN : 978-2-14-027448-0.

[5] MHAMMEDI (M), KARIBI (K), BERDOUZ (K), HARATI (M), ALAGUI (F), An urban planning and modern architecture of the twentieth century in morocco: the construction of a capital, Rabat. International Journal of Civil, Structural, Environmental and Infrastructure Engineering Research and Development (IJCSEIERD), Vol. 8, Issue 5, Dec 2018, (p.p17-36).

[6] PINSON (D), EL HARROUNI (K), Deux Élie et deux patrimoines potentiels pour l'architecture de la seconde moitié du XXème siècle au Maroc? African and Mediterranean Journal of Architecture and Urbanism, Ecole Nationale d'Architecture, Rabat, 2021, pp.9-21.

[7] Karima BERDOUZ La construction de l'image de la ville, le rôle des acteurs locaux, cas de la ville de Rabat, Rabat 2015 mémoire de fin d'études pour l'obtention du diplôme des études supérieures en aménagement et urbanisme INAU, Rabat 2017.

[8] les archives des autorisations de construire aux services des archives, Commune de Hassan, Rabat.

[9] MHAMMEDI (M), KARIBI (K), BERDOUZ (K), Le patrimoine colonial marocain à l'épreuve du façadisme (1912-2012), Transversale n°3, ENSAP Bordeaux, 2018.

[10] RIVET (D.), Le Maroc de Lyautey à Mohammed V. Le double visage du protectorat, Paris, 1999 Ed. Denoel.

Materials Research Proceedings 47 (2025) 103-116
https://doi.org/10.21741/9781644903391-12

Architecture in Morocco, between tradition and modernity the things of forms and the forms of things

AMHAMDI Rimane[1], LAMZAH Assia[1], KHARMICH Hassane[1]

[1]Ecole Nationale d'Architecture de Rabat, Morocco

Keywords: Vernacular Architecture, Traditional Architecture, Modern and/or Contemporary Architecture, Identity Anchoring, Forms of Things, Things of Forms

Abstract. Modern architecture is defined as a search for a new model, an alternative to the traditional model deemed outdated and incapable of meeting the evolving needs and requirements of modern man. In Morocco, more so than anywhere else, vernacular and traditional architecture has proven, especially since the colonial period, to be a reference point in the development of a new architecture with a strong Moroccan identity. This architecture aims to be a fusion of traditional, Berber, and Arab-Muslim architecture, both in terms of form and aesthetics. Revealing the process through which contemporary architecture in Morocco claims its Moroccan specificity and modernity whether in terms of concept, aesthetics, form, function, sustainability, energy efficiency, or well-being is not about contrasting old/modern or present / past. Rather, it is about unveiling: the "things of forms" that define conceptual modes, and the "forms of things" that define formal and aesthetic particularities. The objective of this communication, stemming from a portion (currently in writing) of my thesis research, is to make a modest and partial contribution to the definition of contemporary Moroccan architecture. This architecture, while modern, increasingly asserts its identity by drawing from traditional Moroccan architecture in various ways. The goal is also to show how this architecture, in its constant search for cultural and identity inscription, breaks with the anonymous international style, creating new forms enriched by the contributions of traditional architecture and craftsmanship.

Introduction

Moroccan architecture, rich in a diversity of forms and concepts specific to Arab-Muslim and Amazigh art, has its roots in Hispano-Moorish art and is a synthesis of several artistic contributions and expressions of different sensibilities, forming an integral part of Muslim civilization.

The term "Moorish" means both Spanish Muslim and North African. The double meaning corresponds to historical reality: Hispano-Moorish art is the art of a civilization born in Spain and which emigrated to the cities of the Maghreb, where it found prosperity and acquired a specifically Maghrebi form, a synthesis of Berber and Umayyad-style architecture.

To give greater clarity to our study of contemporary Moroccan architecture, it's worth recalling that the major distinction between the Arab and Amazigh styles is that the former is characterized by specific forms, arabesques, interlacing, and the aesthetics of the horror of emptiness (1), while the latter is characterized by the rigor and simplicity of its formal and decorative geometry.

It is clear that it was the fusion of these two concepts that gave rise to the Moorish style, which covers the forms taken by Islamic art in the Iberian Peninsula under the rule of the Umayyad Caliphs, then under the domination of the conquering Berber civilizations of the Almoravids and Almohads. (2)

Morocco is widely regarded as the crossroads of the Arab East and the Western Mediterranean. Its architectural art is strongly represented by Andalusia. As a result, it remains the stronghold of virtually all Arab-Muslim art forms. An art that flourishes here and is widely applied.

That said, with the French colonization of Morocco, architecture and town planning underwent a series of changes and experiments. A new, or mixed, type of architecture gave rise to what is known as modern colonial architecture, inspired by local culture.

Later, in the twentieth century, numerous movements appeared everywhere, sharing technical and stylistic characteristics such as abstraction, mass production, industrialization and, above all, the rejection of tradition... In short, a widespread belief in the concept of "form follows function". (3)

Meanwhile, in Morocco, as in the rest of the world, so-called modern architecture, in opposition to traditional architecture, began to link the architectural object to the contexts that gave rise to it. New contexts, new needs, new aspirations. And it is to the ways of building, living and looking that it has bent.

Three key periods that are to cite. The following periods in the kingdom's history are key to understanding the birth of contemporary Moroccan architecture.
- A period largely dominated by the ideas of Marshal Lyautey, Prost and Marchisio (1915-1945);
- Period marked by the Athens Charter (1950-1970);
- Today's heterogeneous and diverse period, divided between anonymous and strongly Moroccan, with concerns for cultural anchoring, ecological and environmental imperatives, and so on.

In view of all this (4), is it possible to speak of contemporary Moroccan architecture today? If so, in what way and how?
- Is current Moroccan architecture a continuation or a (partial / total) break with so-called traditional architecture?
- What forms of vernacular or traditional Moroccan architecture does contemporary Moroccan architecture draw on? (Concepts, identity, materials, techniques, sustainability, energy savings, etc.)
- What are the forms of things specific to traditional Moroccan architecture that contemporary Moroccan architecture draws on? (Geometry, simplicity, aesthetics, lighting, shapes...)

The hypothesis that Moroccan architecture, with its strong national identity, is gaining ground in a wide variety of forms, is proving to be a reality. In this paper, we'll highlight some of these, highlighting their sources of inspiration, namely the forms of traditional architecture (concepts, identity, shapes, materials, techniques, sustainability, energy savings...). We will then attempt to reveal, in the same way, the specific forms of this same architecture, from which this so-called new Moroccan architecture also draws (geometry, aesthetic simplicity, lighting, shapes...).

Having said that, this study is deliberately restricted to the analysis of certain contemporary projects whose architectural style claims a typically Moroccan cachet; how and why are they so?

To get to the bottom of this, a historiographical synthesis of the genesis of architecture in Morocco, as we shall see below, is essential.

---

(1) In response to Democritus, for whom the universe was filled with atoms and emptiness, Aristotle was convinced that emptiness did not exist and that the universe was everywhere filled with something. His Latin translations contain the phrase "Horror vacui" or horror of the void.
In the visual arts, the "horror of the void" aesthetic refers to the act of filling the entire surface of a work with finely detailed details, as in the case of the Zellij in traditional Moroccan architecture.
Cf. Wikipedia: https://fr.wikipedia.org/wiki/Horror_vacui
(2) Marçais (G.), Manuel d'Art musulman, Tome I, l'art du Maghreb et de l'Espagne aux XI et XII siècle, p.306-421.
(3) Form follows function: is a principle put forward by American architect Louis de Sullivan, by which twentieth-century architecture suggests that the expected utility of a building should be the main factor in its design.
(4.) - For further information see my doctoral thesis (in progress)

Materials Research Proceedings 47 (2025) 103-116 | https://doi.org/10.21741/9781644903391-12

1. Architecture in Morocco: tradition, modernism, contemporaneity

1.1 Amazigh vernacular architecture

Amazigh vernacular architecture (5) is the highly varied rural architecture of the Amazigh regions of North Africa, particularly Morocco. This architecture is characterized by a diversity and heterogeneity of concepts and forms, which are also found in current Moroccan architecture. It is distinguished by the simplicity of its forms and concepts. It is often based on locally available materials. Its materials, construction methods and organization make it an architecture with its own forms and organization, it remains contextual, sustainable, energy-efficient, practical in its simplicity and economical.

This authentic architecture has never been influenced by the Arab conquest or the expansion of Hispano-Moorish art, and therefore has its own forms of things and things of forms. Certainly, the kasbahs of the south remain the most significant expression of this type of architecture. The forms of things and the things of forms of this type of architecture obey different rules. It is often built on stone foundations with thick adobe walls. A construction that insulates against summer heat and winter cold. It is also often organized around a patio (mrah), a veritable well of light and refreshment for all the surrounding spaces.

What's more, in mountainous areas, adaptations have always been necessary to better cope with the vagaries of the seasons. To protect against the cold, the basic layout of the whole dwelling has to be modified, resulting in a significant reduction in openings, or even their elimination, or at least a sharp reduction in the central patios and their opening to the east. On the plains of the Saharan regions, the adaptation consists in protecting people and livestock from excessive summer sunshine. This de facto means procuring cast shadows from ramparts, ensuring that buildings are high and terraced, with thick walls and terrace floors, ventilation and daylight chimneys, adjustable openings, brise soleil, moucharabiehs, vertical domestic migration day/night and summer/winter, etc. To protect against the sun and sandy winds, ramparts are built on both sides. These are the precious elements that anticipate the architectural future in social, economic and eco-energy terms, from which contemporary Moroccan architecture, among others, naturally draws.

2. Shapes of things and things of shapes in traditional Moroccan architecture

Traditional Moroccan architecture, however, in the case of important buildings, is highly valued for its symmetrical construction, the golden ratio (6) and the contributions of a wide variety of geographically localized arts and crafts. Indeed, a number of these crafts are directly linked to architecture, both past and present, even if they are introduced into the latter in different ways and according to new concepts, as we shall discover below. Among the various elements that lend authenticity and cultural anchorage to this type of architecture are:

(5.) - Ben El Khadir (M.), Lahbabi (A.); Architectures régionales un parcours à travers le nord marocain; ed. Najah el Jadida, Casablanca, 1989.
- MM. El Kasri (M.), Squalli (H.), la grande encyclopédie du Maroc, Culture arts et traditions, Vol. I (architecture section)
- Khermich (H.), Sedreddine (M.), El Gharbi (S.); Les matériaux de construction locaux, un appui pour une architecture contextuelle autosuffisante en énergie. AMJAU, 2019.

(6) Menway (P.), Le code secret, la formule mystérieuse qui régit les arts, la nature et les sciences, p.14-15
- The golden ratio = (1√5)/2 or 1.61803, which corresponds to the Greek symbol φ, equivalent to the sound “Phi”. It is used to represent the proportion described by Euclid in the fifth chapter of his Elements, known as the golden ratio, also known as the golden proportion, the golden mean or the golden ratio.
- For Vitruvius, architecture is “the balanced adjustment of details to the whole, the arrangement of proportion with the objective of a symmetrical result.

Materials Research Proceedings 47 (2025) 103-116 https://doi.org/10.21741/9781644903391-12

2.1 Ceramics and zellij

Ceramics and zellij play a key role in traditional architecture, whether religious or civil (mosques, medersas, sanctuaries, palaces and wealthy homes). Their use can be multiple and varied, either in the form of walls entirely covered with patterns and colors, according to a unique aesthetic of its own kind known, by analogy with Aristotle's idea that "nature abhors a vacuum" as "the aesthetic of the abhorrence of emptiness"; or in the form of panels covered with geometric patterns or excised zellij, revealing floral forms or calligraphic inscriptions...

Zellij is used in 2D form in traditional architecture in the most classical way possible, covering almost all interior and exterior spaces, whereas in so-called contemporary Moroccan architecture, it's used in new and extrapolated ways.

2.2 The wood

Woodworking, an art that has always thrived in traditional Moroccan architecture, whether as an integral part of the architecture or simply as a decorative element. Medersas and Fondouqs (7) are fine examples of woodworking. When wood isn't used as a frame, as in the case of Andalusian-style residences, there's a profusion of cladding and various types of work and sculpture, particularly turned wood for clerestory panels and openwork windows (the moucharabieh).

There are also a number of decorative elements, often made possible by the woodworking technique, such as acanthus, pine cones, shells, "muqarnas" (8), particularly kufic calligraphic inscriptions, and motifs similar to those found on zellij. Wood chasing is applied to ceilings, doors, courtyard windows and furniture. The most frequently used decorative motifs are either "Testirs" in the form of geometric interlacing made up of straight lines with a star at their center; or "Touriqs", consisting of interlacing foliage and rosettes made up of curved lines. As will be explained below, wood is being reintroduced into contemporary Moroccan architecture in a variety of ways, both in its classic form and in extrapolated, innovative forms.

2.3. Chased plaster

The art of chiselled or sculpted plaster is of unrivalled importance in traditional Moroccan architecture, a veritable lacework of extremely varied motifs adorning the interior walls of mosques, shrines, palaces and wealthy homes.

As in other branches of decoration, these are "Testirs" and "Touriqs", calligraphic inscriptions, divided into small panels (stars and sun) with motifs curiously reminiscent of Moorish decoration, also known as "Chmamech". But the art of the plasterers' craftsmen includes not only the ornamentation of flat surfaces, but also the carving of capitals, ogives and vaults. Added to this is the workmanship of "Muqarnas" in the form of honeycombs, a refined, top-of-the-range work.
All these techniques and skills can be found in the kingdom's various medersas, and are extrapolated and applied differently in Morocco's current architecture.

(7). Fondouq: The word derives from the Greek πανδοχείον "pandokeion" meaning inn. In 10th-century Arabic texts, it refers to places intended to shelter travelers and their mounts on the first floor.

(8). Muqarnas: These are ornamental honeycomb motifs in painted stucco, wood, stone or brick in Islamic architecture. They tumble down like stalactites or line the vaults or interior of domes.

Materials Research Proceedings 47 (2025) 103-116 https://doi.org/10.21741/9781644903391-12

3. The birth of modern colonial architecture in Morocco: Lyautey and crossbreeding

Lyautey's policy on urban planning and architecture in colonized territories was to seek a new and appropriate form of integration. Indeed, French architects were required to draw inspiration and formal and aesthetic elements from local architecture and urban fabric, for any new architectural development.

In Lyautey's view, each colony had its own specific indigenous architecture, which should be extrapolated, inspired and adapted. *"Neither a copy of the native habitat, nor the importation of a classic style, but adapted and lived with the local craftsmen, the "Maallems" (master craftsmen), guardians of the "Qaïda" (*tradition)(9).

These were the criteria on which the day-to-day development of what later came to be known as colonial-style modern architecture, a fusion of modern and traditional local architecture, was based.

The local style, the forms of things and the things of forms, served as references and sources of inspiration for architects who made what they considered necessary modifications and improvements to adapt them to the environment and the people. And it was to the traditional urban architecture of Rabat and Salé that the architect Albert Laprade, one of Henri Prost's closest collaborators in Morocco, turned to draw architectural and aesthetic solutions that are now found throughout colonial architecture in Morocco.

Under the protectorate, architects not only had at their disposal a heritage of rare diversity, but above all they had at their disposal expert master masons and craftsmen, the same ones who had worked on the sophisticated construction of medersas and palaces, and thus inherited the advanced know-how once applied to the Alhambra in Granada or the Saadian tombs in Marrakech. What's more, these architects also found an abundance of materials on site: wood of various species, hard or soft stone, various marbles, as well as ancestral techniques and know-how and an "art industry", particularly those associated with architecture.

4. Modern and contemporary architecture, the forms of things and the things of forms

Modern architecture is defined as the search for a new model, an alternative to the traditional model, which is considered outdated and hardly meets the diverse needs and requirements of modern man.

While architectural modernism encompasses several twentieth-century movements with shared technical and stylistic characteristics - abstraction, industrialization, rejection of tradition and, above all, the unifying concept of "form follows function" - modern architecture should not be confused with contemporary architecture. Indeed, as explained above, modernism is a set of ideas and conceptions of the world that rejects the history of the traditional by siding with progress. It favors simplicity, functionality, geometric forms, open plans and the use of modern materials such as reinforced concrete, glass and steel. Contemporary architecture, on the other hand, while drawing its essence from modern architecture, aims for free creative expression and timeless adaptability. It is a force for innovative, uncluttered and minimalist solutions. In short, etymologically speaking, modern is characterized by its opposition to the traditional, and is in line with progress.

Inscribing one's project in the traditional and artisanal repertoire and producing contemporary Moroccan architecture, according to new concepts, another vision, other modes, that new technologies allow, is becoming more and more, a major concern for contemporary architects. The resulting creations take a variety of forms, from a traditional-contemporary mix of forms and materials, to the reproduction of modern forms combined with traditional ones. And so we speak sometimes of harmony, sometimes of synthesis, sometimes of a modern/traditional mix.

(9). Rigollet, cited by J. Marrast, in « L'œuvre de Henri Prost », Académie d'architecture, p.5

Materials Research Proceedings 47 (2025) 103-116 https://doi.org/10.21741/9781644903391-12

4.1. Formal and aesthetic properties

4.1.1. Symmetry :

Symmetry is one of the hallmarks of classical architecture. Symmetry is considered the most important element in the architectural geometry of large-scale projects, with the main entrance to buildings always placed in the center according to a certain number of columns or decorative elements. Virtually all Neoclassical architects after the Renaissance often employed symmetry as a design language, as many monuments testify. In Morocco, a number of traditional buildings such as palaces, medersas and gateways in various cities - Bab Rouah and Bab El Kébir in Rabat, Medersa Al Attarine, Bab Boujloud, Bab Lahdime in Fez, Bab Agnaou, Medersa Ben Youssef in Marrakech, Medersa Bou Inania, Bab Mansour in Meknes - remain important examples of the power of symmetry in traditional architecture. But symmetry is also common in recent Moroccan architecture.

4.1.2. Aesthetic proportions: the golden ratio (divine proportion)

Aesthetics, or the science of form, plays an important role in stimulating sensory perception, in our case visual perception, and in triggering emotions and aesthetic perception.

In architecture, ancient buildings were constructed according to relationships governing different dimensions, such as the spacing between different components and parts of the façade. Such construction is governed by relationships similar to the different intervals in music, which condense a perfect measure corresponding to a guessed proportion, known as the golden ratio, necessary to trigger aesthetic perception.

The aesthetics of proportions has always been studied through the golden ratio, which characterizes the ratio of two quantities X and Y, identical to that of the greater of the two to their sum: if X is smaller than Y, $X/Y= Y(X+Y)$, i.e. $X/Y= 1 \sqrt{(5)}/2$ = (0.618 around).

The "code" of the system of proportions used to build some of Morocco's great traditional monuments calls for in-depth analysis, in order to reveal the reasons behind the great secret of the beauty of traditional Moroccan architecture, which is often wrongly linked solely to the work of craftsmen (zellij, plaster, wood, stone...), and to ascertain how this has or has not been taken up in so-called contemporary architecture.

4.1.3. Ornamentation

By definition, ornament is that which is added to something to give it beauty and grace. In architecture, ornament is decoration in the form of motifs, carvings and moldings used to decorate the various parts of a building. Contemporary Moroccan architecture has its own ornamental motifs, such as worked, cut, chiselled and assembled zellij, carved and/or sculpted stone, the many motifs obtained with plaster or wood, the multiple motifs in wrought iron, etc., all of which were once used for their own purposes.

They have a special place in traditional architecture. Today, contemporary Moroccan architecture, particularly in large-scale projects, makes use of them as a reference point for identity, either in the classical way, or by extrapolating them in different ways. This makes them a source of inspiration for creating complete or partial building envelopes in the form of skins, for an obvious cultural anchorage, but also to act as a brise soleil, reducing heat and therefore cooling the premises thus sheltered.

4.1.4. Eco architecture

Contemporary Moroccan architecture, while claiming its place in the authenticity of traditional architecture, which is in essence sustainable and green, is it, too? and how?

As we know, towards the end of the 1960s, with the critique of modernist dogma claiming that the miracles of technology would solve everything, came the oil crisis in 1973, heralding the finiteness of fossil fuels. Architecture was therefore one of the sectors where a new approach was

Materials Research Proceedings 47 (2025) 103-116 | https://doi.org/10.21741/9781644903391-12

needed to counteract this crisis. Design is therefore seeking to become part of green architecture, which encompasses a number of key strategies. First and foremost, the reduction of energy consumption, the passive exploitation of solar energy for heating and temperature maintenance through reinforced insulation in cold zones, and, in the case of hot zones, an emphasis on cooling and the use of passive ventilation to reduce dependence on air conditioning.

This contemporary Moroccan architecture which, unlike that of other places, nevertheless seeks in its majority for the important buildings, not only a cultural anchoring, by using and the forms of the things and the things of the forms of Moroccan architecture and traditional art, but also to draw its ecological and eco-energetic principles from it.

5. Studies and analyses of contemporary Moroccan architecture

In order to provide some answers to our questions, we will, in the light of the above detailed development, subject three major contemporary achievements, which seem to us to be heterogeneous in design but representative, to an in-depth analysis through their forms of things and things of their forms as described above. This will enable us to reveal the elements borrowed and/or extrapolated in the process of their genesis, in order to confirm or refute our hypothesis as to the birth of contemporary Moroccan architecture.

5.1. Rabat Agdal high-speed train station

The Rabat Agdal LGV station is a recent creation (2018) by architect Y. Melihi. It is a bridge station, linking the Agdal, Akkari and Y. El Mansour riverbanks. The architecture of this station is resolutely modern and modernist in the forms of its things, justifying its design and its inscription in the modernist style, materials, technology, concrete, steel, and glass, on the one hand, on the other hand, Melihi has shown himself to be concerned with the cultural context and the ancestral alliance between modernity and tradition, his project has not escaped the seduction of traditional art and an ingenious extrapolation of a simple motif from craftsmanship, namely to one of its important components: the zellij. But unlike many other projects, he drew on a repertoire of simple elements to design his facades - modernism obliges.

1. Main facade of the LGV station. Note that the hexagonal motif is used in different ways here

2. Berber rugs with a simple geometric pattern (Amazigh art and architecture are characterized by their formal simplicity and geometry).

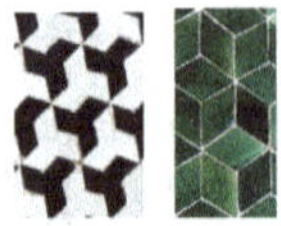

3. Simple Moroccan zellij, source of inspiration for the hexagonal motif used in various ways on the station's facades.

4. Main facade of the LGV station. Note that the hexagonal motif is used here in different ways

5. Detail of the left side of the main façade of the LGV station

6. Detail of the hexagonal pattern borrowed from zellige, skilfully handled and used in different ways, full, empty and reproduced on the station's various facades.

7. Detail of moucharabieh seen from inside. Light infiltration and thermal reduction

8. Detail of the skin in the form of a moucharabieh with solids and voids

9. Detail of the motif reproduced differently on the ceiling

10. Three hexagons Sculpture by V.Vazarely

11. 11. West facade reproduction of the same motif in bas-relief.

12. Reproduction of the motif on glass walls

13. East facade Reproduction of the same motif over a large area of this facade on an enlarged scale.

Materials Research Proceedings 47 (2025) 103-116 https://doi.org/10.21741/9781644903391-12

Through the use of this hexagonal geometric motif, abundant in Amazigh craftsmanship and architecture (ph.2, 3) but also in traditional Moroccan zellij (ph.14, 15) and modern art (ph.10), including in particular the optical works of the famous painter Victor Vasarely, who was strongly influenced by the aesthetics of Amazigh and Muslim art, Y. Mélihi, while not speaking outright about the borrowing of motifs from traditional Moroccan art, but rather about the alternation of solids and voids, emphasizes, in a way, the importance of thingsforms, namely the inclusion of the traditional zellij and the architectural work in question, in the repertoire of optical art. An art which, as we know, indulges in free optical interpretations and visual situations, as in the case of the various applications on the station facades (ph.4,5,6,11,12,13).

14 & 15 Hexagonal pattern may also be found in zellij

16

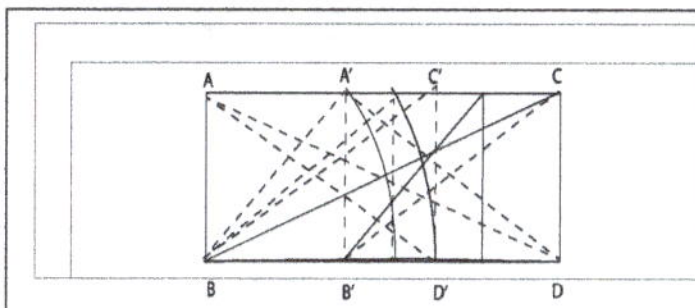

The secret framework of the station's main façade: a revealing layout showing the structuring and distribution of the various elements in a golden section. Such a section is omnipresent in the great monuments of traditional Moroccan architecture.
17

The station's architecture was designed according to contemporary principles and concepts based on transparency, flexibility, simplicity of form, and the use of traditionally-inspired motifs in the cladding and treatment of materials, as well as a concern for environmental integration and prevention. The station's architectural approach is based on sustainability, with self-regulating thermal systems and a predominance of green spaces in the surrounding area, in reference to the city of Rabat, but also and above all to traditional Moroccan architecture, which is often surrounded by green spaces that play a dual role: partial absorption of CO_2 and visual calming.

5.1.1. The shapes of things: diagonal division

The division of the façade's shapes, although often along horizontal lines, can also be achieved by means of diagonals. In the case of the golden section of the station façade, if we trace the diagonals ; having transferred the short side to the long side, starting once at one end and then once at the other, if we trace the diagonals of the resulting squares, we'll find a network of obliques and two parallelograms (AA' DD' and BB' CC'), which, depending on the module and

Materials Research Proceedings 47 (2025) 103-116 https://doi.org/10.21741/9781644903391-12

dimensions of the original rectangle ABCD, are particularly well-suited, on the one hand, to determining oblique divisions with an interesting rhythmic value, and, on the other, to accommodating the most important highlighted elements (see ph. 16 and the layout of the framework, revealing the secret of its aesthetics, ph.17). The façade is thus ordered by the golden ratio with its harmonic regressions. The entire glazed section of the facade is inscribed in a rectangle with the golden ratio, as was the original design intent, or the professional secret of the architect whose aesthetic needs are his second nature.

5.2 Marrakech railway station

Marrakech railway station is also a modern, contemporary station in the heart of Marrakech, rebuilt in 2008 by architect Y. Melihi. A fine example of contemporary architecture firmly rooted in Morocco's traditional architectural heritage. For this architect, "each project is a new beginning, an opportunity to question oneself and to immerse oneself in the new place, because it's not a question of reproducing a predefined style, but of developing a process leading to an architecture that may sometimes go against the grain, but which has the merit of being firmly rooted in its environment" (10). Implicitly, the architect rejects the pre-established rules of modernism and accepts that the modern architecture of this station must be anchored in its environment, i.e. in traditional Moroccan art, for reasons of identity and contextualization.

18. Main facade and entrance to the Marrakech LGV station.
There are many similarities between the building housing this station and Bab Rouah. Only the materials and the interpretation of the sculptures make the difference.

19. Bab Rouah à Rabat

According to the strict, classical rules of traditional Islamic architecture, with rules that are well established in the technical requirements, i.e. geometric simplicity, symmetry, enclosed, private spaces with sober, sculpted decoration on the main facade (ph.10).

Analyzing the volumes, especially the main facades, of Bab Rouah as an ancient traditional Moroccan architecture and of the railway station as a contemporary Moroccan architecture, with the aim of drawing out the forms of their things and the things of their forms, or in other words to reveal the secret of their striking conceptual and aesthetic similarities, many analogies and common points can be noted.

According to the regulating lines that govern the two buildings, it's clear that the partial superimposition of the two golden rectangles ABCD and A'B'C'D' lends a golden suppleness to the compositions of the main façades of Bab Rouah and La Gare. The intersecting diagonals of the two rectangles AD, BC, B'C', A'D' form a small rhombus in the center, with a clock in the middle. This is the visual focal point of the station's main façade. A center that orchestrates the entire building. This arrangement reminds everyone that time is the central element for travellers. Just like the main door in the middle of the imposing Bab Rouah gate, a reminder of the gate's unique entrance. The station, therefore, borrows from Bab Rouah not only the volumetry and aesthetics, but also, to a certain extent, the concept and form; the forms of things and things of forms, in other words.

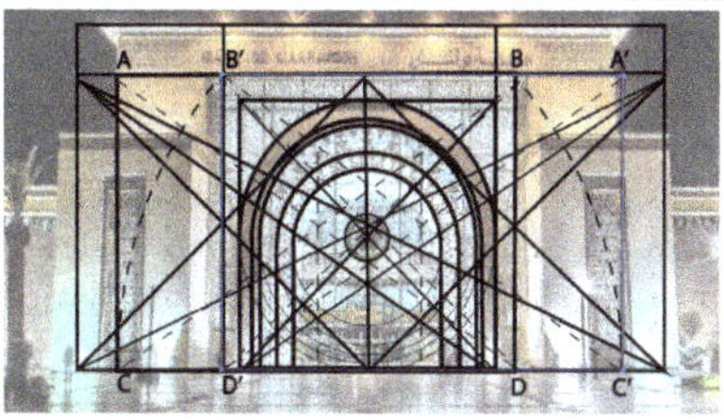

20. Regulatory layout of Marrakech railway station : main facade

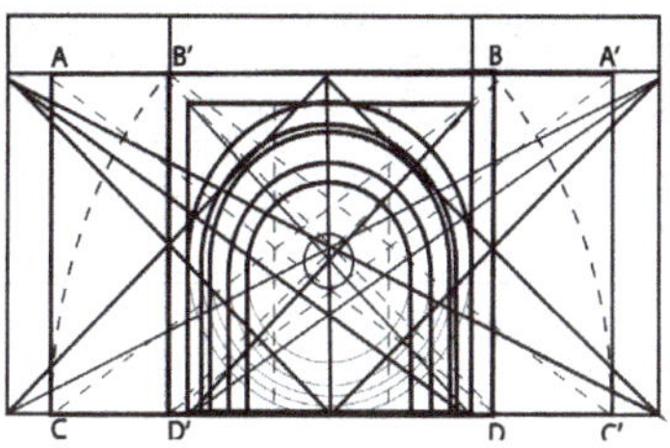

21. Regulatory layout from Bab Rouah to Rabat: Main entrance

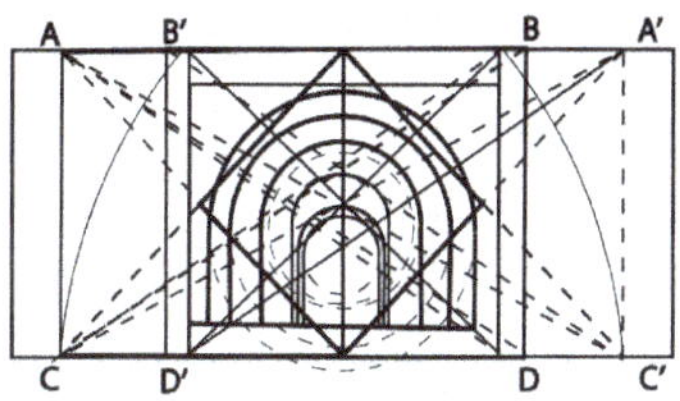

The regulatory layouts of the two buildings reveal a striking similarity. Imposing entrances, symmetry, almost identical elements, treatment of facades animated in similar ways only the materials differ. Successful cultural anchoring.

The folding of the short sides of the rectangles is used here in the simplest form; it is visible that this rigorous diagram makes it possible to inscribe this facade in two rectangles with the Euclidean golden ratio since in these rectangles all the diagonals and the obliques intersect according to this ratio of the golden section and serve to fix, in addition to the elements finding on either side of the entrance doors, the ground, the doors, the clock and the rest of the elements framing these entrances.

5.3. Arribat Center Multifunctional Complex in Rabat

A large-scale project, carried out by N. Basset and his team. This project aims to be an open block integrated into the urban setting of the Agdal district. It seeks to merge forms of things and the things of forms from both traditional and contemporary architecture.

Its design was focused on good integration into the landscape while preserving a "garden city" spirit right in the center of Agdal. Perfectly positioned in its site and offering maximum fluidity,

Materials Research Proceedings 47 (2025) 103-116 https://doi.org/10.21741/9781644903391-12

this project is made up of well-designed spaces where the element of water is present almost everywhere in different forms: swimming pools, ponds, fountains... ensuring good hygrothermal comfort throughout the complex. Moreover, it includes large green spaces aimed at partially absorbing CO2 emissions (photos 25, 26), while the climate systems are equipped with energy-saving devices. The fully glazed facades, mostly oriented to the north and east, are protected by roof overhangs and/or sun-shading devices, allowing for natural light infiltration and thus reducing heat. A significant part of this building consists of an open-air walkway designed to limit energy consumption. The double skin that covers almost the entire building is designed as a "moucharabieh" using the aesthetic technique of the "horror of the void," inspired by a new graphical study borrowed from Zellij (photos 22), and features alternating full and empty spaces, as well as thickening and thinning motifs. Inspired by Zellij, this skin serves a dual purpose: thermal regulation, making the building eco-energy efficient, and providing a Moroccan cultural grounding. As in traditional architecture, these elements are omnipresent for the same reasons.

Moreover, the architect himself acknowledges that his project, through the "open block" concept, is based on the quality of generous spaces inherent to traditional Moroccan architecture, particularly the large square opening onto Ibn Yassine Square. He adds: "The commercial heart of the project, fully connected to the Agdal district, is organized on two levels around pedestrian streets, most of which are covered, recapturing the charm and spirit of the Medina's alleyways." (12)

22. "Contemporary design with concrete, steel, and glass, and cultural anchoring through the covering of the entire building with a skin inspired by Zellij in the form of a moucharabieh, which serves as a thermal regulator."

24

23. "Like traditional constructions, the presence of water in various forms and in different spaces of the building ensures hygrothermal

(12).https://aemagazine.ma/complexe-multifonctionnel-arribat-center-une-nouvelle-centralite-urbaine-pour-l-agdal/

Materials Research Proceedings 47 (2025) 103-116 https://doi.org/10.21741/9781644903391-12

24 et 24 bis. "Traditional motifs as a source of inspiration and extrapolation."

25. Vegetation inside the complex".

26. With its 'garden city' design, this complex is equipped with multiple green spaces intended for the partial absorption of CO2.

Conclusion

The social and cultural conditions that characterize Morocco now compel us to reconsider the notions of traditional, modern, and/or contemporary in the field of architecture. This is a delicate domain, as there is already in place an Amazigh architecture, an Arab-Muslim traditional architecture, and a rich artisanal heritage, which itself constitutes the essence of this architecture. We are not facing a fixed and finished field once and for all, but rather processes of contextual construction and architectural adaptability. This study advocates for the recognition of contemporary Moroccan architecture as a product of multiple temporalities. Indeed, the many configurations of contemporary Moroccan architecture show that great importance should be given not only to the notion of time, to the evolution of mentalities, techniques, and technologies, but also to how they fertilize, accumulate, and merge.

The various investigations carried out in this study on architecture in Morocco, between tradition and modernity, reveal that so-called modern and/or contemporary architecture is increasingly distinguished by creativity, techniques, technologies, and new materials, helping it stand apart from another Moroccan architecture, also called modern, but which is often anonymous and eclectic. Contemporary Moroccan architecture, according to our study, innovates and builds its concepts by drawing from traditional forms and identity references, either by borrowing forms, principles, decorative elements from so-called traditional architecture and/or Moroccan artisanal art, or by extrapolating and transposing typical and specific forms and motifs from these domains, or by combining and associating these elements in various ways to create new forms, particularly in the design of significant buildings. The various possibilities used by this architecture can generally be summarized by either the integration of identity elements' forms, such as incorporating elements directly drawn from traditional architecture, like volume, openings, patios, water features, vegetation, etc., or by "the things of forms," such as various and multiple combinatorial principles (Zellij panels, carved wood as an integral part of the construction, carved stone…), or finally by the extrapolation and transposition of "the things of forms" from the same identity elements generating new creative concepts, like the insertion of entire walls inspired by one of the components of traditional art.

The various "forms of things" and "things of forms" in contemporary Moroccan architecture can be classified into different distinct categories, at least at this stage of the study. The three major contemporary architectural projects studied reveal three distinct types of architecture.

The architecture of the LGV Agdal Station is a contemporary Moroccan architecture that has built its concept on the extrapolation of a single motif from traditional art, innovatively and creatively applied in various ways on the different walls of the building. An original skin that also filters daylight in certain places and lets out artificial light at night, while reducing the internal heat of the station. In addition, other types of integration of the same hexagonal shape are sometimes engraved, sometimes reproduced in 2D, sometimes playing with full and empty spaces, even involving the internal lighting system that ensures continuity of another kind.

The architecture of the Marrakech Station is also a contemporary Moroccan architecture whose concept is the complete extrapolation of another traditional building and the integration of artisanal elements and Berber art motifs in a revisited design, as demonstrated by the main facade of the station. Added to this is its symmetry and its inscription in the aesthetic canons of the divine proportion of classical architecture, which curiously can be found in the traditional architecture of Bab Rouah, and in other important buildings, as they all follow almost the same architectural concept.

The architecture of Arribat Center is also a contemporary Moroccan architecture. Its contemporaneity is primarily due to its incorporation into various repertoires of the forms of things and things of forms of traditional architecture and craftsmanship. An envelope entirely borrowed from the motifs of traditional Moroccan art, but also a concept grounded in the fundamentals of traditional architecture, as elaborated earlier.

Through the analysis of these three projects, we also observe that the initial hypothesis, namely that in Morocco, the design of more and more contemporary architectural projects is deeply rooted in Moroccan culture, is confirmed. Of course, the studied projects do not summarize all the different possibilities for drawing from Moroccan traditional architecture and art, as many other types already exist, particularly in Morocco's major cities. That said, traditional architecture and craftsmanship remain an invaluable and inexhaustible source of creativity and inspiration for a hybrid contemporary Moroccan architecture with a strong national identity.

References

[1] Centre (G.P.) ; Des architectures de Terre ou l'Avenir d'une tradition millénaire ; C.C.I. ; Paris, 1981 ;

[2] Choay (F.) ; L'allégorie du patrimoine ; Seuil ; Paris ; 1999 ;

[3] Fathy (H.) ; Construire avec le peuple ; Acte du sud ; Paris ; 1999 ;

[4] Hemenway (P.) ; Le code secret ; la formule mystérieuse qui régit les arts, la nature et les sciences ; Evergreen ; Köln ; 2008 ;

[5] Lafargue (B.) ; Artiste /Artisan, Figure de l'art 7 ; PUP ; Paris ; 2004 ;

[6] La grande encyclopédie du Maroc (Arts tradition) ; V.1 ; C, L'architecture ;

[7] Marc (G.) ; La Médina, modèle urbain pour le XXI s. ; Archimédia ; février ; 2011 ;

[8] Marçais (G.) ; L'architecture musulmane d'occident ; Paris ; 1954 ;

[9] Marçais (G.) ; Manuel d'art musulman, l'architecture (Tunisie, Algérie, Maroc, Sicile, tom : I et II, Edition Auguste Picard, Paris, 1926, 1927 ;

[10] Marrast (J.) ; Maroc, in « L'oeuvre de Henri Prost. Architecture et urbanisme, Académie d'architecture, 1960 ;

[11] Papadopoulo (A.) ; l'Islam et l'art musulman, Citadelle & Mazenod, Paris, 1976 ;

[12] Rudofsky (B.); Architecture sans architects; Chêne ; Paris; 1977;

[13] Théorie de l'architecture, de la renaissance à nos jours ; Taschen ; Köln ; 2015 ;

[14] Umberto (E.) ; Histoire de la beauté ; Flammarion ; Milan ; 2004.

Materials Research Proceedings 47 (2025) 117-123 https://doi.org/10.21741/9781644903391-13

So-called "vernacular" architecture in the Kongo Central province in the Democratic Republic of Congo (DR Congo): Characteristics and dynamics of architectural production in the Boko and Langa villages in Mbanza-Ngungu

Mylène KOKO ZOLA

Faculté d'Architecture, Campus Solbosch – Flagey, Bruxelles, Belgium

mylene.koko.zola@ulb.be

Keywords: Vernacular Architecture, Characteristics, Dynamics, Architectural Production, Villages

Abstract. The architectural production in the province of Kongo Central had already been noticed and reported in Europe by Duarte Lopez and Filippo Pigafetta in 1591 as well as Olfert Dapper in 1687, who subsequently fueled several other stories and missions in the Congo until the 18th century. The information reported by these authors is scant but allows us to get a more or less idea of the construction methods, the materials used, and the shape of the houses. This demonstrates that architecture remains little known in this region, it is there, but we do not see it because there are very few updated studies on so-called "vernacular" architecture in DR Congo and Kongo Central in particular. Our objective is to observe the existing situation with the inhabitants and their ongoing or current construction practices. Study this constructive reality carefully, to better understand it to be able to document the spatial organizations and details of this architecture; and to characterize Kongo constructive practices. Our period of study extends from the first half of the 20th century, with the creation of the first cement factory in Lukala (CILU) in Kongo Central in 1921, until the middle of the 21st century to highlight the relationships between these different constructive cultures. So, to get there, we will start by mainly presenting the architecture in the Kongo Kingdom, as described in the stories of missionaries and travelers, based on the literature consulted. Then, we will report our observations and analyses on the ground today. Without locking ourselves into a more or less forgotten past of this "traditional vernacular architecture", we will study certain elements which remain permanent over time at the moment of the break about colonization but also to tradition.

Introduction

Limited to the north by the Republic of Congo (Brazzaville), to the south by Angola, to the west by the enclave of Cabinda and the Atlantic Ocean, and to the east by the city of Kinshasa, capital of the Democratic Republic of Congo (DRCongo) and the Kwango Province; The Kongo Central region is one of the oldest provinces of the DRCongo.[1] It constituted one of the parts of the Kongo Kingdom, having benefited from written documentation that dates back long before the 21st century. This relates to the history as well as the social and cultural frameworks that characterized the Kongo Kingdom and the surrounding countries. In architectural terms, this province is marked, on the one hand, by the traditional architecture of the Kongo people ("traditional vernacular architecture") which testifies to a historical past known thanks to the stories of travelers and missionaries between the 16th and 18th centuries. centuries which depict the history, life, and customs of the ancient Kongo civilization. Indeed, these authors described and commented on this civilization using their terms and through their conceptions, marked by the thinking of their time. On the other hand, the colonial architecture of Lower Congo ("imported colonial architecture") testifies to an organization of housing responding to a new social structure

Materials Research Proceedings 47 (2025) 117-123 https://doi.org/10.21741/9781644903391-13

with a European style. The loss of this knowledge, the abandonment of certain constructive traditions, and the dynamics of transformation of the local context would be factors in the variation of architectural quality in the contemporary period.

Architecture remains little known in this region, it is there, but we do not see it because there are very few updated studies on so-called "vernacular" architecture in DR Congo and in Kongo Central in particular. The writings consulted come from the oldest explorers, who give some descriptions of the habitat and the landscape which could be exploited from the point of view of architectural research. But a large part of knowledge comes from anthropologists, historians, geographers, ethnologists, archaeologists, farmers, etc. while planning data and constructive know-how are not their responsibility. Referring to these ancient sources requires only preserving useful information, because it is sometimes repetitive, and taking a critical look at their biases. This work involves analyzing the manifestations of the production of "contemporary vernacular architectures" of Kongo constructive cultures in the space of the province.

The area studied is in the territory of Mbanza-Ngungu, which refers to the part of Mbanza-Nsundi and Mbanza-Mpangu, two of the six provinces of the ancient Kongo Kingdom, documented by explorers as resources and on which we can base ourselves on to properly carry out this research.

Literature to define it

This part aims to define the issue of the qualification of "traditional vernacular architecture", by confronting it with "Constructive Cultures". This is not an easy thing because from one era to another, from one author to another, there are consciences that differ. Nevertheless, we will attempt to develop this concept of "traditional vernacular architecture" by focusing on non-colonial studies that we consider relevant, to understand and highlight what characterizes them.

Construction is not separated from design. It involves not only knowledge but also intellectual reflection and research into our society's capacity for production and invention. It remains a social practice of which architecture is a mirror. Architecture and construction are not two autonomous practices, they are inseparable, without one being directly subject to the other. This complex, moving relationship participates in the processes of historical, socio-economic, political, and cultural evolution. Construction must not become a simple repetition of processes. It is not just a series of distinct pieces of knowledge, it is the result of a synthesis using various techniques and is part of "situational logic". Architecture must truly be the expression of scientific and technical knowledge, of mastered know-how, refocused and perfectly located in the historical context of production.[2]

Rudofski.1964 [3], defines it as Architecture without architects, made with "common sense", that which depends on the existing environment which shapes it. It is the expression of the values of popular culture that each country has invested in housing and its extensions according to Guindani and Doepper.1990 [4], "…in primitive societies more than elsewhere, architecture is the affirmation of a universal order. Guidoni.1995[5], Dzou-Tsanga beliefs and rites. 2016 [6]. In addition, the literature shows the cumulative, historical, and anchored dimension in the transmission of traditional architecture which testifies to the know-how developed throughout the world by successive generations for more than 10,000 years "Scientific know-how of civilizations" (Cicat , Unesco 1996). Oliver.1997 [7] takes stock of vernacular architecture around the world in his book Encyclopedia of Vernacular Architecture of the World, where he explores the theories, principles, and philosophies current in the study of vernacular architecture; he defines this architecture as being the architecture of people (as opposed to the architecture of an architect), architecture without an architect, using materials available on site and implementing traditional techniques. Vernacular architecture is still said in the literature without an architect, spontaneous, indigenous, or rural.

Materials Research Proceedings 47 (2025) 117-123 https://doi.org/10.21741/9781644903391-13

Bisson. 2007 [8], designates vernacular architecture as a type of architecture specific to a geographical area, a region, and its inhabitants; an architecture strongly influenced by the local context, cultural traits, and the impact of physical environments. But Bisson's contribution is to examine the notion of vernacular architecture to establish whether modern vernacular architecture is conceivable and what objects it qualifies. Frey. 2010[9] says that: "everything that is built on the periphery or outside of global economic flows, and which comes from manual, artisanal production, is vernacular, which makes it possible to reestablish the links which favor two processes in its inhabitants: they can recognize themselves in it and make it their own. » For him, vernacular architecture is a source of good ecological, social, and cultural practices; he generally assimilates it to the field of self-construction. Yasmine. 2016[10] Vernacular architecture is etymologically defined as an art and science of building construction specific to the country and the community that erected it.

This vernacular concept is usually used to describe architectures of the past, perhaps considered in a modern sense, which makes the link between the idea of tradition and the updating today of architectural questions: How does the architecture of the 20th century come from the tradition and the precise location of the Kongo territory? How can these elements carry a certain topicality about the issues in 2021 of territory, town planning, architecture, sustainable development, and modern vernacular? How can we define this Kongo architecture? Has there been an evolution of this architecture today as Dzou-Tsanga says: "Architectural expressions have thus diversified over the millennia to offer more security, stability and comfort to man. »? This study could help us position ourselves in this perspective.

Several authors provide suggestions for describing the vernacular architecture and its contemporaneity, particularly in the light of current concepts of "sustainable development": Architecture made with materials available on-site and implements traditional techniques (short cycle) Oliver; Architecture strongly influenced by local context, cultural traits and the impact of physical environments. Bisson, Dzou-Tsanga; A source of good ecological, social and cultural practices. Frey. The vernacular, inspired by traditional housing, is in tune with the times. The idea: take into account the climate, geography, and local materials to build Godfrain buildings. 2014 and Zevi.2016[11] return to the relationships between vernacular architecture and scholarly architecture (parallel between vernacular or popular architecture and Italian dialects). The latest work to date – in addition to the valorization of Kéré. 2016 [12] Yasmine, reports on the in-situ discovery by the three young women of these architectures which have once again become inspiring models thanks to the artisanal know-how used, the natural beauty of the materials used, and the absence of technological challenge. Their work highlights on the one hand the need to take the time to look to understand and feel places, and that of building up a sort of "library" of sensations, and on the other hand the formative virtues of travel, sources of meetings, exchanges and energizing experiences What about vernacular (Lajus, Mener, and Rieublac.2023).[13]

Returning to the archives is the method used in this study to present the architecture of the Kongo kingdom, as it is described in the stories of missionaries and travelers. Subsequently, the restitution of our observations and our analyses on the ground today, is based on interviews, surveys, plans, photographs, and sketches. The objective is to define constructive practices (understand, describe, and promote) and to highlight possible dynamics over time.

Travelers' stories and today's terrain

Since prehistory, man has sought to protect himself from natural phenomena by exploiting the resources of his environment to build a home to live in and protect himself. The creation of these constructions demonstrates the emergence of distinct constructive cultures thanks to the use of local materials, the implementation of construction techniques (local knowledge and know-how), and the diversity of forms used. This point presents the relationship of man with his environment (form and function) as well as our studies on materials.

Materials Research Proceedings 47 (2025) 117-123 https://doi.org/10.21741/9781644903391-13

Plan, form, function and size of the house

According to Cocheteux (1889), there were three types of huts. But it turns out that they have almost disappeared today, giving way to the rectangular shape topped by a double-sloped roof, still extending quite well beyond the walls to form a canopy and supported by several piles. The ancient chronicles do not speak of different types of houses, but rather of a single form, as described previously; with an irregular and disorganized layout, except in Mayombe, with rare exceptions, where all the huts are placed in the North-South direction with a slight North-West and South-East inclination, such that the prevailing winds blow in the direction of the walls having no doors or windows according to Diederich (1880). As for the size, it varies depending on whether the hut is large, spacious, low, small with dimensions more or less than (8x3) m, sometimes decorated with drawings on the walls, except for that of the chief whose dimensions are more or less (15x4) m and is especially responsible for the sculptures on the walls, representing human or animal forms, and which give a fetishistic character.[14]

Maes' work on housing in Bas-Congo in 1908 mentions this information. Let us also add, according to Mengho (1980), that the size of the house varies taking into account the size of the family and the financial and physical capacity of the builder to find and/or purchase the materials (choice, processing) and to build his house. He joins Cocheteux in distinguishing three types of dwelling: the first is the men's house, the most comfortable, made up of rooms whose number varies between one and four; the second is the kitchen hut where housework takes place; and finally, the palaver hut, with basic construction, is where the men meet, especially in the evening to discuss, have a meal together, resolve problems, etc.

Figure 1: on the left types of houses according to Cocheteux, plate VII, fig.19 of the Bulletin of the Anthropology Society of Brussels, t.VII, p.97. Royal Library, consulted in 2023. On the right, a model of house in the Boko village Photo taken in the BOKO village, by Mylène Koko Zola, 2021

The composition of the house, with an average of five people [15], is subdivided into two compartments: a room serving as a living room (acting as a bedroom for the children) and another room as a bedroom for the parents, according to Claessens (1905-1919). As for the older ones, especially the boys, they are housed in a separate hut. The kitchen, located outside the house or under the veranda for practical reasons (use of wood during cooking), is an often simple and basic construction attached to or separated from the accommodation. Apart from these spaces, other elements are added outside: a small field, more rarely a granary on stilts, an enclosure for small livestock (chicken coop, pigsty, etc.) and poultry, latrines, the fruit trees (safout tree, mango tree, avocado tree) giving fruit and shade, and the garbage hole. the existence of a WC in the accommodation is a particularly rare exception; Electricity is practically unknown in rural areas of this region; we use the light from the embers on which the meal was cooked.

In the villages surveyed, the shape of the houses has not changed. It is dominated by the rectangle with a gabled roof, overhanging to create a veranda, often supported by pillars called Malunzi (Tree branch which means pillar, the end of which has the shape of a "U". It plays the role of the structure and facilitates anchoring at the purlin level). On the other hand, we notice a diversity in terms of plan and dimensions, which are done according to the needs and means of each inhabitant of the village. As for the furniture in the house, it is not easy to find a table, a chain

Materials Research Proceedings 47 (2025) 117-123 https://doi.org/10.21741/9781644903391-13

or other objects that constitute its furniture, because the house is considered a refuge serving only to shelter at night and to protect against climate variations; this justifies the fact that the inhabitants of the village spend all day outside their homes.

Materials

Claessens (1905-1919), tells us that the Mayombe especially use straw: Nianga imperatu cylindrica, Maranda leaves, and those of certain palm trees (not being specified by the author). They do not know the use of nails, he adds, so instead of these, they use lianas from which they make ropes that will be used to unite everything. It is especially the vine of the native pepper tree that he calls kefu (Liane of the pepper tree that the native uses instead of nails, although he sometimes uses other elements as well. Claessens; lianas in the shape of ropes acting as nails and used to join the walls of the house among the Yombe, in Mayumbe According to Van Overbergh and De Jonghe, 1907, p.150). that the native uses, although he sometimes uses other elements as well. The walls are made of various materials, such as the petals of palm leaves, veins of a type of palm tree, and other woods. The materials at the roof level are thatched with a cannacea leaf called Masisa, growing like a pineapple on the edge and in the middle of the woods. The results of exhaustive agricultural and sample surveys in the Zaire Atlas (demographic study of Western Zaire from 1975-1976), show firstly that the plant box (branches covered with foliage) only remains in the regions of Kimvula and Moanda, but this material is a lot at roof level; then that the manufacture of raw (adobe) or cooked earth bricks and the addition of cement have been widely used here in terms of construction due to contact with missionaries and the proximity of the Lukala cement factory (CILU), first cement company in the province of Kongo Central, created in 1921. Mengho's study (1980) on rural housing in Congo Brazzaville also presents a diversity of local and external materials (sheet metal and cement). He specifies that more and more of the walls of houses are built from fired bricks and raw bricks. But the roofs are always made of straw or plant tiles and the use of sheet metal is common in several regions of the country.

Today, we observe that the use of leaves for the walls and roof has almost disappeared, giving way to raw or fired brick for the walls, to straw and sheet metal as a covering, the use of which has become common in the villages of the studied region. Referring to the parts of a construction (foundation, elevation, and roof), we can as materials found on the ground, at the level of the roof, as structural timber: Mbasa, Nkoko, Tutu, Nguangua, Mfiolongo, Maleka, Ndele , etc. ; as cover: "Nianga" straw, sheet metal. Elevation, like wood: Mbota, Ntiti, Mvulumukina, Mpungala, Mbasa, bamboo, Nsenga, Futi; as a coating and covering: hollow earth, banana and palm leaves, cement; raw brick; baked brick. The foundation, as a base: pieces of cooked bricks, beaten raw earth

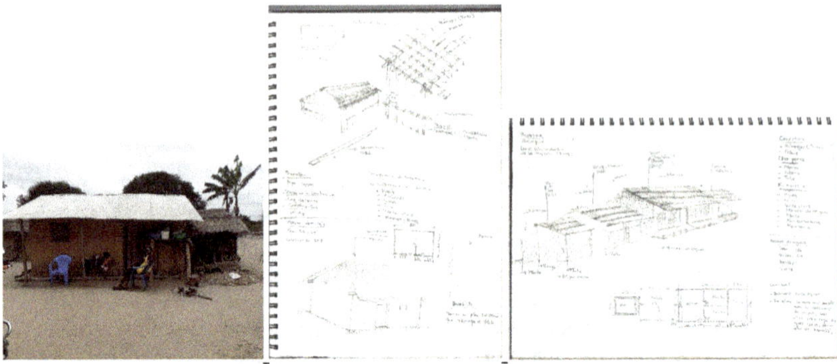

Figure 2: Photo and sketch of houses in the Boko village (in Mbanza-Ngungu), illustrating architectural details (plan, section, facade, implementation, etc.) and the materials used. By Mylène Koko Zola, 2021-2023.

Conclusion

Our conclusion is based on ancient chronicles (16th - 17th centuries) and our observations in the field, but it remains a hypothesis.

(1) The constructive cultures of Kongo vernacular architecture relate to the way of building in a particular historical era. The main holders of this constructive knowledge are the local artisans and inhabitants.

(2) It is an important heritage, which requires conservation, exploitation, and development. As Jean-Louis Michon [16] says, its conservation is essentially a function of the inhabitants themselves, who through their will maintain their tradition. However, this conservation of architectural forms and techniques is linked to the transmission and respect of local traditions and customs. Hence it can only be done when the villager continues to live in, use, and develop his rural environment. To achieve this, this involves thinking about the training of architects and the collaboration of residents in general, and particularly villagers, on the future of these constructive cultures in the region.

(3) Kongo vernacular architecture is architecture from communities, built by people for people, with local techniques and means expressing precise functions, and social, cultural, and economic needs; which proves that it is evolving.

(4) The promotion and perpetuation of these constructive practices within Kongo society according to its lifestyles, would deepen the understanding of the causes which explain their forms and would promote a possible renewal of architectural theories.

References

[1] Special Economic Zones Agency (AZES), Kongo Central technical sheet, https://www.azes-rdc.com

[2] UNESCO Architecture for Education Unit, Architecture and constructive cultures of construction. Elements for a construction teaching center, September 8-17, Paris 1993 (pp. 10-11)

[3] B. Rudofski, Architecture without architects, 1964

[4] S. Guindani and U. Doepper, Vernacular architecture: territory, habitat and productive activities. Architecture Collection. Lausanne: French-speaking polytechnic and university presses. 1990

[5] E. Guidoni, J. d'Yvoire, and B. Levrault, Primitive architecture. World history of architecture 6. Paris: Berger-Levrault, 1980

[6] R. Dzou-Tsanga, Architecture and technical identities in Cameroon, IHMC-Institute of modern and contemporary history (UMR 8066) edition, June 15, 2016

[7] P. Oliver, Encyclopedia of Vernacular Architecture of the World: Band 1 Theories and Principles. Cambridge: Cambridge University Press, ed. 1998

[8] M. F. Bisson, Marie-France, Modern vernacular? : towards an understanding of the notion of vernacular architecture and its links with architectural modernity » Memory. Montreal (Quebec, Canada), University of Quebec in Montreal, Masters in Arts Studies, 2007

[9] P. Frey, Learning from vernacular. Arles: Actes Sud, 2010

[10] M. A. Yasmine, Architectural changes and the quest for development in rural areas Burkjnabe: Supporting the relevant evolution of local constructive cultures for the design of a habitat that is a catalyst for sustainability, 2016

[11] B. Zevi, Architectural Dialects Hardcover – Illustrated, May 11, 2016

[12] B. Kéré, Architecture and constructive cultures of Burkina Faso. Villefontaine, France: CRATerre-EAG, 1995

[13] J. Lajus-Pueyo, A. Medec and M. Rieublanc, What about vernacular, 315 photographs, plans, drawings, sketches, 2023.

[14] J. Maes, housing in Bas-Congo, 1908

[15] Zaire Atlas, Agricultural Survey, 1975-1976

[16] Vernacular architecture confronted with the Venice charter, heritage news, n°61, May, 1995.

Materials Research Proceedings 47 (2025) 124-129 | https://doi.org/10.21741/9781644903391-14

The similarities between Almohad architecture and the vernacular architecture of Marrakech

MAJIDI Salma[1,a*], MHAMMEDI Mouna[1,b]

[1] Ecole Nationale d'Architecture de Rabat, Morocco

[a]majidi.salma@gmail.com, [b]mouna.mhammedi@gmail.com

Keywords: Morocco, Marrakech, Almohad, Vernacular, Monumentality

Abstract. This article aims to gather and emphasize the similarities between Almohad architecture and the vernacular architecture of Marrakech. Hailing from Marrakech, the Almohad Empire, while expanding, integrated local characteristics from the El Haouz region's vernacular style, into its architectural designs. Some of the components of this design are rammed earth, stone, and wood, which are used extensively in structures with thick walls and flat roofs. The city of Marrakech, famous for its key position in the rise of the Almohad Empire in the 12th century, has had a notable impact on its architectural style. Almohad architectural style, influenced by both Amazigh and Andalusian elements, had a great effect on the local architecture of the area.

Introduction

Vernacular architecture generally refers to local and regional building traditions and techniques that have developed over time in response to environmental conditions and cultural practices. One form of vernacular construction in Morocco is the *Qsar,* which is defined as a fortified village made of rammed earth, stone, wood, and local materials. These structures illustrate the adaptation of vernacular architecture to the climate, protecting against heat and cold. The spatial organization of the *Qsour*, characterized by closely spaced houses and interior courtyards, promotes thermal insulation and natural ventilation. Locally available materials, such as rammed earth (pisé) and earthen bricks (adobe), provide different construction techniques, passed down from generation to generation, attest to local craftsmanship, *"the materials used are available on site and are natural, so they are environmentally friendly, renewable, reusable and have very low production and processing costs."* [1].

Almohad architecture developed under the Almohad dynasty, a reformist movement based on the Islamic religion, that ruled over North Africa and parts of the Iberian Peninsula from the 12th to the 13th centuries. Their architectural style can be characterized by monumental buildings, such as mosques, city gates, and fortifications, which served to assert the power and legitimacy of the dynasty. Contrary to the austerity often attributed to the dynasty, Almohad architecture incorporates sophisticated ornamental elements, including arches, Kufic script, and *muqarnas*. The Almohads assimilated and reinterpreted architectural elements from other cultures, particularly Andalusian, thereby creating a distinctive architectural style that combines multiple influences. *"At this point, it seems evident that the Almohades showed a certain taste and inclination for the Andalusian artistic tradition. [...] also in the reclaiming of pieces from some Umayyad monuments in al-Andalus that were in ruins and were even taken to cities like Marrakech to use in its palaces and mosques (Rosser-Owen 2014, pp. 152–98). »* [2].

For instance, *Bab Agnaou*, a monumental gate built in the 12th century, features several decorated arches, entirely built in stones and earth, the monochromatic ornamentation of this massive gate exemplifies the monumental yet modest nature of Almohad architecture [3] which appears mostly in earth and stone colors. Similarly, the *Tinmal Mosque*, constructed in the 12th century, now in ruins because of the earthquake that hit Morocco in 2023, was inspired by the model of the Great Mosque of Córdoba and incorporates elements characteristic of Almohad

Materials Research Proceedings 47 (2025) 124-129 https://doi.org/10.21741/9781644903391-14

architecture. Is this modest and minimalist aspect of Almohad architecture inspired by the local and vernacular architecture of Marrakech? While vernacular architecture focuses on local traditions and environmental adaptation, Almohad architecture is marked by its monumentality and specific ornamental style, but what are the similarities between these types of architecture?

Historic context

Fig 1: Marrakech, in the beginning of the XIII th century [7]

The Almohad dynasty founded by Muhammad Ibn Tumart criticized the religious superficiality of the ruling Almoravids, advocating for a return to a pure and unified Islam based on the concept of *Attawhid*, the oneness of God. This explains the name *Almohades*, derived from Arabic *al-Muwahhidun*, meaning "those who profess the unity of God" [4]. After *Ibn Tumart's* death in 1130, his disciple *Abd al-Mumin* took leadership, conquering Marrakech in 1147 and expanding the empire across the Maghreb and Al-Andalus. By the mid-12th century, the Almohads controlled a vast territory extending from present-day Libya to western Algeria and southern Spain[4]. The peak of Almohad power was reached during the reign of *Abu Yaqub Yusuf* (1163-1184) and his successor *Abu Yusuf Yaqub al-Mansur* (1184-1199), but the empire began to decline following their defeat at the Battle of *Las Navas de Tolosa* in 1212, ultimately succumbing to the *Marinids* in 1269 [5].

Mosques played a central role in the religious and social life under the Almohads, characterized by vast, bright prayer halls often organized around a central courtyard surrounded by porches. The minaret, typically square, is another important element of the Almohad mosque. The Koutoubia mosque, initiated by the Almoravids and completed under the Almohads, features a minaret over 70 meters tall, representing the Almohads influence on the city's religious architecture [6].

Fig 2: Medina of Marrakesh, the Koutoubiya Mosque - Martin Gray © Sacred Sites whc.unesco.org/en/documents/110417

Materials Research Proceedings 47 (2025) 124-129
https://doi.org/10.21741/9781644903391-14

Fig 3: Drawing of clustered home, example of Douar Anammer region of Marrakech-Safi [8]

Fig 4 : Medina of Marrakech, bird eye view - Yann Arthus-Bertrand, 2016

Vernacular architecture in Marrakech, like in different regions in Morocco, reflects a tradition of construction adapted to local conditions, with the use of local materials. Based on examples from the region of Marrakech-Safi, neighboring houses are in a form of clusters called *douar,* and each house is built around a courtyard, it can be assumed that vernacular architecture in Marrakech evolved in response to several factors. The arid climate of Marrakech, characterized by hot, dry summers and mild winters, has influenced construction techniques and materials used. Traditional houses in Marrakech, such as *riads*, are designed to protect against heat while maximizing natural ventilation [8].

Vernacular architecture favors the use of locally sourced building materials, including rammed earth (pisé) and earthen bricks (adobes), palm wood, and reeds. These materials are not only available but also provide thermal insulation. Additionally, vernacular architecture in Marrakech has been influenced by various cultures, since the city was the capital of multiple dynasties, including the Almohad dynasty, before they moved their capital to *Ribat At-Fath*, The Riads, are the form of persistent vernacular and traditional houses, primarily present in the Medina of Marrakech. The Riad is organized around an inner courtyard often adorned with a garden and fountain, creating an oasis of freshness and greenery at the heart of the house, while thick walls and small windows protect the house from heat.

Almohad architecture and vernacular architecture represent two important facets of Marrakech's architectural heritage. Almohad architecture reflects the power and influence of a dynasty through grand monuments, while vernacular architecture embodies the adaptation of inhabitants to local conditions and the transmission of ancestral knowledge. The coexistence of these two architectural styles in Marrakech contributes to the richness and diversity of its urban landscape.

Materials Research Proceedings 47 (2025) 124-129 https://doi.org/10.21741/9781644903391-14

Fig 5 : Tinmal mosque, 2019 [10]

Exploring both vernacular and Almohad architecture

Functionality:

- Vernacular architecture: By definition, vernacular architecture focuses on functionality, meeting the practical needs of its inhabitants using local materials and techniques. In Marrakech, this would translate into dwellings adapted to the hot, dry climate, using materials such as raw earth and techniques such as courtyards for ventilation. [9]
- Almohad architecture: The Almohads, as a powerful dynasty, built buildings with important political and religious functions. The Koutoubia Mosque and the Tinmal Mosque, for example, served as places of worship and assembly, while Bab Agnaou marked the entrance to the Qasbah, the center of Almohad power [6].

Materials and aesthetics

- Vernacular architecture: The materials used in vernacular architecture are generally simple and locally available. Raw earth, stone and wood would be favored, creating a simple aesthetic integrated into the environment.
- Almohad architecture: The Almohads, wishing to assert their power and refinement, used richer, more ostentatious materials, such as marble imported from Al-Andalus, *"[...] Many of these pieces were marble columns and capitals from, for instance, Madīnat al-Zahrā'. This is confirmed in the minaret of the old Almohad mosque in Seville, in the aforementioned courtyard of the Casa de Contratación of the Reales Alcázares of this same city, and even in the mihrab of the Kutubiyya mosque itself."* [2], for columns and decorations. Attention to detail, complex geometric patterns, and the use of calligraphy as a decorative element testify to an elaborate aesthetic quest.

Evolution and adaptation:

- Vernacular architecture is constantly evolving, adapting to the changing needs of the population and environmental conditions. In Marrakech, the influences of different populations (Amazigh, Arab, Andalusian) would have helped shape vernacular architecture over time.
- Almohad architecture: Almohad architecture, though majestic, has remained frozen in time, bearing witness to a particular era and aesthetic. Although some monuments have undergone restoration, they retain their original structure and decoration.

Materials Research Proceedings 47 (2025) 124-129 https://doi.org/10.21741/9781644903391-14

Influence of the Almohad Empire on Vernacular Architecture

The influence of the Almohad and the local architecture of Marrakech on urban planning as well as the city's landscape is deeply rooted in its complex history. The numerous Almohad buildings present in Marrakech such as the Koutoubia Mosque, with its carved details, and the ramparts that surround the historic city, have truly shaped not only Marrakech's visual identity but also its cultural heritage. Both Almohad and vernacular architecture play essential roles in constructing collective identity in Marrakech, shaping its urban landscape and reminding its inhabitants of their historical and cultural heritage.

The way these two types of architecture influence each other can be observed through several convergence points; both architectural styles widely utilize locally available construction materials. For instance, the Almohads used earth bricks and stone to construct the gates of Marrakech, while vernacular architecture in the region of Marrakech-Safi employs raw earth. This similarity likely arises from practical constraints and a desire to integrate with the environment. Additionally, the importance of climate adaptation in both architectural styles, the Qsour of the Marrakech-Safi, which serve as examples of vernacular architecture, are designed to protect inhabitants from heat and sandstorms. Similarly, although Almohad architecture is known for its monumental structures, it incorporates elements that address climate challenges, such as interior courtyards and fountains.

Conclusion

Almohad architecture, while drawing from earlier architectural traditions (such as Umayyad architecture from Andalusia), also developed its architectural forms and solutions. For instance, the use of arches in the Tinmel Mosque showcases a certain level of innovation. However, several elements clearly distinguish Almohad architecture from vernacular architecture; Almohad architecture is primarily a tool of power and religious propaganda. Imposing mosques like the Koutoubia Mosque, as well as monumental gates like Bab Agnaou, were designed to assert the power and legitimacy of the Almohad dynasty. This pursuit of grandeur contrasts with the functional simplicity of vernacular architecture. The Almohads also aimed to use noble materials and even imported marble from the ruins of Umayyad cities (Madinat Az-Zahra) to ornament their monuments. The complexity of geometric patterns, the use of calligraphy, and the attention to detail reveal an aesthetic pursuit different from the sobriety of vernacular architecture. Furthermore, the construction of Almohad monuments required significant technical and artistic expertise; stucco workers, sculptors, and calligraphers...etc. In contrast, vernacular architecture often relies on skills passed down within communities and the participation of the inhabitants themselves.

In conclusion, it is more accurate to describe Almohad architecture as learned architecture, supported by a political and religious elite, rather than vernacular architecture. It is part of a rich and complex architectural and artistic tradition, reflecting the Almohad dynasty's desire to assert its power and grandeur.

Fig 6 : Bab Agnaou, 1889 [3]

References

[1] Sustainable Architecture and Urban Development Sustainable Vernacular Architecture: The Case of the Draa Valley Ksur (Morocco), Eliana Baglioni.

[2] The Almohad Caliphate: A Look at Al-Andalus through Arabic Documentation and Their Artistic Manifestations - Ignacio González Cavero

[3] Archaeological Research at 'Bab Agnaou' in Marrakech (late XIIth century): First Results - Hicham Rguig, Abdallah Fili, Abdellatif Marou, Asmae El Kacimi, 2023

[4] The Almohad Caliphate: A Look at Al-Andalus through Arabic Documentation and Their Artistic Manifestations - Ignacio González Cavero, 2018

[5] Histoire des Almohades / d'Abd el- Wâh'id Merrâkechi - traduite et annotée par E. Fagnan. 1893.

[6] Marrakech – Hamid Triki, Le Prix Aga Khan d'Architectur,e Ceremonie de remise des prix 1986.

[7] La médina de Marrakech : formation des espaces urbains d'une ancienne capitale du Maroc - Quentin Wilbaux, L'Harmattan 2001

[8] Les architectures régionales de Tansift – Youssef Melehi, Aziz Derouich, Ed Okad, 2008

[9] Exposer l'architecture vernaculaire - Nadya Rouizem Labied, Édition électronique URL : https://journals.openedition.org/lha/9700 DOI : 10.4000/lha.9700 ISSN : 1960-5994

[10] Une Parure Almohade - Hammad Berrada, Ed Marsam, 2019

Materials Research Proceedings 47 (2025) 130-137 https://doi.org/10.21741/9781644903391-15

The intellectual representations of vernacular architecture through the surveys of the MNATP, 1941-2001

Frédéric FIRRERI

Laboratoire de recherche en architecture, École nationale supérieure d'architecture de Toulouse (ENSA-Toulouse), 83 rue Aristide Maillol, BP 10629, 31106 TOULOUSE CEDEX 1, France

frederic.firreri@toulouse.archi.fr

Keywords: Folklore, Rural, Popular, Institutions, Housing

Abstract. This article discusses a succession of two national surveys on rural architecture conducted by the National Museum of Arts and Popular Traditions at the Trocadéro in Paris between 1941 and 1948, and then between 1968 and 1988, which gave rise to the publication of a collection named *L'Architecture rurale française*, encompassing twenty-three volumes published from 1977 to 2001[1]. These surveys shed light on the relationship between the peasant in their territory and the scholar in their institution the scholar in the territory and the peasant in the institution. Within this dialectic, one can understand how the production of knowledge about the territory gradually became institutionalized and how the tools of knowledge, work methods, and publication of works can today constitute a part of the history of French rural architecture. Learning from territories therefore means understanding our scientific and institutional mechanisms shaped by scholarly representations. It is through various transformations of these representations that one can detect an intellectual, institutional, and political trajectory. These mutations are revealed by the evolution of the semantic field mobilized by its actors, changes in individuals involved in the surveys, and the shifting economic policies that shaped these studies.

Introduction

The context and results of this extensive collective work have been discussed by several architectural historians. Jean-Claude Vigato, Jacques Lucan, Daniel Le Couédic, Monique Eleb, Lionel Engrand, and Patrick Giromini described the political, methodological, and architectural dimensions, as well as the aims of the intellectual project 1425, but did not establish any connection with the second part of the survey [2]. More recently, at the Collège de France, Jean-Louis Cohen conducted a course entitled "Regionalism and Rurality," followed by a recent article describing both surveys [3]. In 2021, Jacques Lucan also focused its work on the collection as a whole [4]. In the field of humanities, several authors have addressed the first survey and the collection [5]. Lucien Bernot and Gérard Collomb discussed its administrative and methodological aspects, while Pascal Ory, Laurence Bertrand Dorléac, and Christian Faure examined the political and institutional dimensions of the project [6].

In addition to these studies, the actors themselves have produced an enlightening literature. Georges-Henri Rivière and Marcel Maget both contributed to the understanding of the political, institutional, and methodological dimensions of the first survey [7]. Henri Raulin, in 1964, highlighted the entirety of EAR I, followed by Jean Cuisenier who summarized the proceedings of EAR II in 1988 into two articles, one of which appears as a preface in each volume of the collection [8]. Despite the numerous authors who have mentioned or studied the topic, the relationship between EAR I, the related collection of works, and the scholarly representations that emerged had not been examined yet. This article aims to understand and historicize this critical moment in the construction of knowledge about French vernacular architecture.

At the beginning, the Chantier intellectuel 1425

The Chantier intellectuel 1425 is a field survey on rural architecture that was launched in 1941 in the Department of Arts and Popular Traditions housed in the Palais de Chaillot, inaugurated in 1937 during the Paris World Exposition. It should be noted that it was, at that time, an international preoccupation, with folklore museums opening all around Europe at the end of the 19th century, such as the National Museum of Finland and the Norwegian Museum of Cultural History in Oslo (Norsk Folkemuseum) in 1893, the Museum of Ethnology in Vienna in 1895, and the Danish Museum (Dansk Folkmuseum in Copenhagen) in 1897 [9]. It was the government architect Urbain Cassan who, in 1941, suggested to Georges-Henri Rivière, the museum curator, to use the intellectual and artistic projects of the General Delegation for National Equipment to study rural architecture [10]. This opportunity was both generated and taken by a young friend of the avant-garde group of the "Bas-Matérialistes," led by Georges Bataille. The Center 1425 was therefore created under the Vichy regime, where nearly twenty architects collaborated at its creation. By 1945, with the establishment of the Service of Mandatory Labor, nearly fifty new architects would join. This was not the only intellectual project within the institution, as others focusing on traditional furniture (909), traditional crafts (1810), and the organization of documentation of the same museum were launched simultaneously.

The MNATP had already been conducting field surveys since its opening with architects (Guy Pison) under the Front Populaire. Thus, an institutional dynamic was initiated with Léon Blum's government and continued under the Vichy regime. Contemporary French ethnologist Christian Bromberger notes that "the themes (peasantry, craftsmanship, folklore, provincialism) that flourished under Vichy were already well established before the war"[11]. Yet, long before the war, folklore mobilized scholarly societies and their members in commissions, publishing journals and works. However, the subject of architecture was rarely addressed. It was truly the Chantier intellectuel 1425 that introduced this field of study and simultaneously provided it with the epithet of folkloric architecture. Although initially named a Survey of Regional Architecture, it was this term (i.e. folkloric architecture) that came to be used throughout the 1940s. In 1936, in the *Revue du folklore français et du folklore colonial*, Georges-Henri Rivière attempted to define it:

It was introduced in 1848 by the archaeologist William John Thoms, meaning 'knowledge of the people,' from lore 'knowledge' and folk 'people.' Universally accepted today in the scientific community, the term 'folklore' is used to designate both this knowledge of people and the scientific discipline that aims to study it [12].

Folklorists were collectors and accumulators driven by the collection of this material culture. It is within this dynamic that the MNATP created Center 1425, but it sought to go beyond this.

Methodological framework

To establish Chantier 1425, Georges-Henri Rivière relied on two types of sources that provided a certain foundation. The first consisted of various field experiences from survey members, and the second encompassed results from earlier historical surveys. Thus, the initial surveys conducted by the MNATP under Georges-Henri Rivière on Sologne, Lower Brittany, and Provence as early as 1938 were the first collective experience of the future Chantier intellectuel 1425. Georges-Henri Rivière drew upon the experience of André Varagnac, who led the Collective Research Commission for Volume XIV of the *Encyclopédie française*, directed by historian Lucien Febvre [13]. Finally, he collaborated with geographer Albert Demangeon, who led a national survey on folk housing using a questionnaire and on-site surveys of houses. Albert Demangeon and Georges-Henri Rivière teamed up in 1937 to present this work in an exhibition at the Museum of Modern Art entitled "La Maison rurale en France"[14].

Regarding the historical surveys mobilized for Chantier 1425, Georges-Henri Rivière cited the first volume of the Survey on Housing Conditions in France by Alfred de Foville and Jacques Flash, published in 1894, followed by a second volume in 1899. These works gathered around fifty reports drafted by correspondents, primarily teachers spread across the French territory. They were required to answer a questionnaire developed by Alfred de Foville and Jacques Flash, which investigated the definition of a "typical house" or a "characteristic unit" specific to each region [15]. Georges-Henri Rivière also referenced other studies, such as the 1807 survey by the Celtic Academy, the 1885 survey by the Committee for Historical and Scientific Work, and Maurice Vignerot's study from 1939.

At the beginning of the century, the survey method was already well established, and it was already being used to construct knowledge. Georges-Henri Rivière gathered a team to lead the survey, consisting of three architects: Guy Pison (1905-1986), Urbain Cassan (1890-1972), and Pierre Drobecq (1893-1944), along with folklorist Pierre-Louis Duchartre (1894-1983), ethnologist Marcel Maget (1909-1994), and historian Albert Soboul (1914-1982) from 1943[16]. The objectives of the survey were to establish a ground for future scientific research, serve as references for constructing future buildings, and create one of the museum's collections for upcoming exhibitions. Under the direction of Georges-Henri Rivière, the management committee drafted the guidelines, methods, and objectives for the architects. Trained in ethnographic survey methods upon their entry to the MNATP for several weeks, the architects became "investigators." They could choose the department they wished to study and were sent to their designated area. Thus, almost the entirety of the French territory was explored [17].

Instruction for investigator-architects

The management team developed very precise instructions for the investigators, gathered in a booklet that followed them throughout their travels in wartime France. From this small booklet, one can glean the management team's expectations in compiling monographs by buildings. These were to be selected by the on-site investigator, following an inspection of the department they chose. The buildings were required to possess "typical characteristics or characteristic elements that recur with a certain frequency, termed 'constants,' expressing their persistence". These are "specimens". The objective is "to gather, for the main types of houses, and mentally, all these constants to obtain the characteristic houses of the studied region. These characteristic houses are thus imaginative syntheses that you do not encounter in reality." The effort is placed in constructing a fictitious, idealized image that the specimens support. Regarding the method, it states: "It would not be sufficient to record observations; you must classify, refine, and coordinate them to reach a systematic study" [18]. Thus, investigators needed to capture representative buildings and produce standardized raw materials to accumulate results and compare them to decode rural aesthetics, becoming markers of local constructive cultures. This represents a dialectical relationship between the investigators and the management, the field, and the institution.

A national survey

Two architect-investigators were in charge of representing the occupied and free zones of France (i.e. at the time of war), respectively François Beurdeley and Yvon Almairac. One or two investigators were responsible for two French districts. More precisely, the work of the investigators was divided into two phases. The first one involved developing preparatory documents, where each investigator had to keep a daily log to record all the information and events that had marked their workday; a sketchbook to note all the observations of the selected specimen; and a photographic proof book to compile all their photos. These tools formed the very first sources of information, gathered in situ, and allowed for the daily construction of knowledge. The second phase of their work was to finalize the preparatory documents. These were the "definitive" documents, representing an initial form of realization and the first true exploitable source. These documents are specifically detailed in a booklet divided into two parts: descriptive (statistics, precise quantities, etc.) and graphic (with the graphic elements manipulated by architects), and it is separated into three scales: departmental, communal, and architectural. These monographs follow a very precise and rigorous plan developed collectively by the management team[19].

A total of 1,634 monographs were produced between 1942 and 1946 [20]. This empirical and massive practice was, in reality, a process of standardizing the data collected in situ. which allowed to consider folk architecture as a scientific object of study. More than that, it was the beginning of a type of science, a typology (i.e. a collective work of categorizing buildings), allowing for their association or dissociation, tracing a common history with identifiable influences and specificities for each category.

In 1945, at the end of the war, architects were called upon by the government to participate to the reconstruction of France, and the Chantier 1425 faced staffing issues. To address this crisis, the curator established the Centre for Professional Training of Rural Architects and Urban Planners to welcome students for long internships. This renewal took place between 1945 and 1948, and it was truly the 1425 project that continued in this new framework. However, on October 15, 1947, a letter from the Minister of Labor and Social Security of the Provisional Government of General de Gaulle informed Georges-Henri Rivière that the necessary subsidies for the center's proper functioning would not be renewed. This marked the cessation of these field survey activities at the end of that year [21].

From folk architecture to rural architecture

The term "folklore" was definitively abandoned by the end of the Second World War due to its association with the propaganda of the Vichy regime and its past linked to chaotic material accumulation. The museum ceased its field survey activities on rural architecture, but there was still a lot of communication about the 1425 survey in the specialized press, including an entire issue dedicated to it in the January 1947 issue of the journal *Technique et Architecture* [22]. Beginning of 1953, many monographs and articles were published in the museum's journal called *Arts et traditions populaires* [23].

In 1963, Henri Raulin, an ethnologist nearing the end of his thesis work, joined the museum. He began investigations into traditional crafts in the Alps and published an article in 1964 highlighting the high quality and unprecedented nature of the archives from the Chantier 1425[24].

The Corpus of Rural Architecture – 1968 to 1988

In 1968, Jean Cuisenier, an anthropologist and CNRS research director who succeeded Georges-Henri Rivière, reopened the investigation to publish it at the new headquarters of the institution built by Jean Dubuisson. This was made possible by the ongoing efforts of Georges-Henri Rivière's to establish institutional links between the MNATP and the Centre National de Recherches Scientifiques since the 1950s, which was then thriving. Contemporary anthropologist Martine Segalen specified that this institution "functioned with disciplinary commissions deciding on recruitment and career paths of staff, as well as the allocation of resources" [25]. This favorable context allowed for the resumption of the survey through the hiring of scientific staff funded by the CNRS, such as Henri Raulin. The survey took place within the newly established Centre d'Ethnologie Française, created a year earlier in 1967 by a convention between the two institutions, one of the first museum-laboratories in France, although this term had been used since the 1940s. Its experimental nature also explains its fragility, and this museum closed permanently in 2005. The Chantier intellectuel 1425 was named "Corpus of genres, types, and variants." It was directed by Jean Cuisenier, with scientific coordination provided by Henri Raulin, appointed research director at the CNRS. A framework was developed by the latter two, and specialists from each region were contacted to lead the investigation at a regional level.

They were given instructions and the monographs from the Chantier 1425 for each relevant region. Between 1968 and 1988, this second survey was divided into two phases. The first involved

Materials Research Proceedings 47 (2025) 130-137 https://doi.org/10.21741/9781644903391-15

revisiting the field multiple times for each director, using the monographs from the 1940s to supplement, update, and produce new ones [26].

The second phase consisted of gathering the monographs of the analysed territory to produce a general synthesis by region. In this way, the directors traced relationships between monographs while attempting to preserve the density of the original descriptions. It was during this double movement, between completion and rationalization, that a history of types, their variants, and their construction techniques in their cultural, social, and economic contexts gradually emerged. It was during the second survey that the types became apparent, thanks to the connections identified by the directors among the constructions recorded collectively throughout both surveys. These connections were typological-morphological, following the form of the habitat and the materials used; economic, according to the type of activity; social and human, depending on relationships between humans and animals, rituals and religious practices, family structures (nuclear families, root families, paternal or communal families—inheritance); and finally, historical, as the directors attempted to trace the genesis and evolution of the type of habitat. The inquiry policy of the Chantier 1425 was followed by a publishing policy of the second survey. Twenty-seven works were planned at the launch of the collection.

Each volume was structured into two main parts: a textual part, descriptive and historical, written by one or more directors, followed by a graphic presentation of the specimens: plans, sections, elevations, details, stemming from the Chantier 1425, similar to identity cards. They served to illustrate and support the claims made in the first part. From 1977 to 1986, eighteen volumes were published by Berger-Levrault; and in 1988, a volume was released by La Manufacture. Finally, from 1992 to 2001, four volumes were published by A. Die.

The high cost of publication of each volume explains these changes in publishing houses, especially since the collection remains unfinished up to this day. In addition, the volumes on *Périgord Quercy Agenais, Haut Languedoc Rouergue, Maine Anjou Touraine*, and *Limousin* were never published. Due to the decline in the number of buyers, the publishing house A. Die put an end to the editorial project, leaving this puzzle of architectural anthropology incomplete (fig. 1).

Fig. 1: the complete collection of Architecture rurale française.

Conclusion

Several threads guide this institutional history. These surveys are permeated by a growing concern to conserve and enhance rural architectural heritage over the decades through multidisciplinary study. The modalities of knowledge production have evolved within an institutional grouping. While the monographic approach of the 1940s, through in-depth field investigation, was aligned with the desire to transcribe reality in all its complexity, the corpus used regional synthesis by reconfiguring the monographs and made the types of habitats emerge in their historical, cultural, social, and economic contexts. However, the history of the specimens in the monographs from the 1940s appears quite obscure, if not absent, leaning towards a certain mythification of the past.

The directors of the first survey used and claimed a disciplinary crossover of architecture, geography, ethnology, and history to understand an object from all angles—a crossover that had formed during the dark years of the Occupation. The second survey and its publication aimed to push the institutional boundaries further, continuing the multidisciplinary aspiration of the Chantier intellectuel 1425. While architects largely disappeared with the closure of the 1425 Center, over the years, anthropology has asserted itself with a concern for subsequent scientific openness. Only one architect, Daniel Le Couëdic, remained among them, and multidisciplinarity was relegated to the background [27].

The establishment of institutional hierarchies, protocols, and applied investigation devices on a selection of buildings constituted a sampling, an inventory of rural habitats. There was therefore an institutional filter that was collectively generated, although it is the most faithful to reality. At the same time, the MNATP has inserted a dichotomy between scholarly and rural architecture—a form of inequality, a hierarchy opposite to the learned culture of architects represented by the Académie des Beaux-Arts, and more broadly between popular arts (which also encompass furniture, costumes, decor, crafts, dance, song, tales) and fine arts.

From the Popular Front to the Fifth Republic, the surveys persisted. They reinvented themselves according to their political context and evolved with institutional transformations and mandates. Their various initiators maintained the desire to study this material culture scientifically. Finally, we can observe that despite the participation of various ministries in the publication of the collection through subsidies, rural architecture has not been subject to specific protection laws in France. The concerns of the directors of the Chantier 1425 regarding potential destruction of the selected buildings were justified, as many monographs from the 1940s were destroyed at the beginning of the second survey. Questions regarding its legal, financial, and institutional position remain open.

References:

[1] J. Cuisenier (Eds.), L'Architecture rurale française, Corpus des genres, des types et des variantes, Paris, Musée national des arts et traditions populaires, Berger-Levrault, 18 volumes, 1977-1986, p.9.

[2] J.-C. Vigato, L'Architecture régionaliste: France, 1890-1950, Wetteren, Éditions Norma, 1994, p. 324; J. Lucan, Architecture en France, 1940-2000: histoire et théories, Paris, Le Moniteur, 2001, p. 30; D. Le Couédic, La maison ou l'identité galvaudée, Rennes, Presses universitaires de Rennes, 2003, pp.79-80; M. Eleb et L. Engrand, La maison des Français: discours, imaginaires, modèles (1918-1970), Bruxelles, Belgique, Mardaga, 2020; P. Giromini, Transformations silencieuses: étude architecturale du bâti alpin, Genève, MetisPresses, 2022, pp.131-133. https://doi.org/10.37866/0711-15-4

[3] J.-L. Cohen leds the chair « Architecture et forme urbaine » de 2013 à 2021 in: Collège de France. See 22th june 2016 for the Chantier 1425: https://www.college-de-france.fr/fr/agenda/cours/architecture-dans-la-france-de-vichy-1940-1944/regionalisme-et-

ruralite. Also J.-L. Cohen, « Un fonctionnalisme rural: le chantier 1425 des arts et traditions populaires », in: Shahram Abadie, Gauthier Bolle et Amandine Diener (Eds.) Pour une histoire culturelle de l'architecture, essais offerts à Anne-Marie Chatelet, MetisPresses, 2023, pp. 221-228.

[4] J. Lucan, Habiter, ville et architecture, Genève, EPFL Press, 2021, pp. 20-26.

[5] I. Chiva et F. Dubost, L'architecture sans architectes : une esthétique involontaire ?, Études Rurales, n°117, 1990, pp. 9-38, ainsi que J. Christophe, Les journaux de route du Musée national des Arts et Traditions Populaires, n°1, mars 2000, pp. 137-146 et M.-N. Denis, L'enquête d'architecture rurale (1940-1968), une étape dans la construction de l'ethnologie française, in: J. Christophe, D.-M. Boëll et R. Meyran (Eds.) Du folklore à l'ethnologie, Paris, 2009. https://doi.org/10.3406/rural.1990.4647

[6] L. Bernot, Le corpus de l'architecture rurale française, Études rurales, vol. 73, n°1, Paris, 1979, pp. 125-135; G. Collomb, « Logique sociale et composition architecturale : Jean Cuisenier et le corpus de l'architecture rurale française », Bérose, 2020; P. Ory, La belle illusion - Culture et politique sous le signe du Front populaire (1935-1938), Paris, CNRS Éditions, 2016; L. Bertrand Dorléac, Histoire de l'art : Paris, 1940-1944 : ordre national, traditions et modernités, Paris, Publication de la Sorbonne, 1986; C. Faure, Le Projet culturel de Vichy, Éditions du CNRS, Lyon, Presses universitaires de Lyon, 1989.

[7] G.-H. Rivière, « Le chantier 1425: un tour d'horizon, une gerbe de souvenirs », Ethnologie française, vol. 3, n°1/2, Presses Universitaires de France, 1973, pp. 9-14; M. Maget, A propos du Musée des Arts et Traditions Populaires et sa création à la libération (1935-1944), Genèses, n°10, 1993, pp. 90-107. https://doi.org/10.3406/genes.1993.1157

[8] H. Raulin, L'architecture rurale française. Une enquête nationale inédite (1941-1948), Études rurales, vol. 13, n°1, 1964, pp. 96-119; J. Cuisenier, Le corpus d'architecture rurale: logique sociale et composition architecturale, in: I. Chiva *and al.* (Eds.) Habitat et espace dans le monde rural, Éditions de la Maison des sciences de l'homme, 1988. https://doi.org/10.3406/rural.1964.4817

[9] A.-M., Thiesse, La création des identités nationales: Europe, XVIIIe-XXe siècle, Paris, Seuil, 1999.

[10] G.-H. Rivière, 1973, *opus cit.*

[11] C. Bromberger, Introduction. L'ethnologie de la France, du Front populaire à la Libération: entre continuités et ruptures, in: J. Christophe, D.-M. Boëll et R. Meyran (Eds.), Du folklore à l'ethnologie, Paris, Éditions de la Maison des sciences de l'homme, 2009, p.6.

[12] G.-H. Rivière, Les musées de folklore à l'étranger et le futur "Musée français des arts et traditions populaires", in: Revue de Folklore français et de folklore colonial, Paris, Librairei Larose, 1936, tome VII, pp. 58-71.

[13] F. Weber, Marcel Maget, le folklore et la Corporation paysanne : Archiver, publier, enquêter de 1935 à 1944, in: J. Christophe, D.-M. Boëll et R. Meyran (dir.), Du folklore à l'ethnologie, Paris, Éditions de la Maison des sciences de l'homme, 2018, pp. 231-253. https://doi.org/10.4000/books.editionsmsh.10096

[14] A. Demangeon, La Maison rurale en France, catalogue-guide, Jean Gottmann (Eds.), Paris, Denoël, 1937.

[15] J. Flach, Enquête sur les conditions de l'habitation en France : les maisons-type. Tome I, Ministère de l'instruction publique et des beaux-arts, Comité des travaux historiques et scientifiques, section des sciences économiques et sociales, Paris, 1894, and tome II in 1899.

[16] National Archives, Pierrefitte, n°20130277.

[17] F. Firreri, Les Représentations savantes des usagers dans la collection « L'Architecture rurale française »: l'exemple du Dauphiné, in: E. Chapel, G. Bolle (Eds.), Usages et usagers, Les Cahiers de la recherche architecturale urbaine et paysagère, n°19, Ministère de la culture, 20 décembre 2023. https://doi.org/10.4000/craup.13449

[18] Délégation générale à l'équipement national et Service des chantiers intellectuels et artistiques, Enquête sur l'architecture régionale, Paris, Bernard Frères, 1941, National Archives, Pierrefitte, n°20130277.

[19] *Ibid.*

[20] H. Raulin, L'architecture rurale française. Une enquête nationale inédite (1941-1948), in: Études rurales, vol. 13, n° 1, 1964, p.98. https://doi.org/10.3406/rural.1964.4817

[21] National Archives, Pierrefitte, n°20130277.

[22] G. Massé (Eds), Architecture régionale, in Techniques et Architecture, n°1-2, 1947.

[23] Société d'Ethnographie française et Musée National des Arts et des Traditions Populaires (Eds.), Arts et traditions populaires, Paris, Presses universitaires de France, 1953, 72 vol.

[24] H. Raulin, 1964, *opus cit.*

[25] M. Segalen, Vie d'un musée, 1937-2005, Paris, Stock, 2005.

[26] National Archives, Pierrefitte, n°20130277.

[27] D. Le Couëdic, J.-R. Trochet, Bretagne, in: J. Cuisenier (Eds.), L'Architecture rurale française, Corpus des genres, des types et des variantes, Paris, Musée national des arts et traditions populaires, Berger-Levrault, 1985.

Materials Research Proceedings 47 (2025) 138-146 https://doi.org/10.21741/9781644903391-16

Advanced survey technologies for Moroccan traditional earthen architecture: Documenting the Ksar of Tamnougalt

Alberto PETTINEO[1,a*], Giovanni PANCANI[1,b] and Sandro PARRINELLO[1,c]

[1]Department of Architecture, University of Florence, Italy

[a]alberto.pettineo@unifi.it, [b]giovanni.pancani@unifi.it, [c]sandro.parrinello@unifi.it

Keywords: Fast-Survey, Digital Database, 3D Point Cloud, Drâa Valley, Casbah of Caids

Abstract. This research explores the use of fast-survey technologies and digital tools to document and preserve traditional earthen architecture, focusing on the Ksar of Tamnougalt in Morocco's Drâa Valley as a case study. Situated along the historic Trans-Saharan trade route, this village holds significant historical and cultural significance despite its partial abandonment. In this context, the study aims to emphasize the importance of documenting and safeguarding this heritage, given the vulnerability of earthen materials to degradation and destruction. A detailed site survey was conducted using Unmanned Aerial Systems (UAS) and Terrestrial Laser Scanner (TLS) to capture both large-scale morphology and intricate architectural details. The resulting high-resolution, morphometrically rich database enabled the production of accurate 2D drawings, which facilitated detailed graphical analyses and a deeper understanding of the buildings and their components. The research explores techniques to represent traditional architecture's material quality and morphological complexity while preserving the metric reliability of traditional drawings, photogrammetric models, 3D point clouds, and models for innovative descriptive systems. This documentation process not only digitally preserves the architectural peculiarities and traditional building techniques of this historical site, but also provides a valuable resource for future research and conservation efforts. The study highlights the crucial role of digital technologies in heritage preservation, demonstrating the potential of fast-survey techniques in the documentation and management of vernacular earthen structures and contributing to broader discussions on heritage conservation and technological innovation.

Introduction

The fortified village of Tamnougalt is part of the intricate system of Berber settlements in pre-Saharan Morocco, located within the expansive oasis valleys of the Drâa and Ziz, a region known for its *Ksour and Kasbah*. These testimonies, with their ancient roots and inestimable value, represent a unique cultural heritage, which even today, after centuries, continues to captivate with their character and strong symbiosis with the surrounding landscape [1]. Significant geographical factors, such as the Atlas Mountains range, the Desert, and the Ocean, have fostered the flourishing and development of the artistic and technological heritage of earthen architecture in these valley systems with fertile irrigated lands [2].

The traditional architecture of the region, from fortified villages (*Ksour*) to individual residential structures (*Kasbah*), is characterized by its defensive nature. In a context of insecurity and an arid environment, communities needed to protect themselves both from raids and harsh climatic conditions [3][4]. These fortified structures, which share a common defensive architectural organization, developed as individual instances of a linear system along the caravan trade routes that connected sub-Saharan Africa with the Mediterranean [5].

Thanks to its strategic position in the heart of the Drâa Valley, between Ouarzazate and Zagora, Tamnougalt became a vital centre for controlling caravan traffic and a hub for trade between Morocco's interior regions and the Sahara. Originally founded by nomadic Berbers who later settled permanently, Tamnougalt played a central role in the region's economic and political system, also due to its proximity to the irrigated palm groves of the Drâa, essential for local agriculture and livestock farming [6].

The significance of this complex, a valuable example of traditional Berber earthen architecture, remains tangible today in its material remains, which urgently require documentation, conservation, and enhancement efforts to safeguard their integrity.

Fig. 1 - The old settlement of Tamnougalt (bottom), the topic of the documentation, in relation to the new urban development (top).

The Ksar and the Casbah des Caids: Morphological Analysis and Knowledge Actions

The ancient village of Tamnougalt, now bordered to the north by a recently built settlement, is characterized by an irregular arrangement of urban spaces, with a dense cluster of earthen buildings developing around internal courtyards. The architecture is designed to respond to extreme climatic conditions, with narrow, covered alleys protecting from heat and wind, and compact houses optimizing resource use. The village is organized into various districts, each reflecting the social division between the different classes and tribes that inhabited the area. There is the Casbah district, reserved for members of the ruling family, the Tighremt, inhabited by the Casbah's servants, and the Jewish quarter, or Melah. The village's central hub is the Assarag square, around which common-use buildings such as the hammam, bazaar, and marabout can be found. The construction techniques used in Tamnougalt are primarily based on local materials, such as earth, worked into *pisé* or adobe bricks. These materials, in addition to being economical and readily available, offer excellent insulating properties, crucial for keeping interiors cool during the day and retaining heat at night [7].

One of the most representative buildings in Tamnougalt is the Casbah des Caids, the residence of the Caid's family, the local governor. The Casbah is an imposing and fortified structure, symbolising the political and social power of the *Caid* in the region. This building is incorporated within the settlement and developed around several inner courtyards, where most of the daily activities occurred.

Materials Research Proceedings 47 (2025) 138-146 https://doi.org/10.21741/9781644903391-16

Fig. 2 – Above: the morphology of the Casbah within the ancient settlement. Below: Morphological complexity and aggregation of the Casbah's elements seen from the top, and the condition of the structures.

Currently, the Tamnougalt complex and the Casbah itself are in a state of partial abandonment with various issues, due to neglect, seismic events, and the nature of the materials used to construct the buildings. Some areas have been restored and adapted for new uses, such as tourist accommodation, but construction techniques and materials that do not conform to the original characteristics of the structure have often been employed, such as the use of concrete reinforcement elements. This affects the integrity of the traditional earthen architecture, compromising the understanding of its original features.

In this context, in May of this year, thanks to the collaboration between the Department of Architecture at the University of Florence and the Department of Geology, Semlalia Faculty of Science at the Cadi Ayyad University in Marrakech, an international workshop was organized with students from both faculties to initiate exploratory actions and gain insight into the specificities and artisanal construction techniques of this unique architectural heritage. The digital documentation campaign aimed to achieve two objectives: documenting the architectural value of the monument itself, its features, and the structural and conservation issues, focusing on the construction and technological aspects [8] [9].

Through the experimentation of rapid surveying techniques, thanks to the integration of terrestrial LIDAR systems and Structure-from-Motion (SfM) photogrammetry techniques, it was possible to structure and manage a georeferenced digital point cloud database, representing the morphological complexities of the Casbah and the urban development of the village interior, both in terms of Territorial Scale and Big Data [10] [11].

Digital documentation of the complex

The digital documentation campaign was conducted using an integrated survey protocol, applying various methodologies and surveying tools. The choices were made based on the most suitable capacity to interact with the morphology of the complex, ensuring the best results in terms of time efficiency, data processing, and research objectives [12]. Two main survey systems were integrated: Terrestrial Laser Scanner (TLS) measurements and SfM photogrammetry techniques with aerial data acquisition, using UAS systems, and ground-based DSLR cameras. Given the high morphological complexity of the settlement's environments - both in terms of plan articulation and the development of overlapping levels, typical of this architectural type - before proceeding with the survey activities, the complex was divided into spatial units corresponding to areas developed around the courtyards. This subdivision facilitated the surveying operations, which were then conducted on relatively simplified geometric units, with the reassembly of the sections planned for a second phase of processing.

Specifically, the digital documentation activities focused on different levels of resolution:
(i) the entire fortified village, using SfM photogrammetry methodologies with UAS systems to achieve a spatial understanding at the territorial scale; (ii) the interior spaces of the Casbah, through the integrated use of TLS, UAS, and DSLR equipment, for an architectural scale understanding that would facilitate the reading of the intricate internal environments and the material characteristics of the surfaces; (iii) the marabout, the main square, and its connections to the Casbah, combining TLS and UAS systems to understand the spatial relationships between the Casbah complex and the rest of the village, with a focus on maintaining a high degree of accuracy and overlap at crucial junctions. During the measurement operations, specific geometric constraints were defined with control points, through which the high measurement precision allowed for the connection of the environments to a single global reference system.

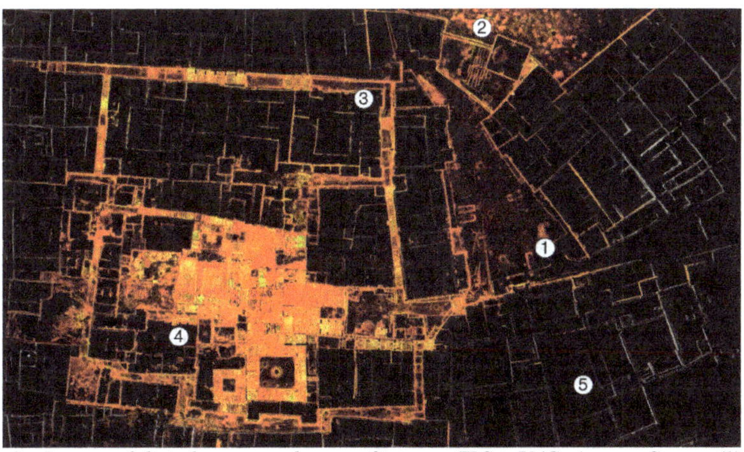

Fig. 3 – Integrated digital survey and areas of interest. TLS + UAS: Assarag Square (1), the Marabout (2), main connections (3), and the spaces of the Casbah (4). The entirety of the village (5) was documented exclusively using UAS equipment.

The laser scanner survey acquired three-dimensional data and measurements through scanning stations developed along polygonal paths, connecting the spatial units, the connections, and the external environments. By distributing the scanning operations across multiple stations, it was

Materials Research Proceedings 47 (2025) 138-146
https://doi.org/10.21741/9781644903391-16

possible to focus on documenting surfaces while avoiding obstacles and addressing challenges posed by the complex spatial arrangement and disturbances like furniture and debris. The TLS survey was carried out using a Leica BLK360, whose features and portability allowed the operator to easily access all spaces, particularly the interiors of the Casbah, making the documentation process efficient and suited to the need for a rapid survey. A total of 650 stations were completed, with an average acquisition time of 2 minutes per scan and a point measurement speed of up to 680,000 points/sec.

In parallel, a dual photogrammetric campaign was conducted: aerial, using UAS equipment, to document the entire architectural complex, capturing surface materials and architectural details to complement the laser survey; and ground-based, using DSLR cameras, to map the main facades and provide detailed documentation.

The aerial SfM acquisition strategy carried out with a DJI Phantom 3 Pro drone, was structured into 5 flight plans, each with 200-250 photos (96dpi resolution and dimensions of 5280x3956). The first phase involved performing grid flight plans (3 plans with nadir and dual 45° angles), enabling an overall acquisition of the Ksar. The second phase consisted of 2 focused flight plans over the Casbah, with concentric multi-altitude trajectories (45° angles), creating a "cone" coverage around the architectural object. The resulting data (a high-density point cloud with an average density of 5 million points/m^2) enabled the development of a single 3D dataset of the urban area. The UAS documentation allowed for the acquisition of elements that were otherwise inaccessible from the ground, such as highly deteriorated areas and roofs. This made it possible, during post-production, to integrate the database and ensure comprehensive and reliable coverage of the morphological and colorimetric information not captured through TLS.

The photographic campaign using DSLR cameras was useful for surface documentation, aimed at mapping any issues within the complex, such as localized degradation and structural problems. This allowed the creation of useful analysis cards to support the tridimensionality point cloud for studying the complex.

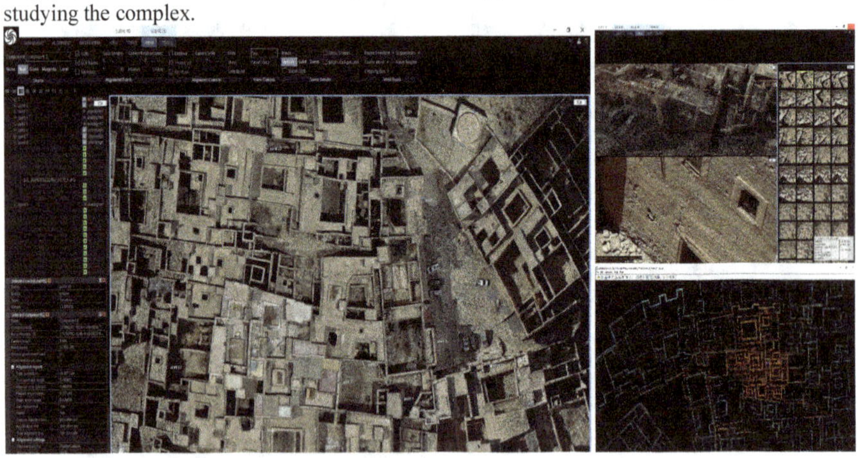

Fig. 4 – Photogrammetric processing (on the right) and subsequent referencing phase to UCS of the dense photogrammetric point cloud (top left) with integration of the data with TLS (bottom left).

Post-Production and structuring of digital database

The post-production phase involved the processing of data from the various instruments and the subsequent referencing of the datasets to a common UCS, to create a single high-resolution

Materials Research Proceedings 47 (2025) 138-146
https://doi.org/10.21741/9781644903391-16

database of the entire complex. The scans obtained from TLS, after undergoing a process of import, filtering, and selection, were aligned within the Leica Cyclone point cloud management software. During the alignment of the TLS data, the scans were divided according to the spatial units established during the survey, considering the relationship of the Casbah with the entire settlement and to the main and secondary connection paths. This subdivision of the acquired areas ensured better control, facilitating database management during registration.

The interaction between the different scans, developed in a cloud-to-cloud mode, was optimal, with average overlap sections of 10m between scans (5m for interior environments and approximately 2m in confined spaces like staircases and internal connections). This resulted in an acceptable mean error of 20mm for the overall data of the Casbah. The first partial result is a high-resolution point cloud (6,211,893,275 points) of the Casbah.

The photographs obtained from the UAS survey were processed through RealityCapture software in a single process, resulting in a dense point cloud (120 million points) of the entire settlement. The processed data, which lacked dimensional value, was then referenced and scaled based on the TLS point cloud by identifying common morphological points in both datasets and integrating the coordinates from the TLS reference system into the photogrammetric point cloud. The integration of databases, both in terms of extensive survey (UAS) and point-based survey (TLS), increased the quantity and quality of the manageable information for the rapid documentation of the complex.

The complementary nature of the instrumentation and the reliability control in the alignment of the metric data allowed for a detailed examination of the morphology of the entire complex, enabling analyses of the architectural context at various scales of investigation. A basic metric network was also established as an organic system adaptable to multiple subsequent actions of discretization and integration. The result is, therefore, a three-dimensional database with information now suited for multiple uses and integrations [13].

Fig. 5 – Partial view of the photogrammetric point cloud of the settlement.

Fig. 6 – Partial view of the integrated digital database.

Conclusion and future development

Starting from the digital database, an in-depth study is underway on the architectural complex to better understand its morphology and construction techniques. Early progress development in this research has resulted in the creation of two-dimensional drawings that highlight the intrinsic complexities of the architectural object, emphasizing technological aspects and potential conservation issues within the structures. The digital documentation, which integrates two-dimensional drawings, photogrammetric models, and point clouds, has accurately conveyed traditional architectures' material quality and complexity while preserving the technical outputs' metric reliability. These digital tools support the conservation of structures and provide a solid foundation for further studies and interventions. From these actions, the goal is to develop complex information systems that collect and organize morphometric data, analysis reports, and two-dimensional drawings. Such systems will make the enjoyment and understanding of the heritage more accessible, facilitating the work of researchers, technicians, and the general public, thereby creating a bridge between the past and the present [14] [15].

Engaging the public in the exploration of this heritage through educational and cultural initiatives can be an opportunity to stimulate interest and awareness of the importance of its conservation. Additionally, initiating targeted recovery paths based on preserving original construction techniques will help maintain the integrity and authenticity of these architectural complexes, avoiding interventions that could alter their essential characteristics. In this way, the architectural heritage of Tamnougalt and other similar structures can not only be preserved but also enhanced and made comprehensible using innovative technologies and the sharing of knowledge, promoting an integrated approach to conservation, research, and cultural enhancement.

Acknowledgement

The documentation activities were carried out during an international workshop held at the village of Tamnougalt, jointly coordinated by the Dep. of Geology of Semlalia Faculty of Science, Cadi Ayyad University (UCA), and the Dep. of Architecture of the University of Florence (UNIFI). The workshop was organized by Professors Mounsif Ibnoussina (UCA), Sandro Parrinello (UNIFI), and Giovanni Pancani (UNIFI), with the collaboration of Professors Admou Khaoula, El Abbassi

Materials Research Proceedings 47 (2025) 138-146 https://doi.org/10.21741/9781644903391-16

Fatima Zahra, Bahammou Younes, Berroug Fatiha (UCA) and Letizia Di Pasquale, Lamia Hadda (UNIFI), Fabio Fratini (CNR Florence), with the participation of Dr. Monica Lusoli (UNIFI). Participants in the workshop included: Cadi Ayyad University - PhDs Khaoula Admou and students Daayout Fatima Ezzahra; Bouhrachi Imane; Hafyan Saadt; Hicham Rihab; Ait Benali Azzdine. University of Florence - PhDs Alberto Pettineo and students Houssem Kouidhi; Francesco Pini; Maria Chiarella; Rachele Parrini; Federico Strazza; Asia Salvestrini; Guglielmo Rocco; Raffaele Vergaro; Mariachiara Sammito; Giacomo Talozzi; Alessia Antonia Nicoletti; Emma Castelli; Davide Vinci; Tommaso Mannari; Lorenzo Gianassi; Jurgen Rina; Andrea Maria Sacchetti; Kawtar El Hamdani; Edoardo Finucci.

References

[1] S.Zerhouni, H. Guillaud, L'architecture de terre au Maroc, ACR Edition, Paris, 2001.

[2] J.Hensens, Qsours et Qasbas du Maroc. Réflexions sur l'Evolution de l'Habitat Rural Traditionnel, Marka Print Pte Ltd, Singapur, 1986.

[3] C. Pietropaolo, The Theme of the Kasbah in Islamic–Mediterranean Architecture. Settlements Along Draâ Valley, Morocco, in Putting Tradition into Practice: Heritage, Place and Design: Proceedings of 5th INTBAU International Annual Event 5. Springer Int. Publishing, (2018). https://doi.org/10.1007/978-3-319-57937-5_45

[4] W.J.R. Curtis, Type and Variation: Berber Collective Dwellings of the Northwestern Sahara in Muqarnas 1983, pp.181-209. https://doi.org/10.2307/1523077

[5] R.A. Austen, Trans-Saharan Africa in world history, Oxford University Press, 2010.

[6] A. Elkarimi. La gestion des ressources en eaux d irrigation dans la vallée du Dra moyen au Maroc. Nouveaux acteurs et nouvelles approches. In: 1st int. research seminar on architectural heritage and sustainable development. 2005, pp. 29-38.

[7] S. Mecca, E. Baglioni, L. Dipasquale, Cultural landscape of the Drâa Valley, Morocco. In: TERRA 2012 proceedings, 11th International Conference, 2012, pp.141-146.

[8] G. Pancani, M. Gentili, Urban Survey of The Mellah of Sefrou (Mo). In Dialogues in Cultural Heritage, 2018, pp. 555-558.

[9] S. Parrinello, F. Picchio, S. La Placa, The Construction of an Informative 3D Model for the Monitoring of City Heritage Risk. In: Reviving Aleppo, Springer Int. Pub., 2024, pp. 243-274. https://doi.org/10.1007/978-3-031-65858-7_11

[10] S.Parrinello, F.Picchio, R. De Marco, Documentation systems for a urban renewal proposal in developing territories: the digitalization project of Bethlehem Historical Center. In: 2018 Metrology for Archaeology and Cultural Heritage (MetroArchaeo). IEEE, 2018. p. 216-221. https://doi.org/10.1109/MetroArchaeo43810.2018.13624

[11] I. Leon, J.J. Pérez, M. Senderos, Advanced techniques for fast and accurate heritage digitisation in multiple case studies. Sustainability, 2020, 12.15: 6068. https://doi.org/10.3390/su12156068

[12] G. Pancani, M.Bigongiari, The Integrated Survey of the Pergamum by Nicola Pisano in the Cathedral of Pisa. Digital Cultural Heritage, 2020, 373-388. https://doi.org/10.1007/978-3-030-15200-0_25

[13] R. De Marco, A. Pettineo, The recognition of Heritage qualities from feature-based digital procedures in the analysis of historical urban contexts. The International Archives of the Photogrammetry, Remote Sensing and Spatial Information Sciences, 2022, 46: 175-182. https://doi.org/10.5194/isprs-archives-XLVI-2-W1-2022-175-2022

[14] A. Mandelli, C. Achille, C. Tommasi, F. Fassi, Integration of 3D models and diagnostic analyses through a conservation-oriented information system. The International Archives of the

Photogrammetry, Remote Sensing and Spatial Information Sciences, 2017, 42, 497-504. https://doi.org/10.5194/isprs-archives-XLII-2-W5-497-2017

[15] S. Parrinello, A. Pettineo, Traditional Architectures Along the Cultural Route of James I of Aragon in the Province of Valencia: Leveraging Laser Scanning and BIM for Heritage Management. Materials Research Proceedings, 2024, 40. https://doi.org/10.21741/9781644903117-36

Materials Research Proceedings 47 (2025) 147-155 https://doi.org/10.21741/9781644903391-17

Vernacular architecture, between the sovereignty of the environment and the restoration requirements: The case of Ksar of Igoulmimen and the medina of Meknes

SEDREDDINE Mouna

Ecole Nationale d'Architecture de Fès, Morocco

m.sedreddine@enarabat.ac.ma

Keywords: Architecture, Vernacular, Environment, Medina, Ksar, Restoration, Contextualization

Abstract. The vernacular represents a homogeneous aggregation of elements coexisting in harmony, wherein nature asserts its dominance. The intelligence of humanity, coupled with the survival instinct and a remarkable capacity for adaptation, results in architecture that serves as a continuation of the surrounding environment. The genius loci—fundamental to architectural production—guides the design process through various distinguishing parameters, including natural, social, cultural, economic, geographical, and occasionally political determinants. This concept is adeptly manifested in the unifying architecture that significantly influences the Moroccan landscape. The medinas, ksour, kasbahs, and rural habitats serve as contemporary testimonials to the alchemical relationship between humanity and its environment. They encapsulate a distinctive craftsmanship that evolves continuously in response to the variability of natural, economic, and cultural factors. This communication will first examine selected national and international human and architectural instances from geographically diverse regions where "vernacular" architecture illustrates the environmental sovereignty over humanity, communities, and the resultant architectural creations. Secondly, it will explore two cases of restoration experiences: the Medina of Meknes and the Ksar of Igoulmimen. This analysis will be grounded in the established literature and prior field research, to assess the actual impact of restoration efforts vis-à-vis a shifting socio-economic framework. Furthermore, the discussion will address various revitalization actions associated with these projects, which occasionally reflect passive rehabilitation attempts. The intention is to transcend the mere aesthetics of heritage that juxtaposes tradition and modernity, thereby interrogating the resilience capital of these human establishments in favor of "intelligent contextualization."

Introduction

The relationship of continuity and communion between humanity and the environment, as manifested in vernacular architecture, is exemplified by the medinas, ksour, kasbahs, and rural housing. These ancestral constructs serve as living testaments to the interdependence of environment and architecture, a union shaped by natural, social, cultural, economic, geographical, and occasionally political determinisms. Such determinisms delineate an "environment" that, when influencing architectural forms, engenders what can be termed a constructive culture unique to a specific region or community. This philosophical perspective continues to resonate with numerous contemporary architects, who endeavor to reconnect their architectural practices with the environment, thereby rejecting the uniformity, standardization, and decontextualization characteristic of certain currents of modern architecture. The genius of the place, foundational to vernacular architectural production, informs the design process through various distinguishing parameters. These historical fabrics exemplify the synthesis between environment and architecture, driven by characteristics derived from the environment itself. They represent an

Materials Research Proceedings 47 (2025) 147-155 https://doi.org/10.21741/9781644903391-17

intelligent adaptation of humans to their surroundings and embody a distinctive know-how that evolves in response to the changing natural, economic, and cultural factors within the environment. The vernacular architecture reflects a cohesive amalgamation of elements existing in harmony, with nature retaining ultimate control. The interplay of human intelligence, survival instinct, and remarkable adaptability results in architectural forms that are a continuation of their surrounding context. This discourse will explore vernacular architectures within an international framework, illustrating the relationship of subordination that architecture maintains to its environment. Furthermore, it will analyze the factors that generate place-specific architectural forms through examples drawn from diverse geographic regions, where the essence of place and regional expertise establish a connection to territory. Subsequently, the discussion will return to instances of vernacular fabrics in Morocco, specifically the Ksar of Igoulmimen and the medina of Meknes, which have undergone "passive" restoration endeavors. An evaluation will be conducted to illuminate the impact of these efforts within a constantly evolving context. Employing field observation as a basis, this approach incorporates documentary analysis facilitated by various prospecting methods, including bibliographic documentation, physical and digital archives, digital documents, iconographic materials, and administrative records. This methodology aims to support, interpret, and verify certain hypotheses regarding the parameters and characteristics that define "vernacular" architecture, while addressing the challenges of active heritage in the context of passive restoration of traditional fabrics in Morocco.

1. Vernacular architecture as a human experience where the built environment resonates with its surroundings

Places differ from one region to another and are part of the landscapes and territories where they are located, which make them unique by giving them an identity through specific natural parameters (sunlight, topography, vegetation, climate, local materials, etc.). Thus, in each region of the world and at different historical times, architectures specific to their anchoring environment have succeeded one another and evolved as the changes that human communities and their environments have in turn experienced.

By physical determinism, and as the history of peoples can attest, humans have for centuries been subjected to the conditions that their environment offers in a relationship of mutual subjection and dominance where architecture becomes a means of affirmation and physical marking (imprint) of the being within its territory but also a way to transcribe its sensitivity to the place and its understanding of the elements that compose it by scaling everything to its level.

Defined by François Varin as [architecture of popular inspiration that has developed and develops its own characteristics in a specific region where it often uses local materials, ways of doing, and traditional forms.][1], it is immediately apparent that the popularity of this architecture is what characterizes it most and differentiates it from those guided by thought currents and internationalized styles. It is the result of an intelligent adaptation of man to his environment and the embodiment of a typical know-how that keeps it in constant evolution according to the variability of the natural, economic, and cultural factors of its environment.

It is commonly accepted that the vernacular automatically refers to the local and by extension to the traditional and authentic, even to self-construction. The vernacular reflects the distinctive historical, cultural, and environmental context of a territory or region. The characteristics of the vernacular are often the result of the determinism of the natural environment and its elements (climate, geography, topography, nature of the soils, etc.). However, all these elements can at any moment change and develop, thus offering new conditions imposing new constraints. In this sense, François Varin will advance that [*Vernacular architecture integrates know-how and models*

[1] François Varin, L'architecture vernaculaire, une définition difficile à cerner. Comité de l'ICOMOS pour l'Architecture Vernaculaire, p. 4.

Materials Research Proceedings 47 (2025) 147-155 https://doi.org/10.21741/9781644903391-17

transmitted from one generation to another, and it evolves according to new needs while assimilating external influences.][2]

The environment with all its components (cultural, religious, economic, political, historical, etc.) can also evolve and transform, thereby giving rise to other societal and thus architectural configurations. That being said, a local architecture is subject to several movements over time, the impact of which can sometimes remove all legitimacy from the vernacular as it is commonly perceived and defined. The case of the village of Gourna, an unprecedented work by Hassan Fathi, remains, as we will see, a revealing example. Faced with this paradigm shift and from the moment the parameter related to temporality comes into play, it remains imprudent to apprehend the vernacular without 'contextualization'.

Thus, and in the face of a certain enthusiasm for the local and the risk of a passive patrimonialization of the vernacular, it is necessary to distinguish and rid oneself of this taste for the picturesque and this tendency towards nostalgia very common among many architects.

In this sense, there is a continuous back-and-forth movement between these two constantly evolving components that are the environment and architecture, within the framework of a relationship of mutual sovereignty and subjection where most often a form of association ends up taking shape and settling.

Examples of human experiences chosen in this communication reflect this continuity that is described in this work as 'association' and account for how some architectural productions have been designed, realized, practiced, and then reinterpreted, thus involving the temporal parameter that persists in view of the evolution of both (environment and architecture).

1.1 The M'Zab of Algeria, an architectural experience whose principles transcend time, cultures, and territories

Located in the north of the Algerian Sahara, the M'Zab valley is a very telling example of traditional human habitat, an image of the Ibadite culture, a culture rooted in time due to its long politico-religious history, and ingenious due to the extremely efficient interactions that Man has managed to maintain with a hostile and inhospitable environment. Here, the physical determinism associated with the ingenuity of the Mozabites and their unwavering devotion to their culture and faith has given rise to a coherent and supportive socio-spatial organization.

The desire to isolate is at the root of the choice of this part of the Algerian territory, whose harsh and inhospitable geographical and climatic conditions make it a shield against enemy invasions of the time. The motivation was also cultural and religious, as the Mozabites saw in this introversion a way to protect and preserve their identity and long-persecuted religious affiliation. Registered in 1982 on the UNESCO World Heritage list, the M'Zab of Algeria has deeply marked the minds of architects of the modern movement who saw in these sites an opportunity to reconcile nature and technology. The rise of industrialization and the international style after World War II sparked a real enthusiasm for a return to nature as the place of the most extreme deprivation, leading man to the truth of the inner experience (1)

[2] Ibid., p.4.

Materials Research Proceedings 47 (2025) 147-155 https://doi.org/10.21741/9781644903391-17

Figure 1 : the city of the M'Zab.
Source : Francis Tack, Copyright : © UNESCO.

What is interesting about this experience is not so much the architecture of the M'Zab as the sensitive approach adopted by André Ravéreau within a culturally and historically different environment from the one he comes from. Through the associative relationship that links M'Zab architecture to its environment, Ravéreau reflects on the possible contribution of this production to new practices. He sees the urban organization of the Ibadites as the perfect model of adaptation to the constraints of the environment, of architecture and urban planning that respect the natural context and local cultures.

Convinced of the impossibility of reproducing it identically due to the differences imposed by other climates and customs, the relationship of association embodied by the Mozabite experience was for Ravéreau a lesson more than a model of regeneration whose principles remain transposable to other territories and other lands, a kind of transcendent association. For Ravéreau, it's a question of regenerating oneself rather than aspiring to mimic. The idea is to learn from living architectural experience to achieve a renewal of modernist dogmas that have wiped the slate clean of the relationship to identity and place. This approach values adaptation to the environment and the purity of the architectural gesture, inscribing the project in the depths of a culture with a rigor devoid of artifice.

Back in France, he devoted himself to experimenting with his thinking in other contexts, at a time when many works in the same vein were appearing in reaction to modernism, encouraging a return to place and its components in favor of an architecture without professional mediation.

Figure 2: Health center Mopti au Mali (1971 et 1974).

Source : Rémi Baudouï et Philippe Potié, L'atelier du désert.

Figure 3: French High School Nouakchotte en Mauritanie (1982 et 1986).

Source : Rémi Baudouï et Philippe Potié, L'atelier du désert.

Ravéreau's sensitive study of the M'Zab resulted in a reconstruction of the self and a renewed appreciation of the real meaning of the vernacular and the relationship with the environment.

Materials Research Proceedings 47 (2025) 147-155 https://doi.org/10.21741/9781644903391-17

1.2 Gourna, a polemical work between perception, design and practice

Highly committed to a nationalist and social approach to architecture, Hassan Fathy (1900-1989) left his mark on a whole generation of young builders with his book "Gourna, a Tale of Two Villages", published in English in Cairo in 1969, a work paying homage to the peasants and in which an urban planning and architectural experience deeply inspired by the vernacular is recounted. Located close to the Luxor archaeological site, the new Gourna village (built between 1945 and 1948, but never completed) is the result of an approach intimately linked to local traditions, cultural authenticity, and the specific geographical features of the rural environment. An approach that goes so far as to involve the local inhabitants in the construction process.

The use of local materials for climatic, cultural, and socio-economic reasons, as well as the involvement of peasants in the construction of their homes as a humanizing spatial approach, particularly marked the architect's work at a time when Egypt was experiencing the industrialization of construction and the internationalization of architectural style.

Figure 4 : Use of Nubian vaults.
Source : website https://www.arquitecturaydiseno.es.

His desire to revive a vernacular building culture that he appreciates for its ecological as well as social and economic virtues has nevertheless resulted in living spaces that will meet with a certain social reluctance, as they somewhat restrict the freedom of their inhabitants. As a result, elements such as the courtyard were transformed into stables, some openings were closed in order to open up others, and elevations were created that were not in keeping with the original spirit of the project.

Figure 5: Damage to houses in New Gourna. Source : https://egyptophile.blogspot.com.

Figure 6 : Damage to houses in New Gourna. Source : New Gourna Village Conservation and Community. World monument Fund.

Thus, although his approach was appreciated by his many disciples throughout the world at different times, Hassan Fathy did not fail to provoke certain criticisms due to this experience, judged by some to be backward-looking or even utopian in a context where poverty and social

Materials Research Proceedings 47 (2025) 147-155 https://doi.org/10.21741/9781644903391-17

inequality were rife, and where a subjugated population, because destitute, had only the looting of ancient tombs to survive.

By analyzing the design process of the architectural work in the case of the village of "Gourna" and the evolution of the dialectical relationship between this architectural production and its socio-cultural and physical environment, we can see that the project, as it evolved, failed to echo its environment. Functional, socio-cultural, and geographical parameters were the pillars of Hassan Fathy's design process, but they were not enough to ensure the success of this mediation, which only the parameter of integration can achieve.

Over and above the obstacles posed by an unstable political and economic climate that is hardly conducive to this kind of profoundly human experiment, Hassan Fathy's very attitude as a designer remains very much conditioned by his nationalistic and paternalistic impulse towards a population for whom he aspires to a decent and prosperous living environment.

The forced adoption of solutions based on traditional know-how was clearly inadequate to deal with problems that go beyond a concern for functionality and identity. A more global approach, taking into account the different aspects of the precarious situation experienced by the Gournis peasants, could have helped to make this graft a success and ensure greater sustainability for the project. Hassan Fathy would later testify to this in his book [*Tradition, he asserts, is not necessarily outdated and synonymous with immobility. What's more, tradition is not necessarily ancient, but may well have been established recently*][3]. Vernacular is not synonymous with secular.

2. Vernacular architecture: the challenge of active heritage preservation versus passive restoration of traditional fabrics

The ksour, the kasbah, and the medina are true works of engineering. Born in an environment where physical determinism has dictated a way of life specific to the environmental characteristics of the place, these human settlements embody the history of regions of Morocco unique in their ethnic diversity, cultural richness, and architectural authenticity.

In these particularly ingenious structures, perfectly integrated into their environment, the genius of the place means that the buildings adapt to their surroundings and form a single body. Construction techniques and materials, spatial morphology, distribution and use of space, orientation, dimensions and volumes, and heights, are all an emanation of their environment. This communion, which has lasted for centuries, has been broken by various factors, both natural (water stress, soil erosion, siltation, global warming) and man-made (migration and abandonment of oases, emergence of new urban centers, cultural changes, disintegration of the family nucleus, etc.).

However, with the renewed interest in the vernacular and its environmental virtues, several restoration projects have focused on these traditional fabrics, whose essence has been shaken and sometimes even profoundly modified.

2.1 Passive rehabilitation of the Ksar in favor of a failed attempt to revive the vernacular

The Ghris area (about Oued Ghris) was chosen because it is characterized by a high concentration of Ksour, mainly located close to the most fertile and water-abundant sites, but more specifically because of its environments, which have undergone a two-speed evolution, from which local architecture is still suffering today. This area includes the Goulmima chain of oases, and more specifically the Ksar Igoulmimen (dating back over 4 centuries), which is the focus of our analysis. The architectural history of this Ksar has seen two great moments when the environment, with its social, economic, and cultural components, evolved in its relationship with its architecture from a relationship based on man's subjection to the determinism imposed by the need to adapt to the harshness of the climate and the various elements of nature, and to the political and ethnic

[3] Hassan Fathy, *Construire avec le peuple*, Paris, Éditions Jérôme Martineau, 1970, p. 59.

environment of yesteryear, towards a relationship where man is sovereign and where dependence on the imperatives of the past is a thing completed. In addition to the political and socio-economic circumstances that strongly influenced the emergence of this type of dwelling, geographical and environmental factors of a material nature, such as climate and the availability of local materials, had a profound impact on the social, and thus the spatial, organization of the ksar. The aridity and harshness of the climate (heat and sandstorms), combined with the need to adapt to socio-economic constraints, revealed the extreme ingenuity of the population in their understanding of their environment. The materials and construction techniques used to build the habitat are elements that give it an originality that is reflected in the way the inhabitants use the space.

Figure 7: Photo of a ruined house in Ksar Igoulmimen.Source : http://goulmima.toubkal.org/ksours.

The arrival of the French protectorate and the establishment of the colonial administration within the new urban center (appearance of new, more attractive housing models, availability of basic equipment, etc.), combined with the decline in the region's productive offer (competition from large mechanized farms, industrialization of production, etc.) and the regression of palm groves (weakening of the ecosystem) were at the root of a profound socio-spatial mutation resulting in the collapse of the social and economic core of the Ksour. The geographical and natural components of the environment no longer play the important role they once did for residents, and the new center has taken over from them.

However, in response to this interest in revitalizing the vernacular, Ksar Igoulmimen has undergone renovation through the development and maintenance of public spaces and facilities (paved roads, rendering, electrification, running water).

The first restoration work was carried out in 2006, but it was limited to aesthetic measures rather than social ones. The aim was to stabilize the population, which was seeking to migrate out of the Ksar and into the city, by creating profitable economic activities and encouraging associations, among other things. According to the testimonies of the Ksar's inhabitants, simple physical restoration is not enough. What's needed is active social rehabilitation in line with the population's current needs. What's more, the rehabilitation work (carried out by Al Omrane) was carried out using modern materials, with no respect for local building methods and techniques.

Figure 8: Rehabilitation actions based on plastering and electrification.

Materials Research Proceedings 47 (2025) 147-155 https://doi.org/10.21741/9781644903391-17

The approach should be active and evolutionary, social and inclusive, involving and committed. Finally, the environment should not be seen solely in terms of its physical dimension. Architecture is a relational system, and the challenge for the architect is to define the political, social, and economic relevance of his project, keeping in mind the importance of the evolutivity that characterizes the environment with all its components.

2.2 Passive rehabilitation of the Ksar in favor of a failed attempt to revive the vernacular

Like the imperial cities of Morocco, Meknes underwent profound changes in its urban morphology. The city's development, however, reached a turning point with the arrival of the Alaouite dynasty in the 17th century. It was then, under the aegis of Moulay Ismail, that Meknes was promoted to the rank of capital of the Kingdom. The city's development under his reign was different from that of past dynasties. One of the most striking features, which has characterized Meknes to the present day, is the pronounced layout between the medina and the royal palace (built on the site of the Marinid Kasbah), which extends southwards. The advent of the Protectorate and its colonial policies marked a real turning point in the city's urban morphology, resulting in a physical separation between natives and foreigners. This division manifested itself in two ways: A new city to the northeast, based on the first urban planning texts not yet applied in France, and a physical barrier materialized by the Oued Boufekrane, which runs parallel to the medina: from the southeast to the northwest.

The emergence of a new urban center, along with regulations concerning the preservation of the architectural elements of the old fabric, condemned the authentic city of Meknes to halt its march through time. Seeing its evolution inhibited and its population deserted, the old center was transformed into makeshift shelters welcoming the massive influx of the rural population who had come to settle in Meknes. This social mobility in both directions (from the medina to the new town, and from the countryside to the medina) has had repercussions on the demographic and socio-economic structure of the medina, as well as on its built environment. As a result, activities that were once essential to neighborhood community life - such as Ferranes, hammams and M'sids - have disappeared from the centers of cohesion, losing the intimacy and independence needed to consolidate social relations. The road system, which had been designed according to a precise hierarchy and well-defined functions, was turned upside down by the new facilities. As a result, alleyways and cul-de-sacs are undergoing real dysfunction due to the presence of activities incompatible with their original functions. As part of this broad policy of upgrading the "traditional heritage" of Morocco's various cities, Meknes has undergone a series of restorations and upgrading attempts. Since 1967, restoration work on historical monuments and public spaces (squares and gardens) has been carried out by various bodies (the prefecture, the municipality, the Department of Culture, the Habous, and the Meknes commune). Doors, walls, towers, Koranic schools, fountains, etc., have all undergone experiments, most of which proved irreversible. The interventions generally carried out to reproduce the original with a sometimes inexperienced workforce (due to the loss of ancestral know-how), have had a real impact on the buildings.

Figure 9 : an alleyway in the medina of Fez / right: one of the alleyways in the medina of Meknes Source : Author's photo

However, far from reflecting their environment (with all its historical, social, cultural, and geographical components) and any attempt at contextualization and revitalization, the various interventions were characterized by their homogeneity, resulting in a uniform landscape where alleyways and neighborhoods looked alike.

This situation, which has turned out to be a strategy by default, has fuelled the impression that, unfortunately, safeguarding is a matter for museums alone.

An analysis of these actions and their results suggests that the overall approach needs to be reconsidered. Once again, the interventions reflect a certain lack of integration of the site's assets and the opportunities it presents. In older areas, tourism is often used as a means to an end, rather than as a tool for revitalizing the old town center and, first and foremost, taking into account the human component that keeps it alive. In the absence of a global, integrated vision, the rehabilitation actions identified are limited to physical interventions aimed at preserving a certain aesthetic appeal to reassure the eye and the conscience (floor and wall coverings, treatment of cracks, consolidation of structures and masonry, etc.). It is often overlooked that the inhabitant and user are the primary factors in the long-term maintenance of a living space. Their practice, when respectful of the place and its values, is the best way to maintain the medina and ensure its longevity. For the inhabitants, the medina is above all a living heritage and a heritage for oneself. It's the model of living and the model of the city and urbanity that interest residents.

References

[1] François Varin, L'architecture vernaculaire, une définition difficile à cerner. Comité de l'ICOMOS pour l'Architecture Vernaculaire, p. 4.

[2] François Varin, L'architecture vernaculaire, une définition difficile à cerner. Comité de l'ICOMOS pour l'Architecture Vernaculaire, p. 4.

[3] Hassan Fathy, Construire avec le peuple, Paris, Éditions Jérôme Martineau, 1970, p. 59.

Tazotats: Vernacular heritage, thermal storage materials, and heat regulation in arid environments

Mohamed FILALI[1,a], EZZAHIR Anass[2,b]

[1] National School of Architecture of Rabat. BP 6372 Av Allal Al Fassi, Rabat, Morocco

[2] Laboratory for Research in Information Systems, Intelligent Systems and Mathematical Modeling, National Institute of Statistics and Applied Economics, Rabat, Morocco

[a]m.filali@enarabat.ac.ma, [b]aezzahir@insea.ac.ma

Keywords: Bio-Based Materials, Vernacular Architecture, Bibliometric Analysis, Risk Management, Seismic Resilience

Abstract. Vernacular architecture, often defined as a traditional building approach that uses local materials and techniques, intimately reflects the available resources, climate, and cultural practices of a specific region. This form of architecture, developed over centuries, illustrates a deep respect for the environment and an intuitive understanding of local conditions. The use of local materials presents numerous advantages. Firstly, these materials are generally accessible in situ, thus reducing construction and transportation costs. Indeed, they are often better adapted to local climatic conditions, offering thermal and structural properties that contribute to the durability and energy efficiency of buildings. In arid regions, the use of raw earth and adobe bricks helps maintain cool interior temperatures due to their thermal regulation capacity. Vernacular construction techniques, such as earth, wattle daub, and dry stone, also reflect a deep knowledge of material properties and the best methods for using them. These techniques, passed down from generation to generation, often incorporate ingenious solutions to maximize the durability and comfort of dwellings. Integrating local materials and construction techniques into modern architecture can offer sustainable and culturally relevant solutions to contemporary construction challenges. By revisiting these traditional practices, it is possible to design buildings that are not only ecologically responsible but also deeply rooted in their cultural and environmental context. This study aims to examine the importance of using local materials and traditional construction techniques in vernacular architecture, highlighting concrete examples and the advantages they offer in terms of sustainability, energy efficiency, and preservation of cultural heritage.

Introduction

Vernacular houses, designed and built in harmony with their natural and cultural environment, offer architectural solutions that meet the specific needs of their inhabitants while minimizing environmental impact. These dwellings, often constructed with local materials and according to traditional techniques passed down from generation to generation, embody a sustainable and resilient approach to housing. At a time when sustainability is at the heart of global concerns, vernacular houses represent a valuable source of inspiration for contemporary architecture, particularly in the search for more environmentally friendly solutions. By leveraging local resources, adapting to specific climatic conditions, and integrating bioclimatic principles, these traditional dwellings offer crucial lessons for modern architectural design.

Vernacular houses are distinguished by their ability to efficiently use local resources while meeting specific climatic and cultural requirements. Unlike standardized modern constructions, they incorporate materials and techniques adapted to the climate and environment, such as thick raw earth walls for thermal insulation in arid regions or sloped roofs in humid areas. These practices reduce energy needs and environmental impacts related to construction. In addition to

Materials Research Proceedings 47 (2025) 156-164 https://doi.org/10.21741/9781644903391-18

extending the lifespan of buildings through regular maintenance and reuse of materials, vernacular architecture proposes a sustainable model, adaptable to contemporary challenges such as urbanization and climate change, while ensuring optimal comfort for inhabitants.

Vernacular houses, reflecting human adaptation to various climatic contexts, are built with local materials and techniques adapted to specific geographical conditions, integrating knowledge transmitted over several generations. They meet the needs of occupants while respecting environmental constraints. In arid regions, these houses use materials like raw earth to maximize interior coolness, while in tropical climates, structures on stilts and lightweight materials ensure ventilation and protection against humidity. In cold climates, they emphasize insulation, such as wooden chalets or igloos, offering thermal comfort despite extreme conditions.

The integration of bioclimatic principles in vernacular architecture demonstrates how traditional societies have harmonized their constructions with local climatic conditions to maximize thermal comfort while minimizing environmental impact. By using techniques such as solar management, natural ventilation, and local materials adapted to geographical conditions, these houses optimize the use of natural resources. Spatial analysis reveals that elements such as room orientation, wall thickness, and the use of interior courtyards respond to specific climatic strategies. The study of these principles offers valuable lessons for contemporary architecture, providing sustainable solutions adapted to the challenges of climate change.

Vernacular architecture, which relies on adaptation to local needs, available materials, and climatic conditions, is the subject of numerous research studies aimed at understanding how traditional cultures have designed sustainable and resilient habitats. Studies explore building typology, construction techniques, and integrated bioclimatic principles, revealing an ingenious design that reflects both cultural values and environmental constraints. For example, research on the use of raw earth in rammed earth houses in the Middle East and North Africa demonstrates efficient heat management, while studies on bamboo dwellings in Asia highlight the lightness and flexibility of this material, suitable for tropical climates and seismic zones. This research shows that material choices in vernacular architecture are not only dictated by their availability but also by a deep understanding of their thermal and physical properties, thus contributing to habitats that are both comfortable and ecologically viable.

Bioclimatic principles are also at the heart of vernacular constructions. Studies on courtyard houses in the Mediterranean have shown their role in natural ventilation and the creation of microclimates, while troglodyte dwellings offer natural insulation against climatic extremes. Furthermore, ethnographic research highlights the importance of cultural and social dimensions of these structures, as demonstrated by studies on tukuls in Ethiopia and Bedouin haimas, which reflect community practices and local lifestyles. Thus, vernacular architecture offers valuable lessons for contemporary architecture, emphasizing the importance of integrating bioclimatic strategies and sustainable materials to address modern challenges such as climate change. Works such as those by Kamalipour [1] on "Contexte socioculturel et morphologie de l'habitat vernaculaire" and Zayn Ali and James Steele [2] on "Earth Architecture" have been fundamental in this understanding, and more recent research such as those by Paul Oliver [3] and Mileto et al. [4] bring contemporary perspectives on the sustainability and resilience of these constructions.

Vernacular architecture in arid regions, such as the Middle East and North Africa, illustrates intelligent management of local resources and climatic constraints. Houses made of raw earth, clay, and rammed earth are commonly used for their ability to store heat during the day and release it at night, thus regulating interior temperature. These constructions minimize openings to reduce sun exposure while promoting natural ventilation. Flat roofs are used as living spaces during cooler nights. This type of design is adapted to water conservation and heat management in desert environments, as noted by Hassan Fathy and Paul Oliver [5, 3]. In tropical regions, bamboo and wooden houses on stilts offer protection against floods and promote air circulation, while steeply sloped roofs facilitate the evacuation of heavy rains, creating a well-ventilated and comfortable space despite humidity, as described by Laurence Keefe [6].

Figure 1: Diverse Typologies of Vernacular Architecture: Regional Adaptations to Climate and Culture

Mediterranean houses are also designed to meet climatic challenges, with thick stone walls that provide thermal insulation and interior courtyards that create refreshing microclimates. These structures, often equipped with pergolas and shutters, combine passive thermal regulation strategies and aesthetic use of local materials. In cold regions, such as Scandinavia, the use of wood and thick walls helps conserve heat, while sloped roofs prevent snow accumulation. Houses are built to maximize heat while resisting rigorous winter conditions [4]. These examples demonstrate the effectiveness of vernacular architecture in various climatic contexts, emphasizing the importance of traditional bioclimatic strategies in creating buildings adapted to local conditions.

Figure 2: Examples of Vernacular Architecture in Morocco

Case Study

The Doukkala region, located in the vast coastal plain of western Morocco, is characterized by a rich historical and cultural heritage, influenced by various civilizations that have succeeded each other there. Geologically, this region rests on a complex stratigraphy dominated by Mesozoic and Cenozoic sedimentary deposits, composed mainly of limestone, marl, and sandstone [7].

Materials Research Proceedings 47 (2025) 156-164 https://doi.org/10.21741/9781644903391-18

Figure 3: Location of Morocco.

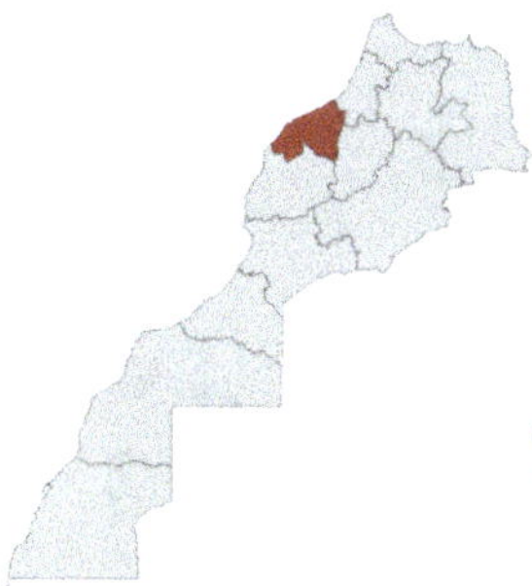

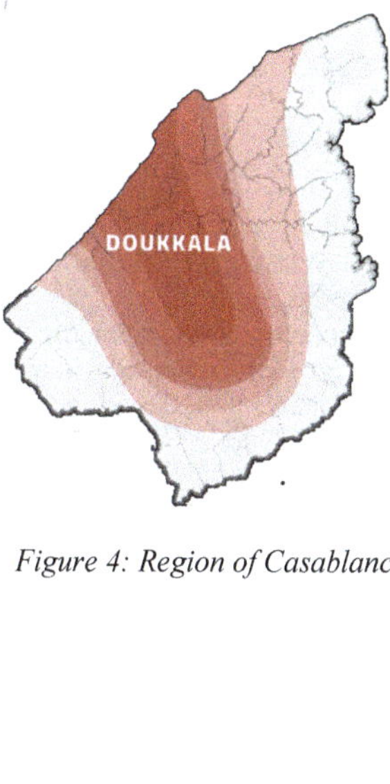

Figure 4: Region of Casablanca *Figure 5: El Jadida & Sidi Bennour Provinces.* *Figure 6: Doukkala.*

Tazotas, dry stone structures characteristic of the Moroccan rural landscape, occupy a central place in vernacular architecture in Morocco. Through numerous studies devoted to these constructions, various aspects related to their origin, materials, construction techniques, socio-cultural role, and contribution to architecture have been explored.

Chahid [8] presents an in-depth historical study of tazotas, highlighting their role in traditional Moroccan architecture, particularly in rural areas. Hammam [9] supports this perspective by showing how these structures have become a symbol of Moroccan vernacular architectural heritage, representative of ancient dry stone construction techniques.

Figure 7: Construction Elements of Tazotats: Traditional Dry Stone Architecture and Its Structural Components.

The studies of Karim Sebti [10] as well as Mohammed Chahid [8] detail the traditional construction techniques of tazotas. These ecological structures are built from local stones, without mortar, which demonstrates remarkable craftsmanship. The durability of these constructions is ensured by the solidity of the materials used and by perfect laying techniques, as shown by Mohammed El Kabir [11] in his analysis of the challenges of tazota restoration.

Steven Coissar [12] explores the role of tazotas in traditional Moroccan society, emphasizing their social function, but also their utilitarian dimension, serving as refuge for peasants, agricultural reserves, or temporary shelters during agricultural work. Maria Genesda [13] broadens this perspective by considering tazotas as reflections of Moroccan material culture, emphasizing the importance of these constructions in defining regional identity.

Tazotas are also considered models of ecological construction. The study by V. Sartiges [14] highlights their contribution to greener architecture, in line with principles of sustainability. These buildings, constructed with local materials and respecting the environment, are part of a bioclimatic approach that reduces energy needs. Heil and Boussaa [15] demonstrate that tazotas adapt perfectly to local climatic conditions, particularly thanks to their design which optimizes thermal regulation. Indeed, tazotas offer natural insulation against high summer temperatures and low winter temperatures, as illustrated by Julien Le Guay [16] in his study on thermal comfort.

Model Development

The development of a Python model for the bioclimatic analysis of tazotas, vernacular architectural structures specific to the Doukala region, is based on an approach integrating both the form and function of these elements. This developed model is based on physical parameters such as the geometry of tazotas, their orientation, and the materials used, to evaluate their impact on thermal comfort and natural ventilation. The model can simulate temperature distribution, incident heat flux, and air circulation inside and around tazotas. The objective is to exploit the architectural characteristics of these forms to maximize energy efficiency in hot and arid environments.

The integration of libraries such as Pythermal comfort for the analysis of thermal comfort indices allows for precise evaluation of the impact of tazota shape on occupant comfort. This model also allows for parametric simulations by modifying the dimensions, materials, or spatial configuration of tazotas (opening dimensions) to explore different bioclimatic strategies and identify optimal solutions for adaptation to local climatic conditions. Through this approach, designers and researchers can better understand and enhance these traditional elements while contributing to the design of sustainable and energy-efficient buildings.

The numerical model developed in this article studies the envelope of the tazota in a bioclimatic and energy approach. The code simulates and analyzes the thermal performance of the envelope based on various physical parameters to evaluate the energy efficiency of construction materials used in the envelope, taking into account their thermal properties, such as conductivity, thermal capacity, and thermal resistance. This model takes into account the dynamic interactions between the envelope and external conditions (temperature, sunlight, wind).

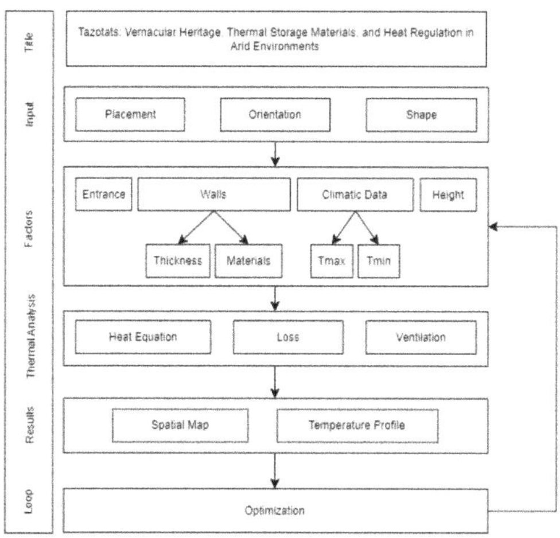

Figure 8: Flow Chart of this Article

The envelope is studied here as an active element, participating in the thermal regulation of the tazota. The model thus allows for a better understanding of how the properties of materials and their thickness influence the thermal performance of the envelope in various climatic contexts.

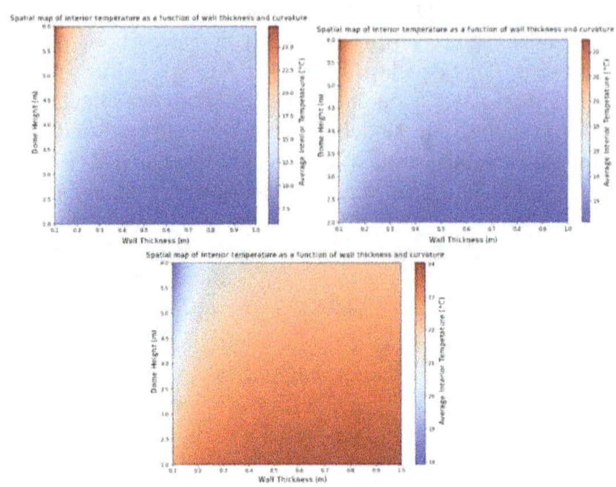

Figure 9: Spatial map of interior temperature as a function of wall thickness and curvature

Changes Based on Materials and Climatic Conditions

Figure 10 shows that:

-The combination of a taller tazota and thicker walls gives the highest interior temperatures (red zone in the upper right). This is probably due to the combined effect of increased air volume (larger dome) and better insulation (thicker walls).

-The combination of a lower dome and thinner walls gives the lowest interior temperatures (blue zone in the lower left), as these configurations likely favor faster heat dissipation and less heat retention inside the structure.

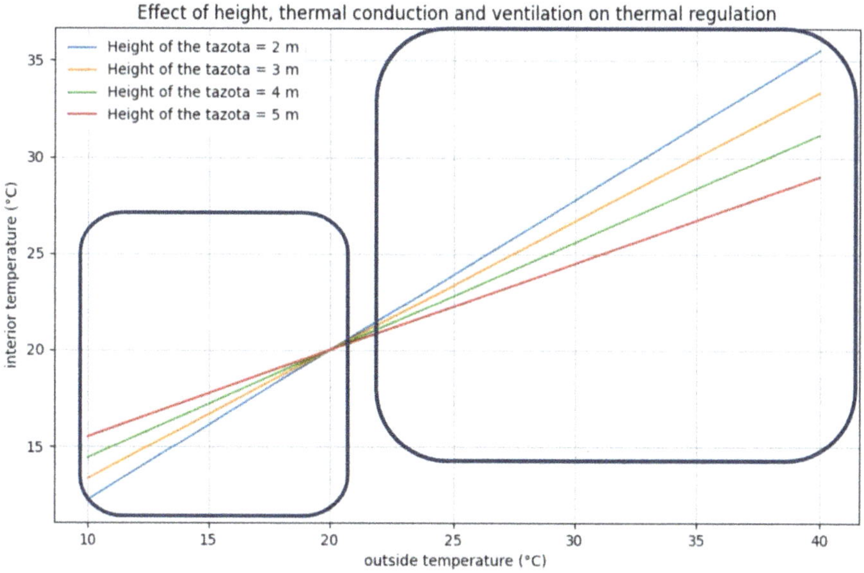

Figure 10: Effect of height on thermal regulation.

The average interior temperature increases almost linearly with the increase in exterior temperature. The higher the exterior temperature, the more the interior temperature increases, regardless of the height of the tazota. The 2m tazota has a higher interior temperature than those of 3m, 4m, and 5m. This shows that lower tazotas are more sensitive to exterior temperature, probably due to lower thermal capacity or less effective ventilation. Taller tazotas (4m and 5m) seem to better maintain a more stable and lower interior temperature, particularly at high exterior temperatures. The height of the dome has a significant impact on internal thermal regulation. Lower tazotas (2m) are more prone to high interior temperatures, while taller tazotas (5m) maintain a lower and more comfortable temperature as the exterior temperature increases. These results could be useful for optimizing the architectural design of domes in hot climates, favoring greater heights to improve thermal comfort.

The spatial map (figure) produced by this model shows temperature gradients based on wall thickness and tazota height. The results show the following:

Wall thickness: Thicker walls offer better thermal insulation, reducing heat transfers between the interior and exterior. Thus, the interior temperature remains more stable when the walls are thick, especially during periods of high heat.

Tazota curvature: The height of the tazota influences how heat is dissipated inside. A higher curvature can contribute to better air circulation, which can help maintain a more comfortable interior temperature.

Interactions between parameters: There is a complex relationship between wall thickness, wall nature, ventilation, and height. For example, a too-low height combined with thin walls could lead to overheating inside, while a high height and thick walls would optimize ventilation and thermal insulation.

This numerical model provides a detailed and visual approach to understanding how different geometric and physical aspects of a tazota influence its internal thermal regulation. The results obtained in the form of a spatial map allow architects and engineers to make informed decisions on tazota design to optimize their thermal comfort.

Acknowledgments
We would like to express our gratitude to Ghita Benjelloun for her valuable contribution to the writing of this article. Her ideas and support have greatly enriched our work.

References

[1] Hesam Kamalipour and Mostafa Zaroudi. "Contexte socioculturel et morphologie de l'habitat vernaculaire : une étude de cas". In: Current Urban Studies 2.3 (Sept. 2014).

[2] James Steele. Une architecture pour les gens - Les œuvres complètes de Hassan Fathy. Thames and Hudson, 1997.

[3] Paul Oliver. Construit pour répondre aux besoins : les enjeux culturels de l'architecture vernaculaire. 2006.

[4] Camilla Mileto et al. Vernacular Architecture: Towards a Sustainable Future. 2014.

[5] H. Fathy. "Energie naturelle et architecture vernaculaire". In: Energie naturelle et architecture vernaculaire. 1973. https://doi.org/10.1201/b17393

[6] Laurence Keefe et al. Earth Building: Methods and Materials, Repair and Conservation. 2005.

[7] Y. A. Laaziz. "Contribution des SIG et de la modélisation volumique à la caractérisation géomorphologique et géologique de la région des Doukkala - Meseta côtière, Maroc - Implication sur les effondrements karstiques". PhD Thesis. 2021.

[8] Ahmed Chahid. Le petit guide touristique d'El Jadida. 2006.

[9] Mohamed Hamam. In: Maalamate al Maghrib. Vol. 6. Salé : Association marocaine de la production, la traduction et l'édition, 1992, pp. 2040-2043.

[10] Karim Sebti. Les Tazotas des Doukkala, p. 114.

[11] Saaidi ElKabir. Guide géologique du Maroc. 2003, p. 121.

[12] Steven Coissar. "Projet de territoire entre recomposition territoriale et cohabitation des activités: L'exemple d'El Jadida au Maroc". PhD Thesis. 2009.

[13] Sergio Gnesda. "Témoins d'architecture en pierre sèche au Maroc: les tazotas et toufri". In: 16 (1996), p. 24.

[14] Vicomte de Sartiges. Les Cabanes en pierres sèches du Sud de la France. 1921. https://doi.org/10.3406/bspf.1921.13326

[15] Julien Heil et al. "Entre-temps. Construction métamorphe en dialogue avec le climat". In: (2023).

[16] Michel Le Guay. Confort thermique dans les lieux de vie. Fondation EFB.

Materials Research Proceedings 47 (2025) 165-171 | https://doi.org/10.21741/9781644903391-19

Cultural revitalization and tourist conversion of historical monuments: The cases of the ancient medinas of Fez and Tunis

BOUKIL Mehdi[1,a]*

[1]Faculty of Letters and Human Sciences, Fez, Morocco

[a]mehdi.boukil@usmba.ac.ma

Keywords: Restoration, Revitalization, Tourism, Cultural Heritage, Medinas, Historical Monuments

Abstract. This article explores the similarities between the medinas of Fes and Tunis, focusing on cultural repurposing projects. Both cities are UNESCO World Heritage Sites and share similar heritage resources, including mosques, mausoleums, madrasas, zawiyas, and traditional houses. As popular tourist and artisanal destinations, both medinas have undertaken multiple heritage rehabilitation initiatives and policies. Significant investments have been made in the cultural repurposing and economic revitalization of restored heritage, avoiding the construction of new cultural spaces by repurposing vacant buildings within the medina. This strategy aims to give contemporary functions to historic buildings while preserving their architectural integrity. By doing so, it ensures the conservation of the historic center, supports its development and breathes new life into unoccupied structures.

Introduction

In the context of preserving built heritage, it is crucial to remember that several international documents, such as the Venice Charter (1964), emphasize the importance of assigning practical functions to historical buildings to ensure their long-term conservation. The repurposing of these buildings can take various forms: it may involve simply integrating new functions while respecting the existing structures, or require more significant architectural interventions, such as additions, extensions, or even demolitions. In any case, and following the recommendations of the Venice Charter, these new uses must preserve the authenticity of the buildings, ensuring that their original layout and ornamentation are not compromised. This approach aims to reconcile the conservation of heritage and historical value with dynamic reuse, transforming these spaces into places for cultural and artistic dissemination. Such an approach not only preserves their integrity but also anchors them in contemporary life, ensuring their transmission to future generations [1].

As part of these efforts, we will study the cultural repurposing of built heritage assets, whether in Fes or Tunis, as an initiative aimed at giving new functional purposes to these historic buildings while preserving their architectural integrity. This approach is driven by the desire to more effectively safeguard these heritage elements while making them available to the local community and visitors for beneficial use. It seeks both to ensure their long-term preservation and to transform them into cultural and educational resources accessible to a broad audience.

The Restoration of Riads and Old Dwellings and Their Transformation into Guesthouses in the Heart of the Medinas of Fez and Tunis:

The restoration of old homes and their transformation into guesthouses is a recurring practice in the two medinas studied. This campaign has led to the restoration and maintenance of several old houses, whether deteriorated or collapsed, contributing to their revitalization and rehabilitation. Investments have been made to turn them into welcoming spaces, with positive effects on the surrounding environment. To convert these houses, it is necessary to bring them up to tourist accommodation standards and undertake upgrading works. If a structural modification is required, an architect is called upon to establish a plan that will be supervised by a commission composed of members from the Municipality, ADER (Agency of Development and Rehabilitation), the Urban Agency, and the Inspection of Historical Monuments (2). *These works generally involve adding bathrooms (sometimes nonexistent), creating a practical kitchen, working on stairs to make them accessible, installing air conditioners and heaters, and covering the patio roof (to use the patio in rainy or cold weather). Ultimately, the main objective is to have a house that appears 'Moroccan' while being comfortable. Preservation also concerns the architectural decor of the house (painted or sculpted wood and plaster, Zelliges...)* (3).

Fig 1: Architectural refinement of a Riad guesthouse in the Medina of Fez, Source: Booking.com

The works must be executed with the utmost care and precisely match the approved plans and drawings regarding their shape, dimensions, quality, appearance, or arrangements (4).

This type of accommodation has opened up the possibility for tourists, especially international ones, to stay in the center of the old city of Fez and Tunis and experience the old lifestyle with new quality standards. In a short time, we have a distinct tourist accommodation, taking advantage of the image, history, and recognition of the Medinas as a UNESCO World Heritage site.

The beauty, refinement, and touch of elegance in traditional architecture and artisanal decoration are defining features of this guesthouses. These elements, often eliciting admiration and wonder from visiting tourists. These houses are designed around a central patio adorned with mosaics, fountains, and plants, featuring carved wooden doors and wrought-iron windows. Rooms and suites are tastefully decorated, creating a friendly atmosphere (Fig 1). This private initiative has also led to a dynamic artisanal and commercial network within the medina. Craft shops, art galleries, restaurants, and businesses have thrived, creating a flourishing local economy.

We can also highlight the role of The Association for the Preservation of the Medina (ASM) in Tunisia which has developed undeniable expertise and unmatched know-how in the preservation of Medina heritage, as evidenced by the many historical monuments it has successfully restored and rehabilitated. Among the houses and palaces converted to host cultural activities, mostly located around the mosque Zitouna and Sidi BenArous areas we can cite the following examples:

- Khair-Eddine Palace, transformed into the Museum of the City of Tunis
- Dar Lasram, transformed into the headquarters of the (ASM) Fig 2
- Dar Bouderbela, transformed into an art gallery
- Dar Ben Abdallah, transformed into the Museum of Arts and Popular Traditions
- Dar Ben Achour, transformed into the National Library of the City of Tunis
- Dar El Jaziri, transformed into the House of Poetry...

Ben Said Imen, in her thesis *(5)*, offers a comprehensive analysis of the processes involved in the restoration and rehabilitation of houses in the old city of Tunis after independence. She highlights the guiding principles of these operations, as well as the role of the various stakeholders, emphasizing the balance between preserving authenticity, fostering innovation, and ensuring comfort for the city's residents.

Fig 2 : Dar Lasram in the Medina of Tunis

The Rehabilitation of Madrasas in the Two Medinas of Fes and Tunis

Madrasas, serving as student residences and Islamic educational institutions, played a central role in the medinas of Tunis and Fes. In Tunis, their development was encouraged during the Hafsid period, from the 13th to the 16th century, while in Fes, they flourished under the Marinid era in the 13th century. These madrasas, often located near major mosques, attracted both local and foreign students, contributing to the intellectual prominence of these cities. Today, they have

Materials Research Proceedings 47 (2025) 165-171 https://doi.org/10.21741/9781644903391-19

become historic tourist monuments, drawing visitors from around the world to admire their architecture and discover their rich educational and cultural heritage.

The Benjelloun Meziane Foundation for Science, Culture, and Art in Fez contributed significantly to this preservation campaign by launching a major restoration project for the Bouanania Madrasa, which lasted eight years (from 1995 to 2004), to restore this architectural masterpiece to its former glory. Skilled traditional craftsmen, selected for their experience, talent, and precision, were chosen to perfectly replicate the work of their predecessors, aiming to restore the madrasa to its original state. This artistic endeavor required perfection and a keen sense of detail.

The ancient madrasas Cherratine and Attarine also underwent complete restoration work. The Cherratine Madrasa benefited from a partnership agreement between the Ministries of Culture and Housing and Urban Planning in 2008, with the intervention of the Al Omran Group. This intervention included the restoration of vertical and horizontal structures, architectural elements, and decorations, using the same original materials: terracotta bricks, earth, lime mortar, wood, plaster, zellij tiles, etc., and the same traditional construction techniques. This attention to detail breathed new life into this prestigious institution, turning it into a beautiful space for exhibitions and education in heritage and architecture (Fig 3).

The Attarine Madrasa benefited from a restoration program funded by a grant from the Arab Bank for Economic and Social Development, amounting to 8.5 million dirhams between 2006 and 2008. Both projects involved the expertise of the Fez Development and Rehabilitation Agency for project management. Between 2013 and 2017, another royal rehabilitation program was launched, covering 27 historical monuments, including the restoration of five old madrasas: Madrasa Al-Sahrij and Sbaîine in the Andalus neighborhood, Madrasa Muhammadiya, Seffarine, and Misbahiya north of the Al-Qarawiyyine Mosque.

Fig 3: Exterior facade treatment, Terrace, and interior facade of the Cherratine Madrasa patio (Source: Al Omran Group)

Fig 4: Sahrij Madrasa dedicated for calligraphy courses Source : Field work

The management of these schools was transferred to the Ministry of Habous and Islamic Affairs, as most of them will once again be converted back to their original educational and residential functions, but with modern conveniences (computers, Wi-Fi, etc.). The Sbaîine Madrasa will focus on teaching Arabic calligraphy, while the Sahrij Madrasa will reopen to tourists, also offering calligraphy classes (Fig 4).

In Tunisia, we highlight the leading role occupied by the National Heritage Institute (INP) in the preservation and promotion of the rich cultural and architectural heritage of the Medina of Tunis. As an entity dedicated to safeguarding this precious heritage, the INP assumes crucial responsibilities to ensure that the historical remnants continue to resonate through future generations. Among these monuments are the medersas, which are abundant in the Medina and have undergone various conversion works, as shown in the following table:

Table 1: Repurposed Madrasas in the Medina of Tunis

Medersa Chammaiya	Training center
Medersa Bachia	Training center
Medersa Slimaniya	House for Medical associations
Medersa Mouradiya	Center of handicrafts
Medersa Bir Lahjar	Culturel Center
Medersa Achouria	Center for culturel associations

Source : Eya Mahroug (6)

Materials Research Proceedings 47 (2025) 165-171 https://doi.org/10.21741/9781644903391-19

Fig 5: Rehabilitation of the Slimaniya Medersa

Source: INP

Integration of Madrasas into Tourist Circuits

In the context of cultural tourism, the madrasas hold particular importance as witnesses to the intellectual and artistic history. They are often included in guided tourist circuits, allowing visitors to discover their scientific and cultural value. During these tours, tourists have the opportunity to explore the madrasas, admire their magnificent architecture, and learn about their role in traditional education. Guides provide historical and contextual information on each madrasa, highlighting its importance in the intellectual development of the region.

Fig 6: Group of tourists at the Attarine Madrasa and entrance fees (Source: Fieldwork)

Tourists are captivated by the remarkable architecture and fascinating history of the madrasas. Every detail of these institutions is a work of art, capturing the imagination of visitors and transporting them to a bygone era of knowledge and spirituality. They flock to admire the richly decorated study rooms, carved columns, and intricate geometric patterns.

Conclusion

The adaptive reuse of vacant spaces within the Medinas offers a sustainable solution that respects and preserves the architectural and cultural heritage of historic centers. By assigning contemporary functions to these historical buildings, this strategy ensures their conservation while promoting economic and cultural development. It not only protects the integrity of the Medina but also revitalizes unused structures, breathing new life into the area without compromising its historical value. This approach successfully merges heritage preservation with modern needs, fostering a dynamic and thriving urban environment.

For example, the madrasas in the Medinas of Fez and Tunis represent invaluable heritage. Their preservation and promotion not only highlight local culture but also attract tourists from around the world. These structures are more than just buildings; they are living symbols of a commitment to beauty and architectural excellence, transporting visitors into a world of aesthetics and inspiration.

Moreover, riads, with their traditional architecture, artisanal decoration, and warm hospitality, serve as guesthouses in the old Medinas of Fez and Tunis, offering visitors an authentic setting and a unique experience that immerses them in local culture and lifestyle. In this article, we aimed to shed light on the enhancement of heritage and tourism in the Medinas of Fez and Tunis through the acquisition and transformation of riads and old residences into guesthouses by both foreign and Moroccan private actors. This qualitative shift has diversified and expanded the tourist accommodation capacity in these cities. Additionally, these houses are being sold for various purposes, such as commercial spaces, cafes, restaurants, craft bazaars, art exhibition galleries... This transformation supports the enhancement of the old urban fabric, which still contains hundreds of old houses that could be converted to accommodate future tourism projects.

References

[1] ICOMOS. (1964). Charte internationale sur la conservation et la restauration des monuments et des sites.

[2] https://www.icomos.org/images/DOCUMENTS/Charters/Venice_Charter_FR_2023.pdf

[3] M. Istasse 2011, Politique patrimoniale à Fez et rôle de l'Inspection des monuments historiques 'Entretien avec Mouhcine El Idrissi El Omari', https://books.openedition.org/cjb/329?lang=fr.

[4] M. ISTASSE 2013, Dynamique de requalification des medinas et préservation du patrimoine : étude en acte dans la Medina de Fès. In medinas immuables ? gentrification et changement dans les villes historiques marocains (1996-2010), sous la direction d'Elsa Casldo et d'autres. Centre Jacques Berque, Rabat.

[5] A. Brahim, M. Laaouane, Piéro D.Galloro 2007 , les maisons d'hôtes à Fez : évolution et étude d'impacts en Médina. Actes du colloque Dynamiques territoriales : des potentialités au développement durable (revue du centre d'Etudes géographiques de l'Université de Metz). P 256

[6] Ben Said, I. (2016). Restauration des demeures dans la Médina de Tunis après l'indépendance, principes et acteurs: entre restitution, innovation et confort. (Thèse de doctorat, École Nationale d'Architecture et d'Urbanisme de Tunis).

[7] Eya Mahroug. (2017). Ambiance patrimoniale dans les opérations de reconversion des bâtiments de la médina de Tunis. (Thèse de doctorat, Université de Carthage, Tunisie). p. 130.

Advanced object detection for heritage preservation: A case study of Ouaouizerth and the Palace of Sultan Moulay Ismail

KHLIFATI Oumaima[1,a*], BABA Khadija[1,b], MASROUR Ilham[1,c], EL HARROUNI Rime[2,3,d], NOUNAH Abderrahman[1,e], EL HARROUNI Khalid[4,f]

[1]Civil and Environmental Engineering Laboratory (LGCE), Mohammadia Engineering School, Mohammed V University in Rabat, Morocco

[2]Euro-Mediterranean School of Architecture, Design and Urbanism, Euro-Mediterranean University of Fez Institution, Morocco

[3]Faculty of Architecture and Urbanism, University of Mons, Belgium

[4]Ecole Nationale d'Architecture de Rabat, BP 6372, Rabat Instituts, Rabat, Morocco

[a]oumaima_khlifati@um5.ac.ma, [b]Khadija.baba@est.um5.ac.ma, [c]ilham_masrour@um5.ac.ma [d]rimaelharrouni@gmail.com, [e]abderrahman.nounah@est.um5.ac.ma, [f]k.elharrouni@enarabat.ac.ma

Keywords: Historical Monuments, Heritage Preservation, Object Detection, Degradation Detection, Architectural Heritage

Abstract. Ouaouizerth, a historic town located about 80 kilometers southwest of Béni Mellal, stands as a testament to Morocco's rich and diverse history. Founded centuries ago, the town is home to landmarks like the Palace of Sultan Moulay Ismail, commissioned around 1710 and also known as Kasbah de Boulaouane. This palace, built during the reign of Sultan Alaouite Ismail Ibn Sharif, showcases a blend of grandeur and architectural skill. Positioned strategically overlooking the Oum Er Rabia river and inspired by Byzantine fortifications, it offers a fascinating glimpse into historical architecture. Preserving these historical monuments is vital, yet traditional methods of documentation and categorization are often labor-intensive and prone to inaccuracies. This study tackles these challenges by developing an advanced object detection model to improve the efficiency and accuracy of degradation detection in these historical monuments. Utilizing the object detection method, a robust model was created for detecting and classifying various forms of deterioration. The dataset included meticulously captured and annotated images from Ouaouizerth and the Palace of Sultan Moulay Ismail, featuring a range of degradations. Annotations were performed using Label Studio software to ensure precise identification and categorization. Data augmentation techniques were employed to create a diverse and representative dataset, essential for effective model training. The model's performance was rigorously evaluated using metrics such as precision and recall, demonstrating its high accuracy in detecting specific deteriorations. The successful implementation of this model holds significant potential for aiding conservation efforts, providing a valuable resource for heritage preservation, urban planning, and tourism promotion, thus ensuring the enduring legacy of Morocco's historical treasures.

Introduction

Ouaouizerth, a historic town nestled approximately 80 kilometers southwest of Béni Mellal in the Middle Atlas region of Morocco, stands as a significant example of the country's rich architectural heritage. Founded centuries ago, Ouaouizerth is renowned for its traditional buildings constructed using the ancient rammed earth technique, a method that employs raw earth mixed with straw or other plant fibers. These rammed earth structures, adorned with intricate geometric patterns,

Materials Research Proceedings 47 (2025) 172-179 https://doi.org/10.21741/9781644903391-20

represent a unique blend of environmental sustainability and cultural craftsmanship, reflecting the deep-rooted architectural traditions of the region [1], [2].

Despite their inherent durability, these historic buildings face ongoing threats from environmental factors, material degradation, and the passage of time. Visible signs of deterioration such as cracks, material displacements, and erosion are prevalent, compromising both the structural integrity and the aesthetic value of these important cultural landmarks. Given Ouaouizerth's importance not only as a symbol of Morocco's architectural past but also as a key contributor to the region's heritage tourism, the preservation of these structures is of critical importance [3], [4], [5].

Among the town's most notable landmarks is the Palace of Sultan Moulay Ismail, also known as the Kasbah of Boulaouane. Built around 1710 during the reign of Sultan Ismail Ibn Sharif, this palace was both a military fortress and an administrative center, strategically positioned overlooking the Oum Er Rabia River. Inspired by Byzantine fortifications, the palace is an architectural masterpiece of its time, offering a glimpse into Morocco's historical grandeur [6]. However, like many other historical structures, the palace has not been spared from the effects of aging and environmental stressors. Issues such as rising dampness, partial collapses, and material erosion are evident, further exacerbating the challenges of conservation.

Currently, surface damage detection methods for historical heritage sites predominantly depend on on-site visual inspection, supported by specialized equipment [7], [8], [9], [10]. Despite the operational benefits of this approach, inexperienced inspectors may inaccurately assess the damage, leading to significant consequences for subsequent structural safety evaluations and repairs. Moreover, large-scale inspections using this method are both inefficient and labor-intensive [11], [12]. These methods, although valuable, are often prone to human error and inconsistencies, creating a demand for more efficient and accurate solutions. The need for automated, reliable methods to detect and classify structural distress is becoming increasingly urgent as efforts to preserve these monuments intensify [13], [14] [13].

In response to this need, this study presents a novel approach for the automatic detection of structural degradations in historical monuments using an object detection model based on the YOLO (You Only Look Once) framework [15]. By leveraging annotated image datasets of the Ouaouizerth monuments, including the Palace of Sultan Moulay Ismail, and utilizing advanced deep learning techniques, this model is designed to detect and categorize various forms of degradation with high precision.

Through the application of this model, the goal is to improve the efficiency and accuracy of heritage conservation efforts, ensuring the preservation of Morocco's architectural treasures for future generations while also supporting urban planning and tourism development.

Methodology

The methodology of this study is divided into five main phases: data collection, data augmentation, annotation, development and training of the object detection model, and model evaluation. These steps were essential in creating a robust and efficient system capable of detecting and classifying degradations in historical monuments, such as the Palace of Sultan Moulay Ismail and the traditional buildings of Ouaouizerth.

1. Data Collection

Images were captured at different locations within the Palace of Sultan Moulay Ismail and across various traditional buildings in Ouaouizerth. The focus was on documenting visible structural degradations, such as cracks, partial collapses, material detachment, and rising damp. A total of 656 images were initially collected, capturing a wide range of deterioration under different lighting conditions, angles, and weather circumstances to reflect real-world variability. Fig.1

presents a sample of the images from our dataset, showcasing various forms of degradation on the walls of the Palace of Sultan Moulay Ismail and traditional buildings in Ouaouizerth.

Fig. 1: *Example images of the walls of Sultan Moulay Ismail's Palace (a) and traditional buildings in Ouaouizerth (b).*

2. Data Augmentation

After collecting the initial set of images, data augmentation techniques were applied to increase the diversity and volume of the dataset. This step is critical for improving the model's robustness and generalization capabilities when detecting degradations in varying conditions.

The following augmentation techniques were implemented:

- Brightness Adjustment: To simulate different lighting conditions, brightness levels in the images were altered. This improves the model's capacity to detect degradations under both overexposed and underexposed scenarios.
- Contrast Adjustment: Adjustments to contrast were made to improve the distinction between the degraded areas and their surroundings, allowing the model to better detect subtle structural flaws.
- Image Rotation: The images were turned at multiple angles to depict varying orientations of the structures. This allowed the model to learn how to recognize degradations from multiple perspectives.
- Blurring: Blur effects were applied to simulate poor atmospheric conditions or optical defects, which could occur in real-world scenarios, further diversifying the dataset.
- Zooming: Zoom-in and zoom-out transformations were applied to simulate variations in the distance and scale of the images, helping the model recognize features from different perspectives and distances.

As a result of these augmentation techniques, the dataset size increased significantly, from 536 to 801 images for the Palace of Sultan Moulay Ismail and from 120 to 500 images for the traditional buildings of Ouaouizerth. This expansion enhanced both the dataset's diversity and its capacity to train an accurate and reliable AI model.

3. Data Annotation

Following the augmentation, the images were annotated to mark various types of degradation. The annotation was performed using Label Studio, a versatile open-source tool for labeling and extracting information from raw data. For each image, specific categories of degradation were identified and labeled, including:

- Cracks
- Partial Collapses
- Spalling (Material Detachment)
- Delamination
- Rising Damp

Each annotation was reviewed and validated to ensure the quality and consistency of the labeled data. This step was crucial for training a model capable of recognizing and accurately classifying different types of deterioration.

4. Object Detection Model

For the automatic detection of degradation, the YOLOV8 object detection framework was employed. YOLO was chosen for its ability to detect multiple objects in a single pass, allowing for real-time detection with high accuracy. Given its efficiency, YOLO was well-suited for identifying the various types of deterioration present in historical monuments.

The model was trained to detect and classify the categories of degradation previously annotated. The training process involved the following settings:

- Input Resolution: 416x416 pixels, to balance detection accuracy and processing speed.
- Loss Function: A multi-part loss function was used, considering both classification and localization errors, as well as confidence scores.
- Training Parameters: The model underwent training for 50 epochs with a learning rate set to 0.001 and a batch size of 32.
- Data Split: The dataset was split into 70% for training, 30 % for testing ensuring that the model was evaluated on data it had not previously encountered during training.

5. Model Evaluation

The performance of the model was evaluated using standard object detection metrics, which are crucial for assessing its effectiveness in detecting and classifying structural degradations. Two key metrics were employed: Precision, which measures the ratio of correctly identified degradations to the total number of detected degradations (Eq. 1), and Recall, which indicates the proportion of actual degradations correctly identified by the model (Eq. 2). This comprehensive evaluation ensured that the model not only detected degradations with high accuracy but also correctly localized them within the images, thereby improving the reliability of the detection process.

$$\text{Precision} = \frac{\text{True Positives}}{\text{False positives}} \tag{1}$$

$$\text{Recall} = \frac{\text{True Positives}}{\text{True Positives+False positives}} \tag{2}$$

Results

Recall-Confidence and Precision-Confidence curves

The Recall-Confidence curve (Fig. 3a) illustrates the model's effectiveness in detecting different defects—cracks, delamination, rising damp, partial collapse, and spalling—across varying confidence levels. As confidence increases, recall tends to decrease for all defect classes, highlighting a trade-off where the model becomes more selective, sacrificing recall (true positive rate) for higher precision. Cracks maintain a higher recall across confidence thresholds, while rising damp and partial collapse experience a sharper decline, suggesting weaker performance at higher confidence levels for these more challenging defects. The "all classes" curve provides an overall assessment, balancing recall and confidence across all defect types. This analysis indicates that in critical applications, especially where detecting rarer or subtler defects is essential, a lower confidence threshold may be necessary to preserve higher recall.

The Precision-Confidence curve (Fig. 3b) shows how the model's precision improves as confidence thresholds increase for various structural defects. Cracks achieve the highest precision performance, with fewer false positives even at moderate confidence levels. In contrast, rising damp exhibits greater variability and the lowest precision, indicating that detecting this defect without generating false positives remains a challenge. The "all classes" curve reflects the average precision across all defects, demonstrating a steady improvement as confidence rises. These insights help determine the optimal confidence level that balances precision and minimizes false positives for each defect type.

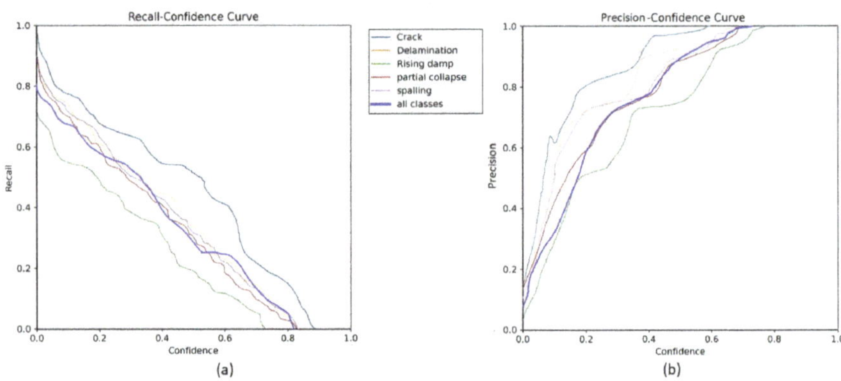

Fig. 2: Recall-Confidence (a) and Precision-Confidence (b) Curves for Defect Detection

YOLOv8 Detection of Structural Degradations

The YOLOv8 model proved effective in detecting multiple structural degradations in the Palace of Sultan Moulay Ismail and traditional buildings in Ouaouizerth (Fig. 3), accurately generating bounding boxes around identified defects. Cracks, in particular, were consistently detected with high confidence scores, ranging from 0.85 to 0.98, showcasing the model's precision across different materials and surfaces. Other degradation types, such as delamination and spalling, were detected with confidence scores reaching 0.89, while partial collapse ranged between 0.85 and 0.88. Rising dampness was identified with a lower confidence score of 0.72, indicating some difficulty in detecting moisture-related issues, suggesting a need for further model refinement.

Materials Research Proceedings 47 (2025) 172-179 | https://doi.org/10.21741/9781644903391-20

Overall, these results confirm the robustness of the YOLOv8 model in accurately identifying and localizing structural defects, making it a reliable tool for heritage conservation initiatives.

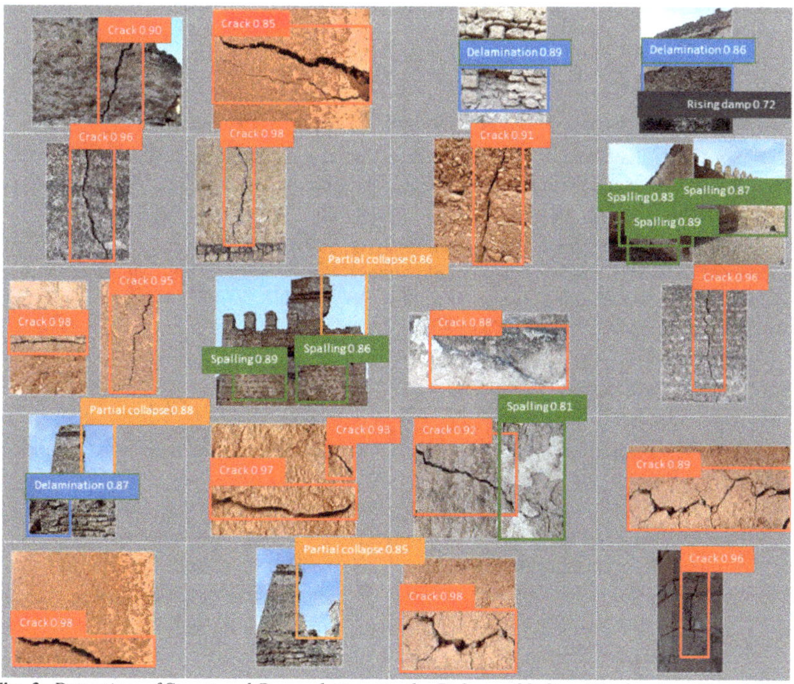

***Fig. 3**: Detection of Structural Degradations in the Palace of Sultan Moulay Ismail and the traditional buildings of Ouaouizerth using the YOLOv8 Model*

Conclusion

The development and application of the advanced object detection model based on the YOLOv8 framework demonstrate a significant advancement in the preservation of heritage monuments, particularly for detecting structural degradations in historical sites such as the Palace of Sultan Moulay Ismail and the traditional buildings of Ouaouizerth. The model effectively detected and classified various degradation types, including cracks, delamination, rising damp, spalling, and partial collapses, achieving high precision and recall metrics. Results from the Recall-Confidence Curve showed that cracks were detected with the highest recall across confidence levels, while defects like rising damp and partial collapse showed a steeper drop in recall. Similarly, the Precision-Confidence Curve indicated that cracks had the highest precision at moderate to high confidence levels, while rising damp showed the most variability, with a lower precision due to false positives. The model achieved confidence scores as high as 0.98 for cracks and up to 0.89 for delamination and spalling. These results confirm the robustness of the YOLOv8 model in accurately identifying structural degradations across a variety of surfaces and conditions, proving it to be an essential tool for enhancing conservation efforts. Further refinement and application of the model to other heritage sites may improve its effectiveness and broaden its utility in urban planning and tourism development.

References

[1] I. Masrour, K. Baba, and S. Simou, "Naturally Strengthening Rammed Earth: The Promising Potential of Biopolymers," presented at the Materials Research Proceedings, 2024, pp. 233–239. https://doi.org/10.21741/9781644903117-25

[2] I. Masrour, K. Baba, and K. Doughmi, "Geopolymers: An Eco-Friendly Approach to Enhancing the Stability of Earthen Constructions," presented at the Materials Research Proceedings, 2024, pp. 226–232. https://doi.org/10.21741/9781644903117-24

[3] S. Simou, K. Baba, and A. Nounah, "Preserving Historic Structures: Advancing Conservation Practices Through Building Material Analysis and Mapping," *Civil Engineering and Architecture*, vol. 12, no. 1, pp. 312–325, 2024. https://doi.org/10.13189/cea.2024.120124

[4] S. Simou, "Investigating Cultural Dimensions in Sustainable Tourism: Carrying Capacity Assessment and GIS-based Management Strategies for Historic Buildings," presented at the Mediterranean Architectural Heritage, Apr. 2024, pp. 311–318. https://doi.org/10.21741/9781644903117-33

[5] O. Khlifati, "Object Detection Method for Automated Classification of Distress in Rabat's Built Heritage," presented at the Mediterranean Architectural Heritage, Apr. 2024, pp. 358–362. https://doi.org/10.21741/9781644903117-37

[6] M. Bouzahzah and Y. El Menyari, "International tourism and economic growth: the case of Morocco and Tunisia," *The Journal of North African Studies*, vol. 18, no. 4, pp. 592–607, Sep. 2013. https://doi.org/10.1080/13629387.2013.836321

[7] I. Flores-Colen, J. de Brito, and V. de Freitas, "Expected render performance assessment based on impact resistance in situ determination," *Construction and Building Materials*, vol. 23, no. 9, pp. 2997–3004, Sep. 2009. https://doi.org/10.1016/j.conbuildmat.2009.05.003

[8] L. A. Santos, I. Flores-Colen, and M. G. Gomes, "In-situ Techniques for Mechanical Performance and Degradation Analysis of Rendering Walls," *Restoration of Buildings and Monuments*, vol. 19, no. 4, pp. 255–266, Aug. 2013. https://doi.org/10.1515/rbm-2013-6606

[9] T. Reyno, C. Marsden, and D. Wowk, "Surface damage evaluation of honeycomb sandwich aircraft panels using 3D scanning technology," *NDT & E International*, vol. 97, pp. 11–19, Jul. 2018. https://doi.org/10.1016/j.ndteint.2018.03.007

[10] Michael O'Byrne, Franck Schoefs, Bidisha Ghosh, Vikram Pakrashi, "Texture Analysis Based Damage Detection of Ageing Infrastructural Elements - O'Byrne - 2013 - Computer-Aided Civil and Infrastructure Engineering - Wiley Online Library." Accessed: Sep. 25, 2024. [Online]. Available: https://doi.org/10.1111/j.1467-8667.2012.00790.x

[11] I. Flores-Colen, J. Manuel Caliço Lopes de Brito, and V. Peixoto de Freitas, "On-site performance assessment of rendering façades for predictive maintenance," *Structural Survey*, vol. 29, no. 2, pp. 133–146, Jan. 2011. https://doi.org/10.1108/02630801111132812

[12] E. Barreira and V. P. de Freitas, "Evaluation of building materials using infrared thermography," *Construction and Building Materials*, vol. 21, no. 1, pp. 218–224, Jan. 2007. https://doi.org/10.1016/j.conbuildmat.2005.06.049

[13] M. Mishra and P. B. Lourenço, "Artificial intelligence-assisted visual inspection for cultural heritage: State-of-the-art review," *Journal of Cultural Heritage*, vol. 66, pp. 536–550, Mar. 2024. https://doi.org/10.1016/j.culher.2024.01.005

[14] O. Khlifati and K. Baba, "Road Distress Detection and Classification: Harnessing the Synergy of Deep Learning and Transfer Learning Approaches," in *Advances in Research in Geosciences, Geotechnical Engineering, and Environmental Science*, K. Baba, L. Ouadif, A. Nounah, and M. Bouassida, Eds., Cham: Springer Nature Switzerland, 2023, pp. 346–355. https://doi.org/10.1007/978-3-031-49345-4_33

[15] J. Redmon, S. Divvala, R. Girshick, and A. Farhadi, "You Only Look Once: Unified, Real-Time Object Detection," May 09, 2016, *arXiv*: arXiv:1506.02640. https://doi.org/10.48550/arXiv.1506.02640

Restoration of the Kasbah of Boulaouane: An innovative approach using eco-friendly geopolymer and biopolymer composite materials

MASROUR Ilham[1,a *], BABA Khadija [1,b], KHLIFATI Oumaima[1,c], SIMOU Sana[1,d], EL HARROUNI Rime[2,3,e], NOUNAH Abderrahman[1,f], EL HARROUNI Khalid[4,g]

[1] Civil Engineering and Environment Laboratory, Civil Engineering, Water, Environment and Geosciences Centre (CICEEG), Mohammadia School of Engineering, Mohammed V University Rabat, Morocco

[2]Euro-Mediterranean School of Architecture, Design and Urbanism, Euro-Mediterranean University of Fez Institution, Morocco

[3]Faculty of Architecture and Urbanism, University of Mons, Belgium

[4]Ecole Nationale d'Architecture de Rabat, BP 6372, Rabat Instituts, Rabat, Morocco

[a]ilham_masrour@um5.ac.ma, [b]khadija.baba@est.um5.ac.ma, [c]oumaima_khlifati@um5.ac.ma, [d]sana_simou@um5.ac.ma, [e]rimaelharrouni@gmail.com, [f]abderrahman.nounah@est.um5.ac.ma, [g]k.elharrouni@enarabat.ac.ma

Keywords: Rammed Earth, Sustainable Construction, Geopolymer, Biopolymer, Mechanical Properties, Cultural Heritage Preservation, Earthen Architecture, Bio-Reinforcement

Abstract. The Kasbah of Boulaouane, a magnificent 18th-century fortress near El Jadida, Morocco, faces a critical challenge. Built by Sultan Moulay Ismaïl, this historical monument is suffering from cracks and instability due to weathering and neglect. The traditional rammed earth walls, once a symbol of resilience, are now threatened. This study proposes a groundbreaking and ecological solution for the Kasbah's restoration: a composite material combining geopolymer and biopolymer. Geopolymers, formed from eco-friendly alkaline solutions and pozzolans, offer exceptional strength, while biopolymers, derived from fermented wheat straw, provide additional reinforcement. Our research demonstrates that this innovative combination significantly increases the flexural and compressive strength of rammed earth. By incorporating these modern, sustainable materials with time-tested techniques, we can not only restore the Kasbah of Boulaouane to its former glory but also ensure its longevity for future generations. This method has the potential to revolutionize the preservation of rammed earth structures worldwide, safeguarding our cultural heritage for years to come.

Introduction

The Boulaouane Kasbah (32°51'25"N 08°01'13"W) is perched on a rugged elevation composed of Cambrian-era schist, layered with Plio-Quaternary conglomerates [13]. Surrounded on three sides by the meandering Oum er Rbia River, it commands a view from the left bank. Constructed approximately in 1710 under the rule of Sultan Moulay Ismail of the Alaouite dynasty by his Pasha, Abou Othman Said Ibn al-Khayat, the structure is positioned equidistant between the prominent cities of Fez and Marrakech, 64 km southeast of El Jadida in a straight line. Initially erected as a strategic military post along a royal and commercial route, the kasbah was instrumental in overseeing the regions of Doukkala, Chaouia, and Rehamna. It has been designated a National Historic Landmark since 1924 [20, 22, 23].

Davidovits' initial development of geopolymers was in 1978. Their structural characteristics are similar to those of zeolites, and they are part of a group of mineral-based binders [25]. Geopolymerization occurs in highly alkaline environments, where aluminosilicate sources react with alkali metal silicates, and form an alkali aluminosilicate gel, commonly known as the geopolymeric binder phase. Within the gel, a tightly interconnected three-dimensional network of tétraédrique de silicate and tétraédrique d'aluminate units forms, with alkali ions, like sodium, in a tetrahedral arrangement providing stabilization [24]. A key distinction between geopolymeric gels and calcium silicate hydrate gels lies in the production of water, which occurs during the formation of geopolymeric gels but not in the same way during cement hydration [7, 11, 14].

The geopolymerization process is multifaceted, involving several stages such as dissolution, orientation, condensation, and polymerization. It begins when aluminosilicate-rich source materials are broken down in an alkaline solution, stemming in the creation of silicate and aluminate oligomers. Polymerization is a process that these oligomers go through, eventually forming complex structures. In 1991, (a) Poly (Sialate – Si–O–Al–O), (b) Poly (Sialate-Siloxo – Si–O–Al–O–Si–O–), and (c) Poly (Sialate-Disiloxo – Si–O–Al–O–Si–O–Si–O–) represent three types of polymerized structures formed during geopolymerization classified by Davidovits [10].

The production of geopolymers is a simple procedure, typically involving the combination of minerals such as metakaolin, fly ash, red mud, or slag [2], coal gangue [12], and waste glass [4] using either an alkaline or acidic activator solution, followed by curing at temperatures below 100°C. Among these materials, metakaolin has gained popularity due to its high reactivity [18], large specific surface area [8], and stable chemical composition, making it an excellent candidate for geopolymer formation.[5]

In recent years, due to their numerous advantages, geopolymers have become a prominent topic of research. These include cost efficiency, lower carbon emissions [9, 19], straightforward synthesis, high thermal and chemical stability, and strong mechanical properties. Consequently, Geopolymers are increasingly being recognized as a more durable alternative to conventional Portland cement, because of their potential for reducing carbon emissions and reducing environmental impact [21]. Their main applications are in construction materials [1, 3, 6, 16, 17, 26, 27], but they are also used in areas such as nuclear waste management, coatings, adhesives , and even 3D printing [5].

Porous geopolymers are gaining recognition due to their ability to perform unique applications that cannot be done with denser forms. Recent trends in research are highlighted by the Web of Science data, which shows a growing interest in porous geopolymers, particularly those made from metakaolin [5].

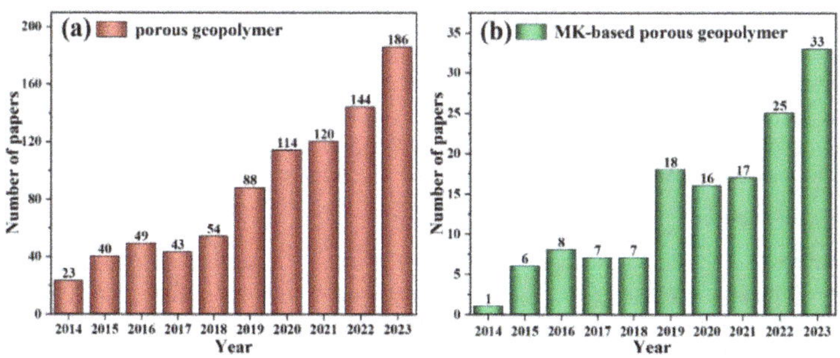

Fig. 1. (a) The evolution of publications utilizing "porous geopolymer" on the Web of Science, (b) The evolution of publications utilizing "metakaolin-based porous geopolymer" on the Web of Science [5]

Biopolymers are naturally occurring polymers, commonly sourced from renewable raw materials such as plants, animals, algae, fungi, or bacteria. Their properties can vary, encompassing plant or animal origins, and exhibiting hydrophilic, hydrophobic, or amphiphilic behaviors[15].

Biopolymers are environmentally friendly, renewable, and low-carbon resources, as they are produced naturally. Polysaccharides have been widely acknowledged for their ability to improve soil properties; biopolymers like xanthan, gellan, guar gum, alginate, and agar have demonstrated notable improvements in the mechanical performance of soils. Additionally, animal-sourced biopolymers have proven effective in enhancing soil. These biopolymers have been applied in various geotechnical fields, including controlling surface erosion, stabilizing slopes, reinforcing soil, creating hydraulic barriers, and improving subgrades[15].

The Kasbah of Boulaouane, an 18th-century fortress constructed by Sultan Moulay Ismaïl near El Jadida, Morocco, is experiencing significant deterioration. This damage has been exacerbated by prolonged exposure to harsh environmental conditions and years of neglect, the monument's traditional rammed earth walls have developed cracks and structural instability. This deterioration threatens the integrity of the historical site, posing a significant challenge to its preservation.

Fig. 2. Structural Degradations in the Kasbah of Boulaouane

This study aims to provide an innovative and sustainable restoration solution for the Kasbah by developing a composite material composed of geopolymer and biopolymer. The proposed solution combines the strength of geopolymers, derived from eco-friendly pozzolanic materials and alkaline solutions, with the reinforcing properties of biopolymers made from fermented wheat straw. The aim is to demonstrate the effectiveness of this composite material in enhancing the compression and flexural strength of rammed earth walls, ensure that the Kasbah is stable and preserved for future generations with long-term stability and conservation.

Materials and methods

Raw materials

Natural pozzolana (PZ) sourced from the Aguelmouss quarry, located in the Khenifra Province, as a key aluminosilicate material. Upon delivery to the laboratory, the pozzolana was dehydrated in an oven at 100°C for 24 hours. To ensure proper granulation, it was ground with a jaw crusher and sieved through a 2 mm mesh.

For the soil used as the matrix in the samples, it was prepared by combining specific percentages of silt, clay, sand, and gravel, following seismic regulations for earthen constructions. The treated soil was next heated at 60°C for 24 hours and sieved through a 5 mm mesh to achieve the desired consistency.

Materials Research Proceedings 47 (2025) 180-187 https://doi.org/10.21741/9781644903391-21

For the alkaline solution was prepared by combining 8 mol/L NaOH with Na_2SiO_3 ensuring between the two components a mass ratio of 1.2.

A biopolymer solution was also used, consisting of a mixture of wheat husk, yeast, and sugar. The preparation involved mixing 5% wheat husk (by mass) with water (95%), along with 0.25% yeast and 0.125% sugar, the latter two being added to accelerate the fermentation process to 48 hours. All components were placed in a beaker and manually stirred with a spatula to create a homogeneous suspension with an even distribution of ingredients.

The fermentation process for the resulting mixture was completed at 40°C and 80% humidity in an environment with limited oxygen for 48 hours. The beaker was merely covered with plastic wrap. At this stage, the solution emitted a characteristic foul odor, indicating successful fermentation.

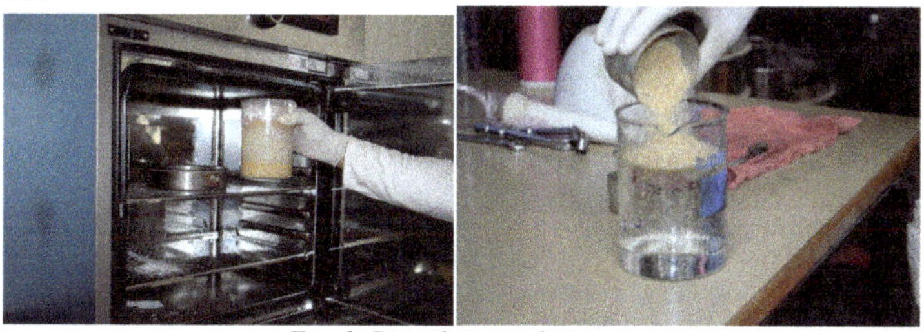

Fig. 3. Biopolymer solution

Experimental Configurations

We investigated four distinct configurations:

1. Reference Soil Material: Soil mixed only with water.
2. Soil Mixed with Geopolymer: Two configurations—soil mixed with a 20% geopolymer solution and soil mixed with a 30% geopolymer solution.
3. Soil Mixed with Biopolymer Solution: Soil combined with a biopolymer solution.
4. Soil Mixed with Geopolymer and Biopolymer: This configuration was further divided into two cases:
    - 20% Pozzolana: Soil mixed with a 20% geopolymer solution, followed by the addition of the biopolymer solution until a paste was formed.
    - 30% Pozzolana: Soil mixed with a 30% geopolymer solution, then the biopolymer solution was added until a paste was obtained.

Rectangular prismatic samples measuring 4x4x16 cm^3 were prepared for the flexural tensile tests, and after the flexural test, compression tests were performed on half of the specimens. Each sample was heated at 70°C for 48 hours, and then maintained at ambient temperature, the testing was conducted after 90 days.

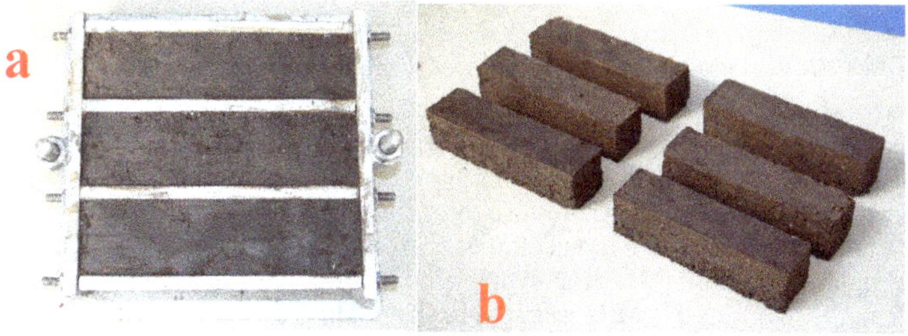

Fig. 4. The specimens before demolding (a) and after demolding (b).

Results and discussion

Geotechnical identification and classification

Following French standards, Multiple laboratory tests were conducted. Based on the granulometric analysis, it can be concluded that the soil comprises a mix of 9% gravel, 52.5% sand, 22.5% clay, and a substantial silt fraction of around 16%. For the Atterberg limits used to evaluate consistency, we found that's: plastic limit Wp=15%, liquid limit WL=28%, and the difference between these limits is used to calculate the plasticity index, Ip=13%. The VBS value was measured at 0.60. This material is suitable for the intended use, as confirmed by its grain characteristics and information from the seismic regulations for earth structures (RPCT 2013).

Flexural strength

Using a UTest machine, the specimens were tested at three points and their maximum load capacity was 250 kN.

Figure 3 demonstrates the flexural strength of different soil compositions, showing that the addition of 20% geopolymer provides the highest flexural strength at slightly over 1.4 MPa, making it the most effective in enhancing the soil's resistance to bending forces. Soil mixed with biopolymer follows closely with a strength near 1.3 MPa, indicating biopolymers as a strong reinforcement agent. When both 20% geopolymer and biopolymer are combined, the flexural strength reaches around 1.2 MPa, suggesting that while this combination is beneficial, the interaction between the materials may not provide the same reinforcement as when used individually. Soil with 30% geopolymer shows a slightly lower strength of around 1.1 MPa, indicating diminishing returns beyond the 20% concentration. The combination of 30% geopolymer and biopolymer yields the weakest result of the modified samples, with only 0.6 MPa, possibly due to reduced compatibility at higher concentrations. Untreated soil shows the lowest flexural strength, around 0.2 MPa, highlighting the significant impact of these additives in improving the mechanical properties of soil.

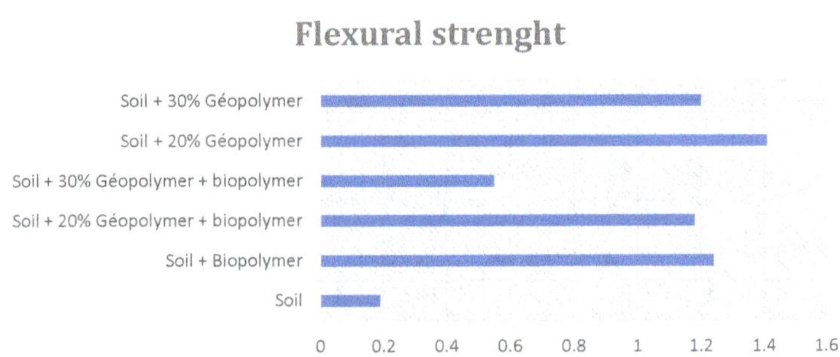

Fig. 5. The flexural strength of adobe material with diverse geopolymer and biopolymer ratios

Compressive strength

Figure 4 presents the compressive strength of different soil compositions, with the addition of 30% geopolymer achieving the highest strength at approximately 2.3 MPa, indicating its effectiveness in improving soil's resistance to compressive forces. The soil mixed with biopolymer also shows a notable improvement, reaching around 2.0 MPa, demonstrating its reinforcing potential. Both the 20% geopolymer and the 20% geopolymer combined with biopolymer configurations exhibit similar strengths, close to 2.0 MPa, suggesting that adding biopolymer does not significantly enhance compressive strength when mixed with 20% geopolymer. However, the combination of

30% geopolymer and biopolymer shows a lower compressive strength, around 1.5 MPa, indicating a possible reduction in the combined effect at higher geopolymer concentrations. Finally, untreated soil has the weakest compressive strength, just below 0.5 MPa, highlighting the significant improvement brought by the addition of geopolymer and biopolymer.

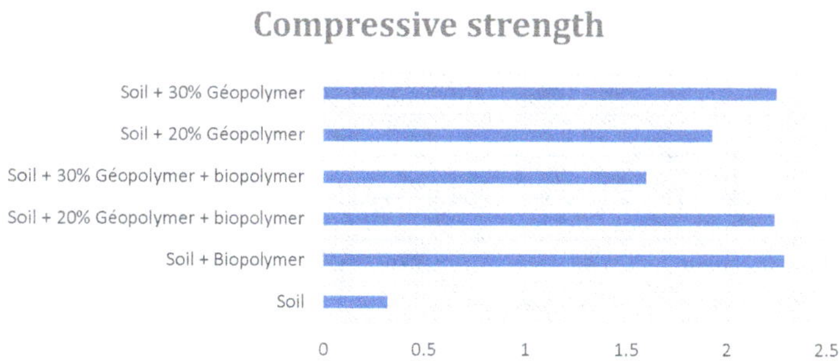

Fig. 6. Varying quantities of geopolymer and biopolymer influence the compressive strength of adobe material.

Conclusion

In conclusion, adding geopolymer and biopolymer significantly enhances the mechanical properties of soil. The greatest compressive strength is obtained with 30% geopolymer, while the best improvement in flexural strength is observed with 20% geopolymer. While biopolymer also demonstrates strong reinforcement potential, especially in flexural tests, the combination of geopolymer and biopolymer shows mixed results, particularly at higher geopolymer concentrations, where the interaction between the materials appears to be less effective. Untreated soil consistently exhibits the weakest mechanical performance, underscoring the importance of these additives in strengthening and stabilizing soil for construction purposes.

References

[1] Almutairi, A.L., Tayeh, B.A., Adesina, A., Isleem, H.F. and Zeyad, A.M. 2021. Potential applications of geopolymer concrete in construction: A review. *Case Studies in Construction Materials*. 15, (Dec. 2021), e00733. https://doi.org/10.1016/j.cscm.2021.e00733

[2] Ascensão, G., Seabra, M.P., Aguiar, J.B. and Labrincha, J.A. 2017. Red mud-based geopolymers with tailored alkali diffusion properties and pH buffering ability. *Journal of Cleaner Production*. 148, (Apr. 2017), 23–30. https://doi.org/10.1016/j.jclepro.2017.01.150

[3] Ayeni, O., Onwualu, A.P. and Boakye, E. 2021. Characterization and mechanical performance of metakaolin-based geopolymer for sustainable building applications. *Construction and Building Materials*. 272, (Feb. 2021), 121938. https://doi.org/10.1016/j.conbuildmat.2020.121938

[4] Bai, C., Li, H., Bernardo, E. and Colombo, P. 2019. Waste-to-resource preparation of glass-containing foams from geopolymers. *Ceramics International*. 45, 6 (Apr. 2019), 7196–7202. https://doi.org/10.1016/j.ceramint.2018.12.227

[5] Bai, C., Zheng, K., Sun, F., Wang, X., Zhang, L., Zheng, T., Colombo, P. and Wang, B. 2024. A review on metakaolin-based porous geopolymers. *Applied Clay Science*. 258, (Sep. 2024), 107490. https://doi.org/10.1016/j.clay.2024.107490

[6] Bajare, D., Bumanis, G. and Korjakins, A. 2014. New Porous Material Made from Industrial and Municipal Waste for Building Application. *Materials Science*. 20, 3 (Sep. 2014), 333–338. https://doi.org/10.5755/j01.ms.20.3.4330

[7] Bourzik, O., Baba, K. and Akkouri, N. 2023. Eco-friendly geopolymer mortar prepared from geopolymer waste: Mechanical and thermal properties. *Environmental Quality Management*. (2023). https://doi.org/10.1002/tqem.22132

[8] Chen, L., Wang, Z., Wang, Y. and Feng, J. 2016. Preparation and Properties of Alkali Activated Metakaolin-Based Geopolymer. *Materials*. 9, 9 (Sep. 2016), 767. https://doi.org/10.3390/ma9090767

[9] Costs and carbon emissions for geopolymer pastes in comparison to ordinary portland cement - ScienceDirect: *https://www.sciencedirect.com/science/article/abs/pii/S0959652611000680?via%3Dihub*. Accessed: 2024-09-27.

[10] Davidovits, J. 2008. *Geopolymer Chemistry & Applications*. Institut géopolymère.

[11] Hanani Ismail, A., Kusbiantoro, A., Tajunnisa, Y., Jaya Ekaputri, J. and Laory, I. 2024. A review of aluminosilicate sources from inorganic waste for geopolymer production: Sustainable approach for hydrocarbon waste disposal. *Cleaner Materials*. 13, (Sep. 2024), 100259. https://doi.org/10.1016/j.clema.2024.100259

[12] Li, X., Liu, L., Bai, C., Yang, K., Zheng, T., Lu, S., Li, H., Qiao, Y. and Colombo, P. 2023. Porous alkali-activated material from hypergolic coal gangue by microwave foaming for methylene blue removal. *Journal of the American Ceramic Society*. 106, 2 (2023), 1473–1489. https://doi.org/10.1111/jace.18812

[13] M., C., Errami, E., Ennih, N., N., G., Zitoune, I., Enniouar, A., A., C. and A., E. 2011. *Valorisation du Géosite delLa Kasbah de Boulaouane (Province d'El Jadida, Maroc).*

[14] Masrour, I., Baba, K. and Doughmi, K. 2024. Geopolymers: An Eco-Friendly Approach to Enhancing the Stability of Earthen Constructions. *Materials Research Proceedings*. 40, (2024), 388. https://doi.org/10.21741/9781644903117

[15] Masrour, I., Baba, K. and Simou, S. 2024. Naturally Strengthening Rammed Earth: The Promising Potential of Biopolymers. *Materials Research Proceedings*. 40, (2024), 388. https://doi.org/10.21741/9781644903117

[16] Nawaz, M., Heitor, A. and Sivakumar, M. 2020. Geopolymers in construction - recent developments. *Construction and Building Materials*. 260, (Nov. 2020), 120472. https://doi.org/10.1016/j.conbuildmat.2020.120472

[17] Ng, C., Alengaram, U.J., Wong, L.S., Mo, K.H., Jumaat, M.Z. and Ramesh, S. 2018. A review on microstructural study and compressive strength of geopolymer mortar, paste and concrete. *Construction and Building Materials*. 186, (Oct. 2018), 550–576. https://doi.org/10.1016/j.conbuildmat.2018.07.075

[18] Panagiotopoulou, Ch., Kontori, E., Perraki, Th. and Kakali, G. 2007. Dissolution of aluminosilicate minerals and by-products in alkaline media. *Journal of Materials Science*. 42, 9 (May 2007), 2967–2973. https://doi.org/10.1007/s10853-006-0531-8

[19] Parathi, S., Nagarajan, P. and Pallikkara, S.A. 2021. Ecofriendly geopolymer concrete: a comprehensive review. *Clean Technologies and Environmental Policy*. 23, 6 (Aug. 2021), 1701–1713. https://doi.org/10.1007/s10098-021-02085-0

[20] Qninba, A., Rihane, A., Bousadik, H. and Agbani, M.A.E. 2020. La colonie mixte de Faucons crécerelles et crécerellettes (Falco tinnunculus - F. naumanni) de la Kasbah de Boulaouane (Centre Atlantique marocain). (2020).

[21] R.A. Sá Ribeiro 2017. A Review of Particle- and Fiber-Reinforced Metakaolin-Based Geopolymer Composites. 03 (2017). https://doi.org/10.4416/JCST2017-00048

[22] Simou, S., Baba, K., Akkouri, N., Lamrani, M., Tajayout, M. and Nounah, A. 2020. Mechanical Characterization of the Adobe Material of the Archaeological Site of Chellah. *Sustainable Civil Infrastructures*. (2020), 118–130. https://doi.org/10.1007/978-3-030-34199-2_8

[23] Simou, S., Baba, K., Tajayouti, M., Jemmal, M., Nounah, A. and Aarab, A. 2020. 3d reconstruction of the Marinids site located at the Chellah archaeological area. (2020), 2807–2814.

[24] Singh, N.B. and Middendorf, B. 2020. Geopolymers as an alternative to Portland cement: An overview. *Construction and Building Materials*. 237, (Mar. 2020), 117455. https://doi.org/10.1016/j.conbuildmat.2019.117455

[25] Tungsten mine waste geopolymeric binder: Preliminary hydration products investigations - ScienceDirect: *https://www.sciencedirect.com/science/article/abs/pii/S0950061808000123*. Accessed: 2024-09-27.

[26] Xie, T., Visintin, P., Zhao, X. and Gravina, R. 2020. Mix design and mechanical properties of geopolymer and alkali activated concrete: Review of the state-of-the-art and the development of a new unified approach. *Construction and Building Materials*. 256, (Sep. 2020), 119380. https://doi.org/10.1016/j.conbuildmat.2020.119380

[27] Yuan, Q., Huang, Y., Huang, T., Yao, H. and Wu, Q. 2022. Effect of activator on rheological properties of alkali-activated slag-fly ash pastes. *Journal of Central South University*. 29, 1 (Jan. 2022), 282–295. https://doi.org/10.1007/s11771-022-4913-0

Hydrography as a guide to reading the urban form of the Medina of Fez

MOUJOUD Lamiae[1,2 a *], CHAHBI Mariame[3,b]

[1]Architecture at Ecole Nationale d'Architecture de Rabat, Morocco

[2]Architecture School of the International University of Rabat, Morocco

[3]Architect and Associate Professor, Architecture School of the International University of Rabat, Affiliated Researcher at UIH/UIR Rabat, Morocco

[a]Moujoudlamiae@gmail.com, [b]mariame.chahbi@uir.ac.ma

Keywords: Adaptation, Comparative Analysis, History, Morphology, The Medina of Fez, Water Management

Abstract. The urban morphological evolution of the Medina of Fez is examined through the lens of hydrography, from the Idrisid period to the contemporary era. The analysis reveals how hydraulic systems have shaped urban development, influencing spatial organization, socio-economic structures, and cultural practices. The research highlights the symbiotic relationship between water management and urban planning, demonstrating a diachronic narrative of adaptation and resilience. The exploration of transformations in water management practices over the centuries illustrates how hydrological considerations have facilitated or hindered urban morphological evolution. The importance of hydrography as a determining factor in the development of Fez is emphasized, offering perspectives on the adaptability of urban spaces in the face of environmental and sociopolitical changes.

Introduction

The medina of Fez, founded in the 8th century, represents a remarkable example of a medieval Islamic city that has preserved its authenticity through the centuries (UNESCO, 1981). Inscribed on the UNESCO World Heritage List in 1981, it is considered one of the largest still-inhabited medinas in the world. At the heart of its development and longevity lies a sophisticated hydraulic system, the result of remarkable engineering and fine adaptation to local geographical and climatic conditions (El Faïz, 2005). Exploring the dynamic relationship between hydrography and urban morphology reveals crucial perspectives on the evolution of cities over time. The Medina of Fez, a site of immense historical importance, perfectly illustrates this interaction, particularly from the Idrisid period to the contemporary era. Hydrographic elements such as rivers, streams, and artificial hydraulic systems have not only shaped the physical layout of the city but have also influenced socio-economic activities, cultural practices, and settlement patterns. Researchers such as Leveau (2014) and Abdessemed (2021) highlight how past water management systems have influenced urban planning decisions, thus revealing a diachronic narrative of adaptation and resilience. This hydraulic system, developed over centuries, is not limited to a simple water distribution network; it constitutes the backbone around which the entire urban, social, and economic life of the city has been organized (Hamdouni Alami, 2011). The main objective of this study is to analyze in detail how water has influenced and structured the urban form of Fez, creating a unique morphology adapted to its context. This research is part of a broader framework for understanding historical Islamic cities and their relationship with their environment. It aims to contribute to the current debate on sustainable water resource management in World Heritage sites, a crucial issue in the face of climate change challenges (IPCC, 2021).

Several research questions guide this analysis:

Materials Research Proceedings 47 (2025) 188-196 https://doi.org/10.21741/9781644903391-22

1. How did the hydraulic system of Fez influence the macro-structure of the city, its street layout and urban fabric?
2. What was the impact of water on the creation and organization of public spaces and domestic architecture?
3. How did economic activities and green spaces develop in relation to the hydraulic network?
4. What are the contemporary adaptations of this system in the face of the challenges of modern urbanization and climate change?

This study is based on an interdisciplinary approach, combining historical analysis, urban morphological study, and examination of traditional hydraulic practices. It builds upon the work of researchers such as Bianca (2000)[1] on urban form in the Arab world, and El Faïz (2005) on the history of Arab hydraulics. The in-depth understanding of water's role in structuring Fez offers important perspectives for the preservation and sustainable adaptation of historic cities. In a context where water management is becoming a major global issue, the lessons drawn from the study of Fez can inform contemporary approaches to urban planning and water resource management.

Methodology

This research on the influence of the hydraulic system on the urban form of the medina of Fez adopts a mixed methodological approach, combining qualitative and quantitative methods:

- Comparative Analysis

A comparative analysis with other historical Islamic cities (Marrakech, Cairo, Damascus) to situate the case of Fez in a broader context and identify the specificities of its hydraulic system.

- Historical Analysis

A review of historical literature to understand the evolution of Fez's hydraulic system and its impact on urban development. This analysis included:

  - The study of historical treatises on Islamic hydraulics, particularly the works of Ibn al-Razzaz al-Jazari (12th century) and Al-Karaji (11th century) (El Faïz, 2005).
  - Examination of documents related to water management in Fez, including records of habous (pious foundations) that managed numerous hydraulic infrastructures (Hamdouni Alami, 2011)[2].
  - Analysis of accounts from medieval Arab travelers and geographers who described Fez and its hydraulic system.

- Morphological Analysis

A morphological analysis of the urban structure of Fez, to identify patterns and characteristics related to the hydraulic system:

Study of the street layout and its relationship with the hydraulic network.

Analysis of building typology and their adaptation to hydraulic constraints.

Examination of the configuration of public spaces around water points.

Historical Analysis

The evolution of Fez's hydraulic system, spanning over a millennium, reflects the remarkable development of this Moroccan imperial city. Founded in the 8th century, Fez was able to exploit

[1] Bianca (2000)[1]: Bianca, Stefano. "Urban Form in the Arab World - Past and Present" (2000) est un ouvrage détaillé sur les structures urbaines traditionnelles dans les pays arabo-islamiques. Ce livre de 347 pages, publié par vdf Hochschulverlag AG, analyse les problèmes auxquels sont confrontées les villes historiques face au développement moderne et aux technologies occidentales, offrant des études de cas sur des villes comme La Mecque, Bagdad, Fès et Alep

[2] Hamdouni Alami, 2011 : Mohammed Hamdouni Alami is an architect and researcher who studied at the School of Architecture in Grenoble. His 2011 work likely focuses on architecture and urbanism in the Islamic world

its advantageous geographical location, benefiting from the abundant waters of the Fez River and numerous natural springs. Over the centuries, the city developed a sophisticated hydraulic system, demonstrating the technical ingenuity of its builders and playing a crucial role in its economic, cultural, and social growth.

This hydraulic history is divided into several distinct periods, each marked by technical innovations, political changes, and significant urban transformations. From the establishment of the first rudimentary facilities at the time of its foundation, through the golden age of the Almoravid and Almohad dynasties, to contemporary rehabilitation efforts, Fez's hydraulic system has undergone constant evolution.

Historical Periodization

1. Idrisid Period (8th-10th centuries):

Foundation and establishment of the initial system

The foundation of Fez by Idris I in 789, followed by its development under Idris II, marks the beginning of the city's hydraulic history. The choice of site, at the confluence of the Fez River and the Boufekrane River, testifies to the importance given to water resources from the very beginning (Madani, 1999)[3].

The initial hydraulic system was structured around:

The capture of natural springs, notably Aïn Ras el-Ma and Aïn Azliten

The construction of open-air canals (seguias) for irrigation and urban supply

The development of retention and distribution basins (sahridj)

The waqfiyyas (foundation deeds) of this period, although rare, already mention the importance of public fountains (siqāyāt) in the nascent urban planning (Sebti, 1987)[4].

2. Almoravid and Almohad Period (11th-13th centuries):

Expansion and development

This period saw a significant expansion of the hydraulic network, linked to urban growth.

The main innovations include The development of the khettaras system (underground drainage galleries) to capture groundwater; the construction of large reservoirs such as the Sahridj Sbaa (Lions' Basin) under the Almoravids; the extension of the underground ceramic pipeline network (qanawāt).

The "Kitab al-Istibsar" by Al-Muqaddasi, written in the 10th century but describing Fez up to the 12th century, provides valuable details on the hydraulic system of this era, particularly highlighting the multiplicity of hydraulic mills along the Fez River (Al-Muqaddasi, 1963)[5].

3. Marinid Period (13th-15th centuries):

Peak and refinement

Under the Marinids, Fez reached its urban and hydraulic peak: Improvement of water capture techniques with the development of new springs such as Aïn Ameir; construction of large hydraulic complexes like the noria of Fez Jdid (14th century); multiplication of hammams and monumental public fountains; the archives of the corporations of master hydraulic engineers

[3] Madani, 1999: Although specific details are not provided, this reference is likely a study on urbanism or the history of Fez, given the context of the discussion on the city's hydraulic system.

[4] Sebti, 1987. Abdelahad Sebti is a Moroccan historian, professor at Mohammed V University in Rabat. His 1987 work could be about urban or social history of Morocco, possibly related to hydraulic systems or urbanism in Fez.

[5] Al-Muqaddasi, 1963 : This 1963 edition presents the work of the 10th-century Arab geographer, Al-Muqaddasi. His work "Ahsan al-Taqasim fi Ma'rifat al-Aqalim" (The Best Divisions for Knowledge of the Regions) offers detailed descriptions of medieval Islamic cities, including their hydraulic systems.

(mâalams) of Fez, some fragments of which date back to this period, reveal a sophisticated social organization around water management (El Faïz, 2005)[6].

4. Saadian and Alaouite Period (16th-19th centuries):
Maintenance and adaptations
This long period is marked by the maintenance and renovation of existing infrastructures; the adaptation of the system to new urban demands, particularly with the rise of tanneries; and the occasional innovations such as the introduction of lead pipes under the Alaouites. Leo Africanus, in his "Description of Africa" (1526), offers a detailed account of Fez's hydraulic system at the beginning of the 16th century, highlighting the persistence of medieval developments (Leo Africanus, 1956)[7].

5. Protectorate and Post-Independence Period (20th-21st centuries):
Modernization and challenges
This period was characterized by a profound transformation of the hydraulic system: Introduction of running water and a modern sanitation network under the Protectorate; progressive abandonment of traditional systems such as khettaras; challenges related to the preservation of hydraulic heritage in the face of rapid urbanization; the plans and maps produced by the Antiquities Service of the Protectorate provide valuable documentation on the state of the traditional hydraulic system at the beginning of the 20th century (Revault et al., 1985).

Comparative Analysis: Fez, Marrakech, Cairo, Damascus

Medina of Fes Medina of Cairo Medina of Damas Medina of Marrakech

Figure1: Source: https://www.geospatialworld.net/blogs/top-10-maps-showing-worlds-oldest-cities

The comparative study of hydraulic systems in Fez and other historic Islamic cities such as Marrakech, Cairo, and Damascus is of paramount importance in this research for several fundamental reasons. Firstly, these cities share a common Islamic heritage and have played a significant role as urban centers at different periods in history. As Boucheron (2017) points out, this comparison allows us to highlight similarities and differences in water management approaches within the medieval Islamic world, thus providing valuable insight into the hydraulic practices of the time.

The geographical and climatic diversity of these cities constitutes a second crucial aspect of this comparative analysis. Each city presents a unique environmental context, which, according to

[6] Faïz, 2005 : Mohammed El Faïz, a Moroccan historian specializing in Arab hydraulics, published a work in 2005 on irrigation systems in the medieval Islamic world. His work sheds light on the sophisticated hydraulic techniques developed in cities like Fez.
[7] Leo Africanus, 1956 : This 1956 edition presents the "Description of Africa" by Leo Africanus, a 16th-century diplomat and explorer. His account offers valuable testimony about Fez in the early modern period, including observations on its hydraulic system.

Materials Research Proceedings 47 (2025) 188-196 https://doi.org/10.21741/9781644903391-22

Tresse's work (2019), allows us to examine how specific natural constraints have influenced the development and adaptation of local hydraulic systems.

This comparative approach also allows us to put into perspective the particular ingenuity of Fez's hydraulic system. As demonstrated by El Faïz (2015) in his research, the comparison highlights the specificities of the "Fassi" system, notably its remarkable adaptation to local topography and its judicious use of natural sources.

The identification of best practices is another major advantage of this comparative study. Madani's work (2018) has revealed that examining innovative solutions adopted in other cities can potentially inspire concrete applications, not only in Fez but also in other contemporary urban contexts facing similar challenges.

Furthermore, this approach allows for a better understanding of the evolution of hydraulic systems over time. Harroud's research (2020) has shown that analyzing the adaptation of different cities over the centuries offers valuable lessons on the challenges and opportunities related to the preservation and modernization of these ancestral systems.

Broadening the analytical framework constitutes another significant contribution of this comparative approach. As Akdim (2016) emphasized, situating the case of Fez in the broader context of Islamic hydraulic engineering considerably enriches the analysis and conclusions of the research, allowing for a more comprehensive understanding of the issues.

Finally, this comparative study allows for the identification of common trends and principles in water management in Islamic cities. The work of Navarro Palazón (2021)[8] has demonstrated that this approach contributes to a better understanding of medieval Islamic urbanism, revealing patterns and practices shared across different regions of the Muslim world.

[8] Navarro Palazón (2021)[8] Julio Navarro Palazón is a Spanish researcher specializing in Islamic archaeology and architecture, particularly in Al-Andalus (medieval Islamic Iberia). He is affiliated with the School of Arabic Studies (Escuela de Estudios Árabes) of the Spanish National Research Council (CSIC) in Granada.

City	Hydraulic System Description	Influence on Urban Structure	Cultural Impact	Key Architectural Features
Fès	Fès developed a hydraulic system starting in the 8th century, incorporating elements like qanats and fountains to provide water to the city, crucial in maintaining the urban population in its arid environment.	The hydraulic system's presence significantly influenced urban planning by shaping densely populated areas such as Fes el-Bali, where structures and markets arose around access points to clean water, promoting trade and social interaction.	The hydraulic system fostered community rituals and cultural events, notably the Fes Festival of World Sacred Music, which celebrates the city's rich cultural heritage and promotes tourism.	Fès is known for its riads and hammams, which exhibit intricate tilework and architectural details influenced by various cultures, including Arab, Berber, and Andalusian designs, reflecting its historical significance as a center of trade and culture.
Marrakech	The khettara system in Marrakech is a traditional underground irrigation system that dates back to the Almoravid period in the 11th century, allowing for the efficient distribution of groundwater to the city	The khettaras significantly influenced Marrakech's urban structure by facilitating the development of gardens, mosques, and homes, particularly in neighborhoods like the Medina, which are designed around water access	The hydraulic system has shaped communal practices in Marrakech, most notably illustrated during local festivals that celebrate water's importance in agriculture and spirituality	Architecturally, the khettaras have influenced structures such as the Menara and Agdal gardens, which incorporate water features that reflect Islamic landscape architecture traditions
Damas	Damas features an ancient hydraulic system known as the 'tali', which consists of underwater canals that enabled water distribution across the city and transformed the arid landscape into fertile land, dating from the mid-6th century to the mid-12th century.	The hydraulic system influenced the urban structure by facilitating the development of a network of public fountains, gardens, and bathhouses. The well-preserved canals and water management practices significantly shaped the layout and planning of the city.	The water system greatly impacted cultural aspects of life in Damascus, fostering communal practices centered around water sources, gardens, and public baths. It became a vital part of community identity and social interaction.	Architecturally, the hydraulic system influenced the design of traditional Damascene houses featuring courtyards, water aqueducts, and decorative elements that reflect its historical and cultural significance. These features are integral to the unique character of Damascus.
Cairo	The hydraulic system in Cairo involved ancient technologies for basin irrigation, diverting floodwaters from the Nile to sustain agricultural practices and urban development, tracing back to Predynastic periods (circa 3100 BCE).	Cairo's urban structure evolved symbiotically with water resource management, particularly with the Nile, influencing both the layout and density of neighborhoods, as well as agricultural practices within the city		The architectural features influenced by the hydraulic system include the Pyramid of Djoser, which utilized a hydraulic lift mechanism for construction. This innovative use of hydraulic technology highlights advanced engineering methods in ancient Egypt

Figure 2 Comparison between the Medina of Fez and other historical Islamic cities, to analyze the similarities and differences between the hydraulic system of Fez and those of other important cities in the medieval Islamic world. Source : Author

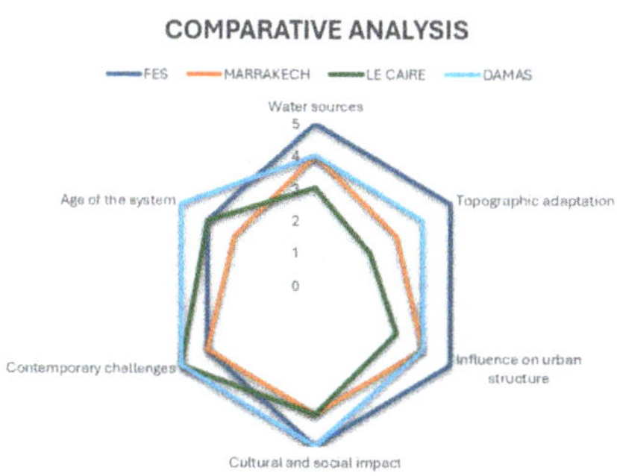

Figure 3 : Multidimensional Evaluation of Hydraulic Systems: Fez and its Historical Counterparts Source : Author

Synthesis: This comparison highlights the richness and diversity of hydraulic approaches in the medieval Islamic world while emphasizing the particular ingenuity of Fez's system in its specific context.

Morphological Analysis: Urban Structure

The medina of Fez presents a unique urban configuration, strongly influenced by its hydraulic system. The city developed on the slopes of two hills separated by the Fez River, creating a natural amphitheater structure (Bianca, 2000). This topography was ingeniously exploited to optimize the gravity distribution of water, a key element in the efficiency of the city's hydraulic system (Navez-Bouchanine, 2003).

Urban duality is a striking characteristic of Fez, with Fez el-Bali on the right bank and Fez el-Jdid on the left bank of the river. This division is not only geographical but also historical and functional. Fez el-Bali, the oldest part, houses the majority of traditional hydraulic infrastructures, while Fez el-Jdid, founded in the 13th century, benefited from a more planned hydraulic system (Hamdouni Alami, 2011).

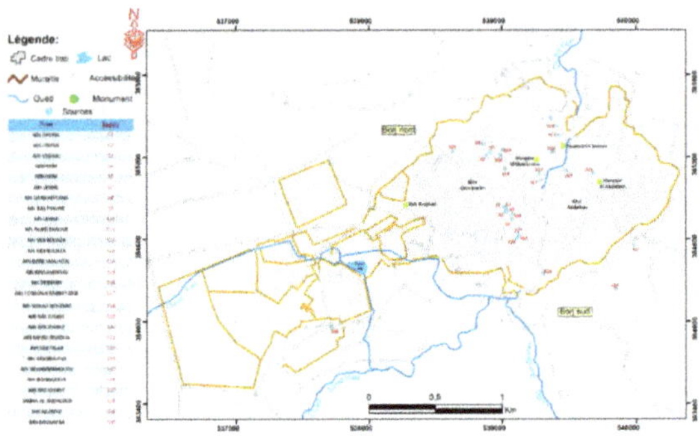

Figure 4 : Map of the distribution of geolocated springs in the Medina of Fez
Source: Ayacine, M., Chanyour, Y., Obda, K., & El Bouchikhi, M. (2023). The springs of the Medina of Fez: Diagnosis for valorization (Fez; Morocco). In M. Amyay & M. Laouane (Eds.), Water springs in the Mediterranean basin: Hydrology, Valorization and preservation (pp. 143-159). Faculty of Letters and Human Sciences Saïs-Fez

Analysis of historical maps and topographical surveys reveals that Fez's urban expansion closely followed the development of its hydraulic network. The oldest neighborhoods are systematically located near the main water sources, particularly around the Ras el-Ma spring, considered the hydraulic heart of the city (El Faïz, 2005).

1. Street layout and urban fabric

The hydraulic network of Fez has profoundly influenced its street layout, creating a complex hierarchy of circulation routes. Our field observations, corroborated by the analysis of old plans, show that the main arteries of the medina often follow the path of major canals (Madani, 2012). This configuration has several advantages:
It facilitates the maintenance and monitoring of main canals.
It allows for efficient water distribution to residential neighborhoods.
It creates natural commercial axes along the waterways.
Secondary streets and dead-ends (derbs) branch off from these main axes, forming a capillary network that reflects the distribution of secondary pipelines. This organic structure has given rise to Fez's famous urban labyrinth, where winding streets are not the result of chance but of meticulous adaptation to topography and the hydraulic system (Geertz, 1972). The density and compactness of Fez's urban fabric are also direct consequences of water management. Our morphological analysis reveals that:
Buildings are tightly packed, creating narrow streets that provide shade and reduce water evaporation.
This compactness minimizes the length of pipelines, thus reducing water losses.

Materials Research Proceedings 47 (2025) 188-196 https://doi.org/10.21741/9781644903391-22

Urban blocks have irregular shapes, adapting to the layout of underground pipelines (Boukhari & Naïma, 2018).

2. Public and architectural spaces

Public fountains (seqqayas) have played a crucial role in structuring Fez's public spaces. Our study identified more than 60 historical fountains in the medina, each generating a unique public space. These fountains create a spatial hierarchy based on their hydraulic and social importance:

Large fountains, often associated with main mosques, have generated important squares that have become centers of community life.

More modest fountains have created micro-public spaces at the heart of residential neighborhoods, serving as daily meeting points (Hamdouni Alami, 2011).

The domestic architecture of Fez has also adapted to water management. The analysis of several traditional houses reveals common features:

The almost systematic presence of a central patio with a basin or fountain serves to collect rainwater and create a microclimate.

Ingenious systems for collecting and storing rainwater, are integrated into the architecture of houses.

An elevation of the ground floor above street level protects against occasional flooding (El Faïz, 2002).

3. Specialized districts and green spaces

Water has also dictated the location of economic activities, creating specialized districts. Tanneries, which require large quantities of water, have settled near the Fez River. Our study of the spatial distribution of artisanal activities shows a clear correlation between their location and proximity to water sources (Serrhini, 2020).

The green spaces of Fez, although limited in the dense medina, are closely linked to the hydraulic system. The Jnan Sbil garden, for example, was created in the 18th century thanks to a sophisticated irrigation system derived from the Fez River. These green spaces play a crucial role in the ecological balance of the city, offering cool areas and contributing to the management of runoff water (Ait Hassou, 2014).

4. Evolution and Contemporary Adaptations

The analysis of Fez's historical evolution reveals that the hydraulic system continued to influence urban development until modern times. However, rapid urbanization and changes in lifestyles have put this system under pressure. Our field observations show that:

Some historical canals have been covered or diverted to make way for new constructions.

The introduction of modern running water has reduced the use of traditional public fountains.

Pollution of the Fez River poses major challenges to water quality and the urban environment.

In response to these challenges, rehabilitation efforts are underway. The Fez River restoration project, for example, aims to clean up the river and create public spaces along its banks, thus reintegrating water into the contemporary urban landscape (ADER-Fès, 2019).

Conclusion

The study of the hydraulic system in the medina of Fez reveals how water has been the determining factor in the formation and evolution of its urban structure. This unique urban form, the result of centuries-old interaction between humans and their environment, constitutes an exceptional heritage that offers valuable lessons for sustainable water management in historic cities.

The detailed analysis of water's influence on the urban macro-structure, street layout, public spaces, and domestic architecture of Fez highlights the ingenuity of medieval builders and their ability to create a hydraulic system perfectly adapted to local constraints. This system, which allowed the city's development and sustainability for centuries, now faces major challenges related to climate change and rapid urbanization.

The preservation and adaptation of this hydraulic heritage to contemporary needs require an integrated approach, as emphasized by Toutain (2016). This approach must combine:

Restoration and modernization of traditional hydraulic infrastructures.

Integration of modern water management technologies.

Adoption of nature-based solutions to enhance urban resilience.

Raising awareness and educating the population about the importance of this heritage and sustainable water management.

Ongoing initiatives in Fez, such as the Fez River restoration project and the development of a climate change adaptation plan, show the way towards more sustainable and resilient water management in historic cities.

This case study of Fez offers valuable lessons for other historic cities facing similar challenges. It underlines the importance of understanding and valuing traditional hydraulic systems, not only as cultural heritage but also as a source of inspiration for sustainable solutions to 21st-century challenges.

In conclusion, the example of Fez demonstrates that preserving hydraulic heritage is not just a matter of conservation, but also a crucial issue for urban sustainability and resilience. It invites us to rethink our relationship with water in cities, drawing inspiration from the wisdom of the past to build a more sustainable future.

References

[1] ADER-Fès. (2019). Rapport annuel sur l'état de conservation de la médina de Fès. Agence pour la Dédensification et la Réhabilitation de la médina de Fès.

[2] Ait Hassou, M. (2014). Le système hydraulique traditionnel de la médina de Fès : patrimoine en péril. In Proceedings of the International Conference on Water and Heritage (pp. 78-89). ICOMOS Netherlands.

[3] Bianca, S. (2000). Urban form in the Arab world: Past and present. Thames & Hudson.

[4] Boukhari, S., & Naïma, R. (2018). Water management in the medina of Fez: A model of sustainability. Procedia Engineering, 209, 135-142.

[5] El Faïz, M. (2002). Jardins de Marrakech. Actes Sud.

[6] El Faïz, M. (2005). Les maîtres de l'eau : histoire de l'hydraulique arabe. Actes Sud.

[7] Geertz, C. (1972). The wet and the dry: Traditional irrigation in Bali and Morocco. Human Ecology, 1(1), 23-39. https://doi.org/10.1007/BF01791279

[8] GIEC. (2021). Climate Change 2021: The Physical Science Basis. Cambridge University Press.

[9] Hamdouni Alami, M. (2011). L'eau dans le monde musulman médiéval : l'exemple de Fès. Presses Universitaires de Rennes.

[10] Madani, T. (2012). L'eau dans le monde musulman médiéval. Presses Universitaires de France.

[11] Navez-Bouchanine, F. (2003). La fragmentation en question : des villes entre fragmentation spatiale et fragmentation sociale ? L'Harmattan.

[12] Serrhini, F. (2020). La gestion de l'eau dans la médina de Fès : enjeux et perspectives. Revue Marocaine des Sciences de l'Eau, 25(3), 45-62.

[13] Toutain, O. (2016). Réhabilitation des médinas au Maroc : défis et opportunités. In Urban Heritage in the Middle East (pp. 203-218). Springer.

[14] UNESCO. (1981). Médina de Fès. Centre du patrimoine mondial.

Materials Research Proceedings 47 (2025) 197-202 https://doi.org/10.21741/9781644903391-23

Natural risks and resilience of vernacular architecture: The case of the Medina of the City of Larache

SALOUI Chaimae[1,a], ADIDI Abdelaziz[1,b]

[1]Institut National d'Aménagement et d'Urbanisme, Rabat, Morocco

[a]chaimaesaloui@gmail.com, [b]a2adidi@yahoo.fr

Keywords: Vernacular Architecture, Natural Risks, Urban Management, Urban Identity

Abstract. Larache, the capital of its province, is a city with remarkable potential in the Tangier-Tetouan-Al Hoceima region. It is situated along the Tangier–Kenitra axis, which has seen significant economic growth due to major international trade routes, including air and port connections. Despite its strategic location, Larache, particularly represented by its vernacular architecture and traditional housing in the medina, is evolving in a direction that prompts further reflection. Currently, the vernacular architecture of Larache's medina is a key element of its cultural heritage, reflecting local traditions and techniques adapted to environmental conditions. Unfortunately, this architectural wealth is often overlooked and threatened by natural risks that diminish its value and functionality. Today, there is a growing need to redirect urban planning towards developing strategies that protect the integrity of these structures while addressing natural risks. In this way, the old medina can become a model for the preservation and enhancement of vernacular architecture, while reinforcing local identity and promoting sustainable tourism.

Introduction

The value of an ancient way of life is often only fully appreciated after its disappearance. In today's context, human lifestyles have changed due to the modification of thoughts, habits, and behaviors. This change has expanded to impact the living environment: housing. Today, we face standardized, uniform, depersonalized housing, losing its identity. In contrast, vernacular housing meets the needs of inhabitants while expressing the values of popular culture, addressing social, cultural, and economic demands: a natural integration that shapes the environment through originality, invention, and character. Vernacular housing is considered functional architecture through the use of local materials and techniques, the enhancement of lived spaces, and the reality of life within the community. Vernacular architecture thus becomes a living expression that transcends walls and partitions, reflecting values, sensitivities, and cultures.

Problem Statement

The city of Larache is experiencing an imbalance between its natural and urban characteristics. This issue becomes apparent when we analyze the region's natural elements, which, due to poor land management, have become natural risks such as floods and forest fires, thereby threatening the city's proper functioning. As for the vernacular architecture of Larache, we observe a rich mosaic of styles influenced by its history and culture (traditional houses, streets and medinas, religious architecture, ramparts, and kasbahs). Given this, we consider Larache a suitable example for evaluating the resilience of the Medina's vernacular architecture in the face of natural risks. It is essential to propose solutions to protect, conserve, and enhance the vernacular architecture of this region. This is, in fact, the primary objective of this article.

Methodology

In this article, our analysis is structured around three key elements to better assess the resilience of Larache's vernacular architecture. First, we will begin with a quantitative diagnosis of the city's physical characteristics, including climate, soil, human factors, the diversity of vernacular

Materials Research Proceedings 47 (2025) 197-202 https://doi.org/10.21741/9781644903391-23

elements, and the way of life of their users. Next, we will address the evolution of vernacular architecture and the resilience it has demonstrated throughout its history. Finally, we will propose recommendations and solutions to ensure the preservation of this vernacular heritage within the Medina of Larache.

1. Vernacular Housing in Morocco:

Vernacular architecture is defined as the relationship that connects three elements: Man, Nature, and Materials. It represents a link between humans and their environment, resulting in a built structure of high cultural value. Humans organize their space based on their economic system, social structure, and the techniques available to them. This space, in turn, supports their activities and reflects their thoughts throughout their journey. That said, vernacular architecture reflects the local traditions, available materials, and cultural practices of a specific region. It is often characterized by its integration into the landscape and the use of traditional construction techniques. Unlike formal architecture, it is usually carried out by local craftsmen rather than certified architects. There are different types of vernacular architecture in Morocco. Examples include the architecture of the ancient medinas in imperial cities (Marrakech, Fez, Chefchaouen, Rabat, etc.), the kasbahs or fortified structures in strategic regions (Kasbah of the Oudayas, Ait Ben Haddou Kasbah, Kasbah of Chellah, etc.), riads, and the earthen architecture of the Saharan regions. In this article, we will focus on the ancient medinas, especially the one in the city of Larache.

2. The Medina of Larache

The term "medina," derived from Arabic, originates from a word meaning "city." Initially, the medina referred to an urban agglomeration without specific distinction but has since evolved to designate the older part of the city, in contrast with modern European-style neighborhoods.

The medina develops along three key aspects: spatial distribution, lived experience, and a sustainable urban model. Firstly, spatial distribution is organized around public spaces and pedestrian pathways. The fabric of the ancient medina is centered around the sacred space: the great mosque, which serves as the central point around which spaces for economic activities and trade are arranged. Residential quarters follow. The ancient medina has evolved to encompass the duality between "modernism" and "tradition," developing as a process of urban growth where two distinct urban systems (the old town and the new city) coexist. Secondly, regarding lived experience, the medina is considered the living heart of Larache, combining craftsmanship, expertise, community life, and social cohesion, which reflect solidarity, sharing, and exchange with neighbors. Thirdly, on an urban level, the medina is divided into three distinct levels: the Derb (narrow street), the Houma (neighborhood), and the traditional house. These elements reinforce the concept of mixed activities and a hierarchy of pathways. The design of the traditional house, in particular, is based on principles and techniques that make it simple, innovative, and functional. The construction takes advantage of direct sunlight, the thermal inertia of materials, natural air circulation, and the maximization of natural light. All of this is aimed at maintaining a pleasant indoor climate, aided by local materials and the integration of courtyards in traditional designs.

Photo. 1 : The Derb (Narrow Street)

Photo 2. The Medina of Larache

Photo 3. The Houma (Neighborhood)

3. Physical Elements: Climate and soils

3.1. Climate

The province of Larache benefits from a Mediterranean oceanic climate characterized by the alternation of a cool, humid season from October to April and a relatively hot, dry season from June to September. The average annual rainfall ranges between 700 and 1,100 mm, concentrated almost entirely between October and May. It is one of the wettest cities in Morocco, thanks to frequent disturbances from the Atlantic in the autumn, winter, and early spring. It is not uncommon for some municipalities in the province to record over 100 mm in a single day (Saloui, 2000).

3.2. Soils

The province of Larache is characterized by the following types of soils:
-Clay soils in the mountainous region;
-Clay-sandy soils in the hills and plateaus of the northern province (R'mel of Ouamra);
-More or less heavy alluvial soils in the Loukous plain (northeast limit of Ksar El Kébir);
-Terraced soils on clay marl in the hills of the southern province (towards Asilah).

The humid climate, combined with the existing soil types, encourages the development of a series of lakes, ponds, and wetlands, which may exacerbate the impacts of flooding. These physical characteristics indicate that, apart from the R'mel plateau of Larache, the province is potentially exposed to flood risks, as seen in the 1960s, 1990s, and 2002 in the city of Ksar Kébir and in the Larache wetland, causing significant damage to property and human lives. Only reasonable urban planning is likely to limit these risks of natural disasters: floods and forest fires.

4. Human Factors

4.1. Demographic Status

The province of Larache has a population of 496,687 inhabitants (H.C.P., 2014), of which 46% live in rural areas. The urban population recorded positive growth between 2004 and 2014, at a rate of 1.1% (Table 1).

Population of the Province of Larache	2004	2014	TAAM (2004-2014)
Urban Area	219 577	265 660	1,1
Rural Area	252 809	231 027	-0,1
Urbanisation Rate	50,4	53,5	--

***[TAAM]: Average Annual Growth Rate**

4.2 Population Dynamics in the Province of Larache

Considering the opportunities available in the city of Larache in terms of employment offerings, and consequently attracting rural labor employed in agriculture and maritime fishing, along with its natural population growth, the city is experiencing a steadily increasing demographic growth

Materials Research Proceedings 47 (2025) 197-202 https://doi.org/10.21741/9781644903391-23

(see Graph 1 below). This situation has led to a rising demand for land acquisition and housing. Consequently, housing prices have risen rapidly. That said, the stakeholders in the province tend to prioritize their own interests over effective urban management.

Since the late 1960s, substandard housing has begun to appear in areas highly exposed to flood risks, as well as modifications to housing within the old medina (deteriorated homes and unauthorized extensions) without prior studies. This has resulted in the exposure of the city of Larache and the deterioration of its medina, an exceptional heritage.

Year	Total population
2014	496,687
2015	499,804
2016	502,883
2017	505,931
2018	508,917
2019	511,815
2020	514,589
2021	517,225
2022	519,707
2023	522,035
2024	524,209

Graph 1. Evolution of the population of Larache's Province

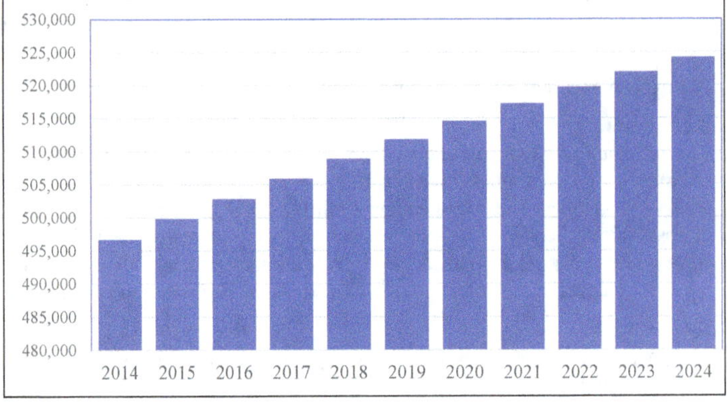

5. The State of Vernacular Housing and Urban Expansion in Larache

Since the 20th century, the medina of the city of Larache has undergone progressive degradation, which can be divided into three major phases. The first phase, that of marginalization, began with the establishment of the French protectorate, which shifted the city center from the traditional medina to new European-style neighborhoods, where colonial administration was situated. Subsequently, impoverishment and ruralization marked the departure of affluent families from the medina to the new city, symbols of modernity, while rural populations occupied these old neighborhoods, transforming the medina into refuge spaces for the less fortunate. This change led to a gradual deterioration of the built environment, characterized by the fragmentation of traditional houses, often divided to accommodate multiple families. The departure of wealthy residents disrupted the socio-economic balance, and the maintenance of homes was neglected, worsening their condition. Real estate speculation also contributed to this deterioration, with the replacement of old houses by modern buildings. Meanwhile, economic and administrative institutions were externalized, and infrastructure modernization altered the hierarchy of activities within the medinas. Finally, the disappearance of green spaces and gardens diminished the quality of life in these neighborhoods, making it more difficult to preserve their architectural and urban heritage. That said, following various developments, the city of Larache has seen an expansion of constructions outside its traditional boundaries. However, it was decided to create buffer zones between the pre-existing fabric and the new constructions. Urban development during the protectorate created an urban continuity with the medina of Larache. Consequently, the layout of a large square, formerly known as Place d'Espagne (now Place de l'Indépendance), was designed (see Photo 5), featuring a set of buildings of high architectural quality (see Photo 3). Subsequently, urbanization followed a radial plan, delineating avenues that organized urban development in well-defined directions. As a result of this dynamic, the city now consists of various complementary fabrics, each presenting unique urban and architectural characteristics specific to its zoning.

Photo 4. City of Larache

Photo 5. Place de l'indépendance (2024)

Photo 6. Urban Planning During the Colonial Era

6. Discussion of Results

Due to the population growth in Larache, following rural depopulation, its attractiveness, and positive growth, the demand for housing is increasing. Consequently, the construction sector is experiencing strong demand from the population. Based on the principle of supply and demand, urban space is under pressure for construction across its various geographical zones, especially in the old medina. In light of this situation, and to prevent housing from being directed toward areas at high natural risk, it is important to raise awareness among the relevant authorities regarding regulations and urban planning documents, as well as to sensitize the population to the importance of choosing their living environments.

Regarding the ancient medina of Larache, it is essential to consider preserving and conserving the city's identity to protect its history, urban territory, and heritage buildings that define Larache today and enhance the identity of its residents within their city.

Conclusions

Larache is a city with rich natural and human potential. However, due to the physical specificities of the city (climate, hydrography, relief, soils, wetlands, etc.), combined with human factors (demography, urban expansion, etc.), the city of Larache, particularly through its ancient medina, remains increasingly exposed to serious natural risks, especially floods, forest fires, degradation of fabric, and marginalization. However, only effective land management involving all concerned stakeholders can limit these risks and their impacts on the environment and people. Therefore, it is wise to thoroughly study the territory, analyze its strengths and weaknesses, and identify a specific action plan for the Larache region. This plan would ensure economic development, user safety in the province, and the preservation of its identity.

References

[1] Saloui, A. (2000). Etude de la variabilité des précipitations et des bilans hydriques dans la partie N-W du Maroc. PhD thesis, Académie des Sciences et Techniques, Faculté des Sciences et Techniques, Lille, France ;

[2] Saloui, A. (2000). Précipitations et bilans hydriques dans les plaines atlantiques marocaines. Publication de la FLSH de Mohammedia, Série: Colloques, n°8 (in Arabic) ;

[3] Haut-Commissariat au Plan, Direction Régionale de Tanger-Tétouan-Al Hoceima. (December 2020). Monographie de la région de Tanger-Tétouan-Al Hoceima ;

[4] Haut-Commissariat au Plan, Direction Régionale de Tanger-Tétouan-Al Hoceima. (2014). Recensement général de la population et de l'habitat: Caractéristiques démographiques et socio-économiques, Région Tanger-Tétouan-Al Hoceima ;

[5] Choukri, B., & Seqqat, N. (January 2019). Élaboration de l'étude de la charte architecturale et paysagère de la ville de Larache, Phase III: Documents finaux, Agence Urbaine de Larache-Ouazzane, Ministère de l'Aménagement du Territoire National, de l'Urbanisme, de l'Habitat et de la Politique de la Ville ;

[6] Ministère de la Santé et de la Protection Sociale. (October 2020). Projection des populations cibles des programmes de santé pour 2014-2019/2020-2026 ;

[7] Ministère de l'Aménagement du Territoire National, de l'Urbanisme, de l'Habitat et de la Politique de la Ville, Agence Urbaine de Larache. (May 2017). Note de présentation du SDAU du Littoral de Larache ;

[8] Guindani, S., & Doepper, U. Architecture Vernaculaire (collection architecture), Territoire, habitat et activités productives, published under the direction of Frédéric Aubry, Mario Bevilacqua, and Plemenka Supic ;

[9] Fougnar, C. (2021). La médina comme modèle urbain marocain : cas de la médina de Salé. Final year project, Ecole Nationale d'Architecture de Rabat, Morocco ;

[10] Rhazouani, M. (2023). De la médina à l'écoquartier : le défi de la durabilité de Rabat. Final year project, Ecole Nationale d'Architecture de Rabat, Morocco ;

[11] Bendaba, Z. (2020). La revalorisation d'un tissu ancien : cas de boulevard Laalou et du quartier de Loubira de la médina de Rabat. Final year project, Ecole Nationale d'Architecture de Rabat, Morocco.

Materials Research Proceedings 47 (2025) 203-215 https://doi.org/10.21741/9781644903391-24

Habitat reinvention: Tented living in waiting for reconstruction

Soumia OUBAH[1,a*], Mohamed TAMIM[2,b*]

[1]Faculté des Lettres et des Sciences Humaines Aïn Chock. BP.: 8507 Hay Inara- Casablanca, Morocco

[2]Institut National d'Aménagement et d'Urbanisme, Madinat Al Irfane, B.P 6215, Rabat -Instituts, Morocco

[a]oubahsoumia921@gmail.com, [b]ytamimoh@yahoo.fr

Keywords: Earthquake, Habitat, Reinvention of Habitat, Reconstruction

Abstract. This study examines « rural habitation » in a mountainous region of the Western High Atlas affected by the earthquake of September 8, 2023. It aims to understand how residents lived before and in the aftermath of the earthquake, as well as the implications of ongoing public reconstruction efforts concerning local and heritage dimensions of rural habitat. Utilizing an ethnographic approach, the research focuses on the case of Ouneine in Taroudant province, highlighting the transition from traditional vernacular architecture to temporary tent structures. It reveals the community's creative adaptations and differing perceptions of their vernacular habitats, emphasizing the need for participatory approaches in reconstruction that respect local knowledge and practices.

Introduction

This communication aims to understand and succinctly convey the changes in habitat and dwelling experienced by populations following a natural disaster, specifically the earthquake of September 8, 2023, in the Western High Atlas [1], with a particular focus on Ouneine (Taroudant Province). Three guiding questions structure this research:

-How did habitat and dwelling present themselves before the September 8, 2023 earthquake?

-What is the situation following the earthquake up to the present day?

-How can we characterize the approach of public action regarding the implementation of reconstruction related to traditional habitat in the area?

The selected field is not unfamiliar to us; Since the late 1970s, researchers from the Institut Agronomique et Vétérinaire Hassan II (IAV-Hassan II) have conducted several action-research projects aimed at development in this area. Since 1997, members of the same team have continued these efforts within the Targa-aide Association, up to the present day.

The results presented in this article are drawn from a Master's thesis in sociology [2], which aimed to document the unfolding of daily life post-earthquake from the perspective of dwelling. Observations were primarily conducted in the douars of Tamsoult and Agard, as well as the sub-douar Talat, which belongs to the Tanjjmcht fraction. The first visit took place from January 2 to 8, followed by a second visit from April 17 to May 5, totaling nearly a month of immersion in the field. During our time in the field, we adopted an ethnographic approach, spending entire days with the residents, both in cracked houses where the fear of aftershocks remains palpable and in tents, to observe the various dwelling practices at play in the area.

A rural habitat of the sedentary peasantry in grouped and compact douars

The earthquake disrupted the daily lives of the inhabitants and highlighted the intimate relationship they maintain with their living environment. These populations are predominantly sedentary mountain peasants living in clustered douars. The rural and peasant dimension is evident through several key characteristics: the limited size of agricultural holdings, the multiplicity of plots within a single holding, the significance of human (particularly family) and animal labor as energy sources, and the importance of off-farm income (from other activities, migration, remittances, etc.) in the average household income[1]. This latter indicator reflects the significance of pluriactivity at the individual or household level, as well as the strong interdependence between urban and rural areas in the lives of these populations.

The douar or lmodaa in Ouneine: Territory of daily life for rural populations: In rural Morocco, daily life unfolds in the douar (or lmodaa in Ouneine), a fundamental and infra-communal territory. According to the 2014 census (RGPH), the Moroccan rural environment comprises 32125 douars, excluding nomads; at times, the notion of douar overlaps with that of "census district." The definition of a douar often emphasizes buildings; however, it only makes sense when several elements are combined and considered: (1) a cluster of more or less grouped dwellings; (2) a common or multiple kinship, whether real or fictive [3] ; (3) an agrarian territory or finage over which the members of the douar have rights (land, water, tree vegetation, etc.); and finally, (4) an ability for collective organization and action on matters of common interest. This capacity for the organization is manifested through the jma'a, which remains a flexible concept depending on whether it involves decision-making or information dissemination. Thus, there will be as many forms of jma'a as there are issues to discuss (management of the mosque, irrigation water distribution, road construction, establishment of a football field for youth, timing for fruit and grain harvesting, etc.), with separate jma'as for information and decision-making. The douar evolves: new douars emerge while others disappear (absorbed by ongoing urbanization). Additionally, since the implementation of the National Initiative for Human Development (INDH) in 2005, many douars have established local development associations to benefit from project calls; these associations provide the douar with a voice that can be heard, as the douar is not institutionalized and the commune remains territorially too extensive. Youth predominantly lead these associations: they are literate, have social needs that sometimes differ from those of their elders, and thus represent a crucial interlocutor for certain public policies.

The clustered nature of the habitat around the douar mosque, typically in areas unsuitable for cultivation, facilitates contact and discussion regarding matters of common interest for the population.

From this definition of the douar, it becomes clear that the dwellings are inseparable from the agro-pastoral spaces that sustain the communities. The home is indeed the center of the peasant agricultural operation. This raises the question of how well these various aspects defining a douar are taken into account in the ongoing reconstruction process in the earthquake-affected areas.

The study region is a mountainous environment characterized by compact housing, where inhabitants collectively construct their homes on uncultivable slopes while cultivating fields in the valleys. In Ouneine, this architectural landscape is distinctly marked by vernacular architecture.

The house (tigmmi), the hearth (Takat), and the family

At the infra-douar level, we encounter lineage and family units. The terms Afouss and Ikhss (the hand ; the bone) are commonly used to indicate a person's lineage affiliation. In the context of clustered housing in Ouneine, each lineage occupies a designated area for construction.

Before the earthquake, the dwellings in Ouneine were primarily characterized by traditional structures that incorporated local materials (earth, stone, wood) and reflected ancestral cultural practices. These homes were designed to meet family, community, and climatic needs, embodying a living heritage where each space held specific significance.

[1] For the entire Morocco, the share of "mixed income from agricultural activities" in the total income of an average rural household was 38.11% in 2019; consequently, just over two-thirds of the income comes from off-farm activities, either locally or in nearby towns and small urban centers. HCP/Social Indicators 2023; p. 245.

They were mainly constructed of rammed earth, featured thick walls, and were generally rectangular in shape, with an interior courtyard known as "Tadwariyt." The roofs, often flat, were made of wood and earth, serving to dry various agricultural products (such as nuts, almonds, and dried fruits) and providing a space for nighttime use by the family, particularly for sleeping during the summer months.

These dwellings, with an area exceeding 100 m², are multifunctional and accommodate both humans and animals, serving as an extension of agricultural fields. In these homes, inhabitants store and dry their harvests, with a dedicated space for agricultural equipment.

The family is organized around the "Takat" (household or hearth) as a structure, meaning that a single dwelling can house multiple families grouped around several "Takatin." Thus, a typical dwelling in Ouneine includes a kitchen ("Anoual"), a living room ("Ahano"), storage and drying spaces for harvests, as well as a space for animals ("agrour"), all arranged around the interior courtyard ("Tadwariyt").

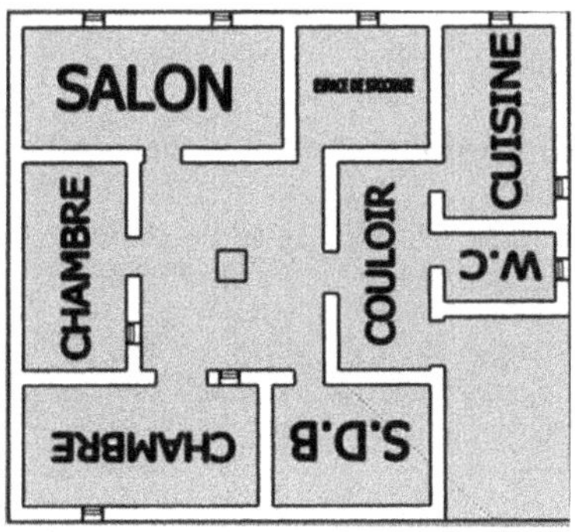

Figure 1: Typical layout of the dwelling prior to the earthquake in Ouneine (created by Soumia Oubah, 2024).

By creating a sketch illustrating the arrangement of homes in the douar with the help of the residents, we noticed a dynamic at play in the way of living: a small number of extended families still live under one roof; indeed, sons tend to separate their households after marriage. Furthermore, residents prefer to physically separate the spaces designated for animals from those reserved for living, maintaining a distance of several meters to mitigate the inconveniences caused by manure odors.

The hearth, or "Takat," represents the central element of the family unit, "Tiguemmi," around which family life is structured. The separation of kitchens indicates a fragmentation of the family. Moreover, Hanae Bekkari [4] showed over 30 years ago that residents placed "Takat" above their homes due to the smoke it produced, but with the modernization of housing, they have begun to isolate it progressively. Today, residents prefer to set up "Takat" outside the dwelling, on the terrace or in an uncovered space to avoid smoke [4]. They also use a gas stove located inside the kitchen arranged in the house. The kitchen is thus a place where women prepare daily meals, and where the family sometimes takes its meals.

"Ahano" served as a multifunctional space intended for living, sleeping, and receiving guests. It provided an ideal setting for relaxation and rejuvenation, offering a panoramic view of the surrounding fields and mountains [4].

The storage spaces, known as "Ahano El Khzin," are always locked. Previously, this room was completely dark, but it is now becoming increasingly illuminated [4]. In addition to food provisions, these spaces contain boxes holding clothing, scrolls, and agricultural tools used for fieldwork.

Before the earthquake, residents began to increasingly incorporate concrete into the construction of their homes. To avoid adding layers of clay each time and to fill gaps in the rammed earth walls, they started covering these walls with cement.

Simultaneously, many families added a floor or a concrete extension, often following the marriage of sons. Some even chose to build a new dwelling, separate from their parents' home, constructed with reinforced concrete rather than traditional local materials. Their homes are characterized by features such as kitchens equipped with cabinets and tiled floors, as well as living rooms with plaster ceilings, reflecting a desire to improve living comfort while adapting to contemporary standards.

Thus, the house, or "tigmmi," is experiencing significant changes in its design, driven not only by family dynamics or the rising urban lifestyles but also by the introduction of new public and collective services previously nonexistent in the douar: road access, electrification, individual connections to drinking water, and the widespread availability of primary and secondary education.

How have people reinvented their habitat in the months following the earthquake to the present day?

The earthquake caused a sudden rupture from the previous daily life, forcing residents to leave their homes or what remained of them, adopting an urgent mode of living in tents while waiting for reconstruction, the duration of which was uncertain. In the weeks and months following the disaster, temporary solutions emerged, often marked by tents and improvised arrangements, reflecting resilience in the face of adversity. In this context, we needed to explore how these new living conditions affected not only the daily lives of residents but also their relationship with their former homes and their culture.

The tent-house : from collective to individual by household

In the aftermath of the earthquake, only a few tents were provided per douar to shelter the population, with efforts made to preserve gender separation as much as possible. Subsequently, there was a need to equip the tents, particularly with electricity, and to restore access to drinking water for the douars.

About two weeks after the earthquake, the Royal Armed Forces delivered collective tents (according to the population size of each douar) and stipulated that they should be grouped in a single location. Residents moved into these tents and organized themselves around them, with variations from one douar to another.

Materials Research Proceedings 47 (2025) 203-215 https://doi.org/10.21741/9781644903391-24

Photo 1: The only collective tent I found still occupied by a family (taken by Soumia Oubah, 2024).

Subsequently, new tents intended for households were distributed by local authorities; each household was assigned one tent (or a maximum of three). The provided tent measures 3.20 meters in width, 3.70 meters in length, and 3.70 meters in height. It consists of a double waterproof fabric, ensuring protection against rain infiltration. Windows are placed at regular intervals of 76 cm on either side, although they do not open fully, being covered with mesh fabric, and they also feature a double layer that allows for closure. At the center of the facade, there is a door, also made of double fabric, equipped with a Velcro closure.

Photo 2: The tent designated for the household (taken by Soumia Oubah, 2024).

These tents proved insufficient, even for households consisting solely of a couple. Residents found themselves compelled to locate a plot for tent installation that was not far from the damaged house and to set up a dedicated space for meal preparation and basic needs. Additionally, with the proliferation of tents, it became necessary to erect a fence to demarcate the boundary between inside and outside. This initiated a genuine process of tinkering and creativity in the living space.

The tent-house: reorganization of space and intensive investment of labor and money by households

As will be evident in the following descriptions, the tent is crucial for addressing immediate needs, but it remains inadequate on its own: it requires modifications and financial expenditures for all the accessories that can transform it into an alternative dwelling to the former home. This resulted in significant costs for the population, particularly in terms of physical exhaustion from having to rebuild everything. However, it should be noted that the state (as well as other humanitarian organizations) demonstrated substantial support for the population to recover and resume activities. Alongside the provision of tents, numerous public and private service operators for water and electricity were present in the field of study. Additionally, the granting of financial aid of 2,500 DH per beneficiary for one year helped alleviate the burden on the population. New products arrived on the local market (souk and municipal center), including plastic materials, wooden beams, wire mesh, tarps, ropes, portable earthen "stoves," and plastic mats.

The delivered tent is used as a bedroom. Residents commonly refer to it as the "sleeping tent" ("Qitoun fin Kan'so"), even though it serves multiple functions, similar to other rooms in the home. Some residents sleep in it, eat there, and watch television, while others engage in cooking, especially during cold weather, or use it as a storage space for provisions or clothing.

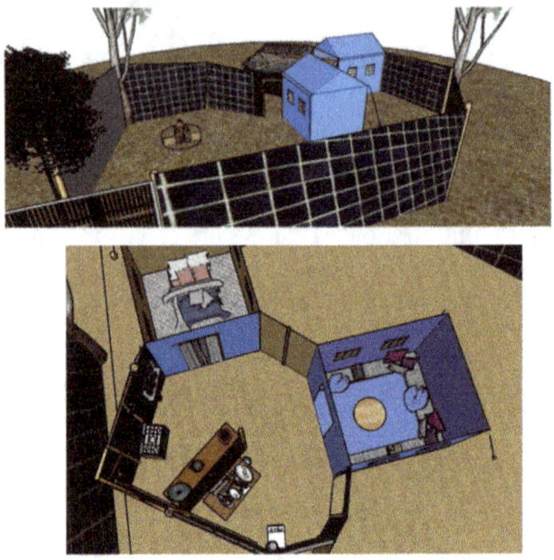

Figures 3 and 4: A 3D layout of the arrangement of tents within a household (created by Soumia Oubah, 2024).

Among the additions are the kitchen, referred to as 'Kozina', and 'Anoual'. The former constitutes a space where one can find, notably, a cooking area in the form of a modern stove and a butane gas bottle. Women hang cooking utensils on hooks fixed to the wooden beams of the room or place them on a work surface known as 'Tabla', made by reusing an old door supported by bricks or stones. It is here that women prepare meals using gas stoves, sitting in the kitchen while washing dishes elsewhere.

Materials Research Proceedings 47 (2025) 203-215 https://doi.org/10.21741/9781644903391-24

Photos 5 and 6: Improvised kitchens: Adaptation to new living spaces after the earthquake (taken by Soumia Oubah, 2024).

In 'Anoual' (kitchen), the space is roofless and consists solely of the 'Takat' oven in the strict sense (that is, a clay oven embedded in the ground, made by women, shaped oval, hollow inside, and featuring a front opening extending to the top). It burns wood to produce flames and generates embers, which are then transferred to a kanoun (mejmer) for baking bread, preparing morning soup, and cooking porridge. It also serves to heat large quantities of water. The reproduction of the 'Takat' in the tents represents a form of cultural resilience, reflecting the desire to maintain habitat-related practices before the earthquake.

Photo 7: Preserved culinary heritage: The "Takat" reproduced in the post-earthquake habitat (taken by Soumia Oubah, 2024).

Paradoxically, at first glance, the rooms appear to be specialized and designated for single use. However, after spending time with their occupants and exploring their various usages in greater depth, it becomes clear that the specialization reflected in the furniture merely indicates a desire to replicate the spatial organization adopted in homes prior to the earthquake.

Materials Research Proceedings 47 (2025) 203-215 https://doi.org/10.21741/9781644903391-24

Photos 8, 9, 10, and 11: Beyond Appearance: Revelations on the Multifunctional Use of Rooms (taken by Soumia Oubah, 2024).

Indeed, the housing conditions after the earthquake force residents to prioritize the multifunctionality of spaces over specialization. For example, bedrooms serve as areas where inhabitants sleep, eat, and watch television. For those without a designated visitor reception area, guests are welcomed in the bedrooms, where parents and children sleep together, often separated by makeshift covers.

There is often an open space, sometimes right after the first entry framing the tents of the same household, or between two sections of the tent. This area is also multifunctional, used for laundry and dishwashing, as well as to receive visitors whom residents prefer not to show their interior space. Additionally, during warm weather, men sometimes sleep in this area. It is noteworthy that only men can use this space for sleeping, as it is visible from outside.

Photo 12: Versatile outdoor space: From dishes to visits, a vital "external" area (taken by Soumia Oubah, 2024).

Regarding the specialization of rooms, two spaces stand out, whether in tents or houses: the first is the kitchen, which is specialized both in terms of use and users, as only women engage in meal preparation there; the second notable space is the storage area, "Bit L Khzin," for food supplies. Indeed, it has been observed that the residents are careful to cover their food reserves, whether in tents or houses. Moreover, during visits to most homes, it was noted that the storage room is generally located far from the main entrance. Aicha, a 75-year-old woman, illuminated this for me by emphasizing that « "Bit L Khzin" should only be seen by four eyes: two belonging to the caretaker and two to the cat that enters to feed." She also specified that "previously, only one woman, usually one who no longer menstruated and was thus considered always purified, held the keys to 'Bit L Khzin.' In the case of guests, she was the only one allowed to access it to provide her daughters with the necessary supplies for preparing meals for guests." » Aicha from Talat. Noted that this is a way to preserve Baraka.

Living also involves managing the boundaries between the outside and the inside, as well as the decision of when to step out and interact with the outside world, and when to remain inside to enjoy moments of intimacy.

After arranging and organizing their tents, the residents drove wooden posts into the ground on either side of the existing opening in the tent, where they subsequently installed doors made from their former homes. Some crafted these doors from wood and plastic, thereby creating a means to regulate access and delineate their living space.

Additionally, to preserve privacy and demarcate the inhabited area, extensions were added to these tents. After a few months of settling in, the residents also planted wooden posts a few meters around the tent, which they covered with a tarp and equipped with a door.
These extensions serve to claim more space, which they use to set up the "Takat," requiring an open area, as well as for washing clothes and utensils. Furthermore, faucets are absent from the kitchens but are installed in these added spaces to prevent water damage to the furnishings.

The tent is not equivalent to a home

Several concerns have been expressed: sounds are not muffled by the fabric or tarp of the tent, human silhouettes are visible at night, fire can easily spread in the plastic, the tent does not provide protection against theft, and it is necessary to always have someone present in the tent, etc.

When examining the aspects that have affected residents' relationship with their habitat, a recurring question during interviews with tent inhabitants is the absence of a private space that preserves their intimacy. They perceive the tent as akin to the street. When they are at home (in the tent), they do not regard this space as intimate or familiar. They feel at home only in the sense that they are not in someone else's place, yet they lack a space to withdraw from others. One resident from Agard expresses this idea as follows:

« As you can see, we are currently living in the street. We are continuously observed by all passersby. Anyone can enter our space at will. The people walking by (she points to passersby near her door) could come in at any time as if it were their own home. Furthermore, she adds: 'Inside a house, if you want to avoid interaction, you simply close the door and act as if you are not home. However, here, everyone knows you are present, so you don't have the option to close your door and enjoy a moment alone (Tsanti l'damk). »

The tent may still endure for a long time; efforts are being made to care for it

For their occupants, the tents served as temporary shelters designed to provide refuge during an emergency following the earthquake. As time passes and residents notice delays in the proposed solutions, they have begun to adapt to these new living conditions. They have started to modify their tents to better reflect their cultural identity and project their aspirations. Personalizing their habitat has become a means of expressing this.

At the entrance, ornamental plants are arranged, not for food but for their aesthetic appeal, creating a green space inside to soften the dark character of the tarps surrounding the tents.

Inside the rooms, handmade decorations invite contemplation. Elaborate rugs made using various techniques and materials, woolen accessories, and cushions adorn the space. These artisanal objects reflect the craftsmanship employed to personalize their homes.

Some residents retrieve family photographs from their former homes to infuse warmth and memories into the cold beams of their tents. Others choose to paint these beams to customize them. Through these personal touches, each inhabitant seeks to claim their new dwelling as a place that feels like home.

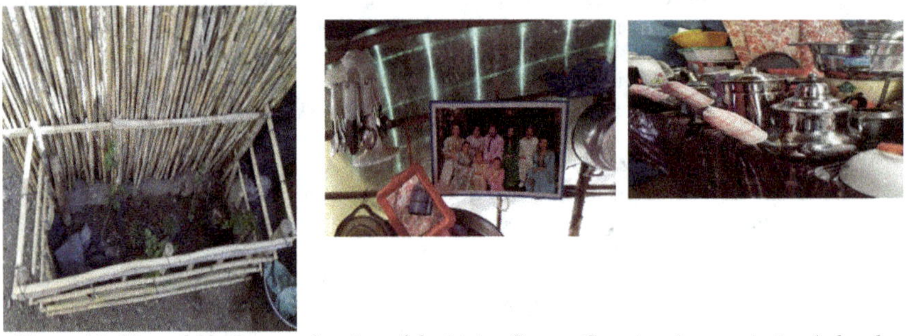

Photos 11, 12, and 13: Personalization of the Living Space: Creative Appropriation (taken by Soumia Oubah, 2024).

From the descriptions provided, we can first deduce that the time of the population is not the same as that of the State. What may seem like a short-term issue for the State can prove to be very lengthy for the population. Consequently, it becomes necessary to reinvent daily life and, particularly, the habitat, while awaiting reconstruction according to the new standards imposed by the State.

In this reinvention of the habitat under the tent, we observe the ingenuity of the population—both women and men—in their use of available materials. This evokes Michel de Certeau's [5] concept of popular arts of making, where subtle "strategies," "tactics" of resistance, and the necessity to "live" come into play.

Reconstruction by the State

It is important to note that the reconstruction undertaken by the State encompasses not only « housing » but also infrastructure such as roads, as well as agriculture and livestock.

Public reconstruction efforts are framed as a central issue in this recovery process. Far from being limited to mere material rehabilitation, reconstruction offers the opportunity to rethink rural housing in light of local and heritage-specific characteristics. This raises questions about how public authorities integrate the aspirations of residents and cultural dimensions into reconstruction projects. What lessons can be drawn from traditional practices to build a future that respects the identity and heritage of this region? This study aims to explore these questions by observing the lived experiences of residents and the ongoing reconstruction policies, to better understand the issues related to housing in the context of Ouneine.

A Top-Down Approach

Overall, what is happening today can be summarized as an administrative "proceduralization" of the population, leaving little room for input from below. This began with the opacity of the housing census, where the population is unaware of the criteria used to select beneficiaries (for financial

Materials Research Proceedings 47 (2025) 203-215 https://doi.org/10.21741/9781644903391-24

aid and reconstruction); followed by the implementation of plans that are imposed from above without local discussion regarding possible plans that would address the rural needs of the population.

On September 20, 2023, the King of Morocco emphasized that the revival of local development in the affected areas must follow a Top-Down approach, listening to the populations and taking their legitimate grievances into account. Two major concerns of this approach should be considered :

• Preserving local heritage, both public and private, as well as human and material resources.

• Listening to the population and engaging in dialogue with them to participate in defining the future of their habitat and aligning with the needs imposed by the situation and the grievances expressed by the residents.

Thus, the government, in undertaking this phase of reconstruction, has allocated an amount of 120 billion dirhams over five years. This funding is intended to benefit 51,300 households, with differentiated amounts based on the condition of the damaged homes. Families whose homes have been destroyed are allocated 60,000 dirhams, while those needing repairs receive 20,000 dirhams [6].

Coordination Problems in the field of study

On the field, the State, through its numerous agents and those from the private sector to whom it delegates tasks, interacts with a large number of 'beneficiary' individuals. The municipality serves as a conduit. Everyone is caught in turmoil, resulting in a loss of energy.

As the implementation of reconstruction principles unfolds, the operation faces several challenges, particularly in coordinating among multiple stakeholders, including local authorities, architects, engineers, and civil society, each operating under distinct logics. This initiative has been centralized in the hands of actors who are distant from the culture and needs of the affected populations, while local authorities have neglected coordination and communication with these residents [6].

The inhabitants live in uncertainty, perceiving the inconsistencies and delays from the authorities as a potential deadlock in resolving their situation. They lack precise information regarding the reconstruction process. Initially, three commissions were established to oversee reconstruction, but testimonies indicate that each commission provides a different assessment of their housing situation compared to the others.

Some residents have begun excavation work without knowing the next steps or when the architect will return to provide further guidance. This uncertainty persists, particularly for those who have not received reconstruction aid and lack building permits. They have no clear perspective on upcoming steps, hindering their ability to plan for the future.

The Fraction, the Douar, the Jma'a, the Lineage, and the Family... Concepts Cast Aside

Numerous questions could be discussed with the population during the local jma'a meetings, and this is within short timeframes, at least to provide information. A multitude of problems arose for families following the State's decisions regarding the reconstruction of housing. First, the distributed plans are essentially 'template plans,' subject to modifications by the architects overseeing the construction sites. However, the margin of maneuver for these architects appears to be very limited, and even the elements of the contract between the architect and the beneficiary of the financial aid for reconstruction are not clearly explained. Moreover, a family member who is a beneficiary may not necessarily be the sole owner of a house or part of a house that has partially or completely collapsed, leading to inextricable conflicts within families.

The government approach creates a dual precariousness among residents, hindering the reconstruction process. Originally designed to help residents regain housing after the earthquake that forced them to leave their homes, this approach has proven paradoxical. The imposed

Materials Research Proceedings 47 (2025) 203-215 https://doi.org/10.21741/9781644903391-24

requirements, such as the obligation to obtain a building permit and the prohibition of beginning reconstruction without the architects' approval, have exacerbated the precarious living conditions of the inhabitants.

For example, the following response from a survey participant encapsulates the current plight of the population: “We suffered from the earthquake of God (Zalzal dyal LLah), and now we are suffering from the earthquake of the government (Zalzal l Houkouma).” In this regard, Hassan from Tamsoult also states: “We don’t want anything from them anymore; we just hope they let us rebuild our own houses ourselves. If they had let us, we would have already returned home.”

Remaining in the Tent or Returning to the Old House Despite the Risk of Aftershocks?

Indeed, the government’s approach to reconstruction views residents as a homogeneous group, while our empirical findings reveal that they have distinct relationships with their habitat. In the face of the tragedy of the earthquake, residents express heterogeneous connections to their living conditions, reflecting varied experiences and feelings. Some women living in tents strongly reject their current situation, feeling unable to adapt to this precariousness, which they perceive as a degradation of their dignity. They nostalgically recall the sturdiness of their former homes, lamenting the comfort and intimacy they provided, while also regretting the lack of support for their reconstruction.

Others, both men and women, demonstrate a quest for safety. Traumatized by the earthquake, they choose to remain in the tents, considering them safer in light of potential aftershocks. Although some have access to aid, their skepticism regarding the durability of new constructions leads them to prefer a solution that allows them to maintain some control over their lives.

Finally, some residents have opted to reintegrate their damaged homes. They conceal the cracks to regain a sense of security and are willing to repair their dwellings while preserving their emotional essence. Their attachment to these places drives them to negotiate solutions with the authorities, seeking to safeguard their homes while ensuring their safety.

Thus, these varied attitudes illustrate the complexity of residents’ relationships with their habitat, intertwining rejection, the search for safety, and attachment, revealing an interesting heterogeneity that should be considered in the reconstruction process.

Conclusion

The situation in Ouneine before the earthquake of September 8, 2023, was characterized by traditional dwellings made from local materials and rooted in ancestral cultural practices. These houses, multifunctional and tailored to the needs of families, represented a living heritage where each space held its own significance. In contrast, following the disaster, residents were abruptly forced to leave their homes to live in tents, symbols of desperate urgency.

The state's response, although initially perceived as support, often appeared disconnected from local realities. The tents, while necessary, were frequently inadequate. However, despite these challenges, the residents demonstrated remarkable creativity in arranging their tents. They developed specific housing practices suited to these conditions, transforming these temporary spaces into functional refuges that reflect their cultural identity.

Meanwhile, reconstruction efforts were hindered by a lack of coordination and communication with the residents. This gap between the expectations of the inhabitants and the actions of the authorities created profound uncertainty, blending hope and disappointment. Families continue to navigate their daily lives, striving to recreate a semblance of normality despite precarious conditions.

In summary, the reconstruction of traditional housing in Ouneine requires not only an approach that respects local cultural practices but also attentive listening to the needs and expectations of the residents. As the future remains uncertain, it is crucial to promote a collaborative approach that incorporates the voices of the community to transform the hope for renewal into a tangible reality.

References

[1] J. Dresch, Recherches sur l'évolution du relief dans le Massif Central du Grand Atlas, le Haouz et le Souss; Tours, Arrault et Cie, Maîtres imprimeurs, 1941.

[2] S. Oubah, "Habiter après le séisme, les différentes facettes des modes d'habiter : Cas de la commune de l'Ouneine après le séisme du Haouz," Master's thesis, Faculty of Letters and Human Sciences Aïn Chock, 2024.

[3] Le Coz, Douar et centre rural : du campement au bourg, In, Revue de Géographie du Maroc (RGM), n°8, 1965.

[4] H. Bekkari, "Au-dedans, au-dehors, au-delà de trois villages de l'Ouneine," Final project for the architect diploma, National School of Architecture, 1990.

[5] M. de Certeau, L'invention du quotidien ; 1. Arts de faire (t1), habiter, cuisiner (t2), ed. Gallimard (folio essais), 1990 (tome 1), 1994 (tome 2).

[6] M. Tozy, "Le séisme du Haut Atlas vu d'Ouneine," Sociétés politiques comparées, vol. 62, Jan.-Apr. 2024, https://fasopo.org/sites/default/files/charivaria1_n62.pdf

Learning from seismic culture in the vernacular architecture of Tulahuén, Chile

Natalia JORQUERA SILVA[1,a*], Jaime ORTEGA PERALES[2,b] and Tomás SEPÚLVEDA SCHWEMBER[3,c]

[1]Departament of Architecture, Universidad de La Serena, Amunategui s/n, La Serena, Chile

[2]Instituto de Investigaciones y Postgrado, IIP - Facultad de Arquitectura, Artes, Diseño y Urbanismo, FAADU - Universidad Mayor de San Andrés, Héroes del Acre 1850, La Paz, Bolivia

[3]Universidad Bernardo O´Higgins, Centro de Estudios Históricos y Humanidades/ Director Ñaupa Estudios, *Avenida Viel 1497, Santiago,* Chile

[a]natalia.jorquera@userena.cl, [b]rk.ortega93@gmail.com, [c]naupa.estudios@gmail.com

Keywords: Adobe Vernacular Architecture, Seismic Threat, Building Cultures, Vernacular Seismic Culture

Abstract. Chile is one of the most seismically active countries in the world, due to the subduction of the Nazca Plate beneath the South American Plate, which causes earthquakes with magnitudes above 7.0 approximately every ten years. This high seismicity has always shaped the characteristics of vernacular architecture and traditional construction techniques, particularly adobe buildings, through a process of trial and error. Seismically vulnerable structures have disappeared, while those that have survived owe their resilience to the development of a seismic culture among the inhabitants and the creation of technical solutions that allow buildings to respond better to seismic events. This is the case of Tulahuén, a village located in the mountains of the Coquimbo region in northern Chile, where vernacular architecture, characterized by the use of adobe, incorporates various types of wooden reinforcements, built by the residents themselves, which have enabled these structures to withstand the intense local seismic activity. This article is the result of two research studies, in which, through direct observation and the collection of oral memories from local inhabitants, the seismic-resistant devices present in Tulahuén were characterized and their role in the resilience of this vernacular architecture was evaluated.

Introduction

In highly seismic regions, where experiencing the consequences of an earthquake is part of the collective memory, societies have developed a seismic culture. This culture consists of knowledge about how to behave during an earthquake and how to prevent damage to buildings. While seismic countries developed building codes during the 20th century to ensure earthquake-resistant designs for new buildings constructed with industrialized materials in urban areas, traditional villages composed of vernacular buildings also have a set of tacit norms. Although these norms are not formally written, they can be observed in the urban morphology (the way buildings are grouped), in the architectural design, and in the construction features of the structures [1].

Some examples of constructive devices adopted in vernacular architecture in areas where earthquakes are a recurrent and well-known phenomenon for the local population [2] include: "counter-arches" (Fig.1-a) placed between two buildings to restrict their lateral displacement; buttresses (Fig.1-b) to reduce the free span of overly long walls; ties (Fig.1-c) that improve the connection between transverse walls; and horizontal wooden "chains" or "keys" (Fig.1-d) that "tie" the walls of a building and reduce the spread of seismic shear cracks. All of these construction solutions enhance the performance of buildings against the lateral forces induced by earthquakes

[3, 4]. Since these strategies represent a "tacit norm," they are widely present throughout the region, identifying them as a "typological" solution.

Figure 1. Different types of earthquake-resistant devices (credits: Natalia Jorquera, 2024)

Chile has a deep-rooted seismic culture, derived from one of the highest levels of seismic activity on the planet, caused by the subduction of the Nazca Plate beneath the South American Plate. This process accumulates energy that is released in the form of large and long-lasting earthquakes [5]. The country holds the unfortunate record for registering the highest magnitude earthquakes in history (such as the 1960 Valdivia earthquake, with a magnitude of 9.5, and the 2010 Cauquenes earthquake, with a magnitude of 8.8, among others) as well as the highest frequency of seismic events, with earthquakes exceeding 7.0 in magnitude occurring on average every 10 years [3], according to data from the National Seismological Center.

In this context, various seismic response strategies have been adopted over the years by the different vernacular architectures that have existed in the country. These strategies were developed using the natural resources available, in a territory characterized by great geographic, climatic, and cultural diversity over 4,300 kilometers.

In the arid regions of northern Chile, dominated by the Atacama Desert, structural systems based on adobe or stone masonry walls have been developed. These systems, having very low tensile resistance to seismic forces, based their earthquake response only on their geometric configuration.

In the central regions of the country, where a temperate climate and moderate rainfall have allowed vegetation growth, ductile wooden elements (horizontal "keys") are incorporated into the vernacular adobe masonry buildings, improving their behavior against the tensile forces caused by earthquakes.

Finally, in the rainy southern regions, where vegetation and wood are abundant, vernacular architecture is based on the use of flexible wooden frameworks with lightweight infill, which perform very well under seismic forces due to their ductile behavior.

The architecture and seismic culture of Tulahuén

The Coquimbo Region belongs to Chile's semi-arid north and is considered a transition zone between the Atacama Desert and Chile's fertile central valley. It is characterized by being the narrowest part of Chile, with an average width of 150 km from the Andes Mountains in the east to the Pacific Ocean in the west, and an altitude range from sea level to 6,216 meters above sea level.

The region's semi-arid nature meant that, until the mid-19th century, vegetation was almost exclusively shrubby, limiting building materials to earth, which was used in various techniques, the main ones being adobe masonry and *quincha* (a vegetative structure filled with local earth).

Like the rest of Chile, the Coquimbo region experiences high seismic activity. The most destructive historical earthquakes include the 1943 Ovalle earthquake, with a magnitude of 8.2;

the 1975 Coquimbo earthquake, with a magnitude of 6.9; the 1997 Punitaqui earthquake, with a magnitude of 7.1; and the 2015 Illapel earthquake, with a magnitude of 8.3 Mw [6, 7].

In this region, at 1,200 meters above sea level, Tulahuén is located (Fig. 2). The vernacular architecture there, as in the entire region, is a mixture of the construction traditions of the indigenous peoples who inhabited the territory—especially the Diaguita culture—and the architecture brought by the Spanish invaders who arrived in the Americas and Chile in the mid-16th century. From the former, the use of earth as the preferred building material was inherited, while from the Spanish, the legacy consists of architecture with rooms organized around interior courtyards and houses grouped to form villages with continuous facades (Fig. 3).

Figure 2. Location of the case study (credits: authors)

Figure 3. General view of the context and the town of Tulahuén (credits: authors).

Tulahuén, along with other villages in the Limarí Valley, was part of large agricultural estates during the Spanish colonial period (16th-19th centuries). Following the first organization of independence from Spain in 1810, the region underwent a transition from agrarian administration to the formation of the current localities of Carén, Monte Patria, Mialqui, Huamalata, Chañaral, Los Molles, Juntas, and Tulahuén. However, there has been a historical continuity to the present day in agricultural vocation, as the region produces wine, fruits, vegetables, and fodder for livestock. This productive activity gave rise to the so-called agricultural communities that still exist today. These traditional collective organizations, based on principles of cooperation, are formed by farmers who manage and share resources such as land and water. These communities have faced challenges such as drought and increasing pressure from intensive agriculture, but they remain a fundamental element of the traditional agricultural system and the social fabric of the region.

Another fundamental activity in the Coquimbo Region has always been mining, and in Tulahuén specifically, the extraction of lapis lazuli stands out, being the only other place in the world besides Afghanistan where this semi-precious mineral is found.

The vernacular architecture of Tulahuén has been conceived to carry out agricultural activities. Therefore, in addition to the housing (Fig. 4-a), livestock pens, agricultural storage facilities (Fig. 4-b), and "pasteras" (Fig. 4-c)—large enclosures for storing animal fodder—are very common [6]. While this architecture shares characteristics with other traditional adobe constructions in the Coquimbo Region and other parts of Chile, it presents two specific features: it consists of large, tall enclosures for storing tools, food, and fodder related to agricultural activities. For this reason, it features long walls (without transverse bracing) and overly slender walls (with low thickness relative to their height). The second particularity is that, being a mountain village, the roofs have steep slopes for the evacuation of snow and rain, which simultaneously generate thrusts on the walls, subjecting them to stresses that could cause them to overturn (Fig. 5). These two conditions render this adobe architecture particularly vulnerable to the thrusts induced by earthquakes. Consequently, it features a series of wooden reinforcements that counteract these seismic thrusts and improve the connections between all the structural parts of the buildings, preventing damage. These reinforcements, constitute the earthquake-resistant devices that reflect the seismic culture of Tulahuén.

Figure 4. Different architectural typologies are present in the town of Tulahuén (credits: authors).

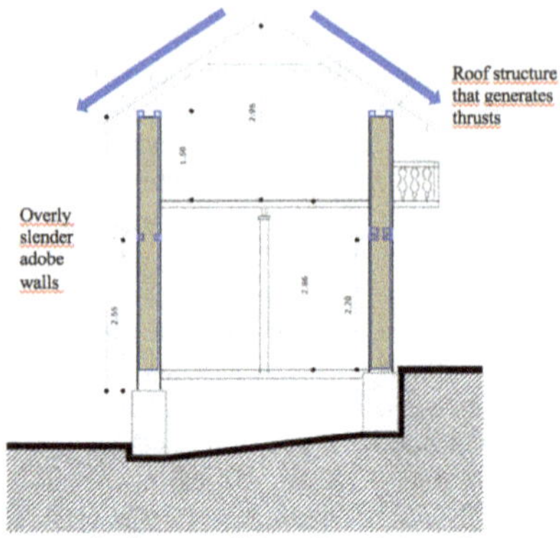

Figure 5. Typological section of a "pastera" (credits: Jaime Ortega y Natalia Jorquera, 2024).

Earthquake-resistant Devices in Tulahuén

In the historic area of Tulahuén, there are around one hundred adobe structures, all of which feature some type of earthquake-resistant device made from wood sourced from Algarrobo (*Neltuma chilensis*), Cordilleran Cypress (*Austrocedrus chilensis*), or White Poplar (*Populus alba*), the latter being a species introduced by the Spanish in this territory.

To characterize the various earthquake-resistant devices, their dimensions, sections (circular, square, or rectangular), geometry, and location within the buildings were recorded. Based on this data, three types of devices were identified [6]:

Device Corner Key.

These are located at the corners of the buildings and primarily aim to ensure the connection between the perpendicular adobe walls, which tend to separate during an earthquake. They consist of two longitudinal wooden pieces and a series of transverse pieces that together form a kind of ladder. They are inserted between the layers of adobe, and at the height of the building, they can appear one or more times (Fig. 6).

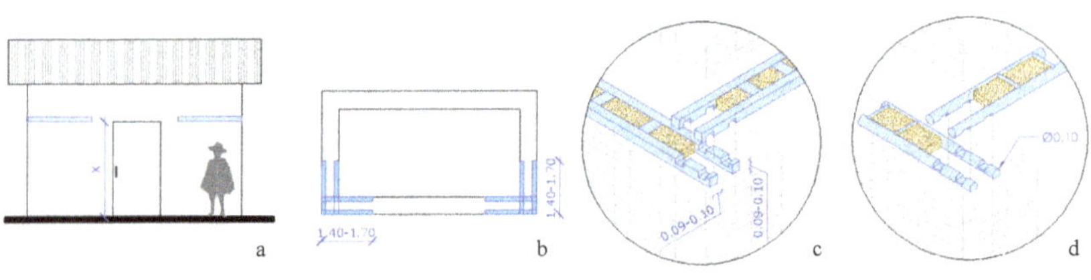

Figure 6. Drawings of the corner device (credits: Jaime Ortega, 2024).

Device Diagonal Corner.

This corresponds to a type of corner reinforcement that, unlike the previous devices, is installed diagonally relative to the adobe walls (Fig. 7) and can be observed from the interior of the buildings. Like the corner device, its purpose is to ensure the connection of the perpendicular adobe walls, preventing their separation and averting any outward tipping.

These diagonal corner devices are generally installed at the height of the lintel of the openings and consist of two essential parts: a wooden piece that crosses the wall diagonally and stakes that secure the brace within it.

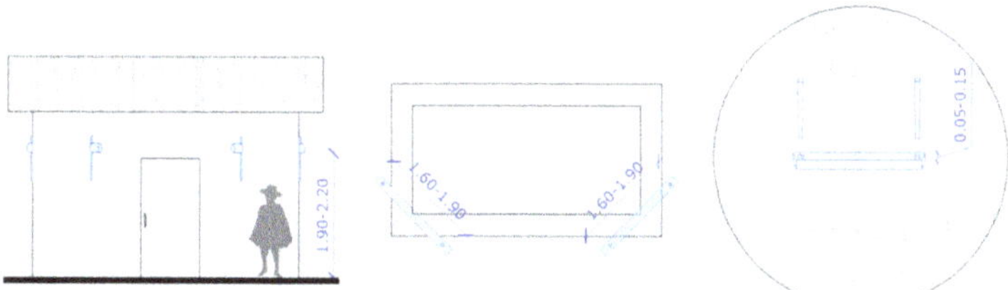

a b c

Figure 7. Drawings of the Diagonal Corner device (credits: Jaime Ortega, 2024).

Device Wall Tie.
A third device present in most traditional adobe buildings in Tulahuén is the wall tie (Fig. 8), referred to as "candados" (padlock) by the local inhabitants. The purpose of this device is to enhance the connection between the horizontal structural elements (roof and floor) and the adobe walls. The device consists of two parts: a beam and a stake perpendicular to it, which appears as a vertical element protruding from the facades of many buildings (Fig. 9), serving as a distinctive feature of Tulahuén's vernacular architecture.

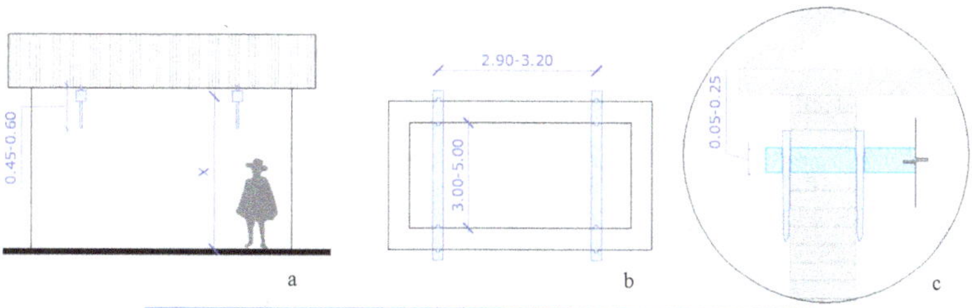

Figure 9. Typical facade with "candados" (credits: authors).

The three types of devices have proven to be very effective, both in improving the connections between the different construction parts and in preventing the tipping of the adobe walls. However, devices made from Poplar wood have encountered issues in recent years; being a softwood, it has become the preferred food for termites, leading to a decrease in its load-bearing section and, consequently, its effectiveness. This local pathology affecting the wood has left buildings in a vulnerable situation, and damages resulting from seismic action have begun to manifest, particularly the separation of the perpendicular adobe walls at their junction, threatening future tipping. This problem has not arisen in devices made from the local woods of Algarrobo and Cypress, which, being more durable, are not subject to attack by wood-destroying insects. This highlights how the use of endemic resources as construction materials is always the best response for a local environment, being one of the essential characteristics of vernacular architecture

Conclusions

Earthquakes, which in the past were faced with knowledge by local communities, have increasingly become more severe disasters due to the loss of traditional seismic cultures. This not only threatens the safety of the population but also the architectural heritage. Recovering this knowledge is essential for any seismic region in the world.

The study and understanding of local seismic cultures constitute an essential tool for generating guidelines that promote appropriate interventions in architectural heritage. These guidelines should rescue and refine the strategies that have allowed for the conservation of historical buildings in areas of high seismic activity, while avoiding incorrect solutions that compromise their durability and authenticity.

The seismic culture of Tulahuén is an outstanding example of how a community has developed a cultural and constructive response to the threat of earthquakes, in a context of extremely vulnerable architecture due to its geometric characteristics. Like other vernacular architectures, the seismic-resistant devices of Tulahuén emerged from a process of trial and error, where the damage caused by earthquakes allowed for the observation of adobe structures behavior and the development of preventive measures.

Despite the effectiveness of wooden reinforcements, constructions using poplar wood present problems due to termite attacks, which weaken the structure. This underscores the importance of using local materials adapted to the environment, ensuring the durability and resilience of vernacular buildings. Local species are more resistant to pests, contributing to the preservation of the structural integrity of seismic-resistant devices.

The researches underpinning this article emphasize the importance of preserving these devices as a fundamental part of local architectural heritage, considering them unique attributes that must be conserved and restored in the event of deterioration. Furthermore, the seismic-resistant devices of Tulahuén can serve as a source of inspiration for other vernacular architectures with similar characteristics.

Finally, one of the most valuable contributions of this research has been to help the inhabitants of Tulahuén recognize the heritage value of these constructive elements. Although residents understood the role of the devices in improving the seismic performance of their buildings, they had not fully realized their importance for heritage conservation or considered them in new constructions.

Acknowledgments

The project that gave rise to this article has been financed by the National Fund for Cultural Development and the Arts-FONDART, a national funding area, Convocatoria 2024, of the Chilean Ministry de las Culturas, Las Artes y el Patrimonio, folio n° 750219.

References

[1] L. Dipasquale, O. Sidik, S. Mecca. Earthquake resistant structures. In: M. Correia, L. Dipasquale, S. Mecca, editors. *VERSUS Heritage for tomorrow. Vernacular knowledge for sustainable architecture.* Florence: Firenze University Press; 2015. p. 232-239.

[2]. P. Pierotti, D. Ulivieri. *Culture sismiche locali.* Pisa: Edizioni Plus; 2001.

[3] N. Jorquera. *Culturas sísmicas: Estrategias vernaculares de sismorresistencia del patrimonio arquitectónico chileno.* Arquitecturas del Sur. 2014;32(46):6-17, http://revistas.ubiobio.cl/index.php/AS/article/view/739

[4] M. Correia, G. Duarte, D. Viana, T. Gomes. *Vernacular seismic culture in Portugal: On-going research.* In: C. Mileto, editor. *Vernacular Architecture: Towards a Sustainable Future.* London: CRC Press; 2014. p. 217-224.

[5] R. Rauld. *Cortical deformation and seismic hazard associated with the San Ramón fault in the mountainous front of Santiago, central Chile (33°S).* Santiago, Chile: PhD thesis in Geological Science, University of Chile; 2011.

[6] J. Ortega. Dispositivos sismorresistentes de la arquitectura en tierra de Tulahuén, región de Coquimbo, Chile. In Congress proceedings 22°SIACOT, La Serena, Chile, 2024.

[7] J. Fernández, C. Pastén, S. Ruiz, F. Leyton. Estudio de efectos de sitio en la Región de Coquimbo durante el terremoto de Illapel Mw 8.3 de 2015. Obras y Proyectos. 2017;21:20-28. https://doi.org/10.4067/S0718-28132017000100003

Vernacular Architecture facing seismic hazards - Al Haouz and Chichaoua Provinces

TLEMÇANI MEKAOUI Nezha[1,a *], KARIBI Khadija[1,b], BERDOUZ Karima[1,c] and MESSOUS Ouafa[1,d]

[1]National School of Architecture of Rabat, Morocco

[a]n.tlemcanimekaoui@enarabat.ac.ma, [b]k.karibi@enarabat.ac.ma, [c]k.berdouz@enarabat.ac.ma, [d]messous.ouafa@gmail.com

Keywords: Earthquake, Al Haouz Mountains, Chichaoua, Vernacular Architecture, Earth and Stone Construction, Reconstruction

Abstract. The architectural and historical authenticity of the Al Haouz and Chichaoua provinces is expressed through earth and stone architecture, showcasing a local expertise that has proven its worth over decades. While it may be considered uncomfortable from the perspective of modern architecture, this simple and "archaic" know-how, if it can be described as such, has demonstrated its effectiveness given the region's climatic conditions. This vernacular architecture, characterized by traditional construction techniques and the use of local materials for buildings, is now threatened by several factors. These structures have shown vulnerability to seismic shocks, as evidenced by the recent earthquake that affected the region. The issue of the historical sites in Al Haouz and Chichaoua represents a major challenge in terms of preserving local vernacular architecture, which demonstrated its vulnerability during the earthquake that struck the region on September 8, 2023. This vulnerability places researchers and professionals before a complex challenge, where the need to preserve heritage intersects with the integration of modern protection techniques and socio-economic valorization. The ability of historical buildings to withstand natural disasters while maintaining their architectural integrity is questioned. Nevertheless, the earthquake that shook the soil of the Al Haouz region also jolted the authorities, professionals, architects, and civil society. It brought to light the isolated parts of the Al Haouz mountain. This article is based on a firsthand experience, post-earthquake, lived on-site, revealing the drawbacks of a vernacular architecture that is deformed, vulnerable, and undergoing transformation.

Introduction

In this article, we will address vernacular architecture facing the challenge of seismic hazards through two temporalities: before and after the earthquake that struck Morocco in September 2023, with a particular focus on the post-earthquake phase: reconstruction.

Vernacular architecture refers to a form of architecture that is inherently tied to the local context, available materials, traditional skills, and the specific climatic conditions of a region.

Our focus will be on Al Haouz and Chichaoua, regions known for their architecture that blends with the mountains, merges with the landscape, and reflects a distinct vernacular identity.

This article is based on a firsthand experience, post-earthquake, lived on-site, questioning the shortcomings of a vernacular architecture that is deforming, vulnerable, and undergoing transformation.

It is grounded in fieldwork, qualitative interviews, observations, comparisons of reconstruction documents, and confrontation with the reality of intervention sites.

Materials Research Proceedings 47 (2025) 224-231 https://doi.org/10.21741/9781644903391-26

1. Definitions and Context

Many authors, such as Amos Rapoport (1969) [1], emphasize that the three main characteristics of vernacular architecture lie in its adaptation to the climate, its use of locally available materials in the immediate environment, and its traditional construction techniques.

In this regard, the definition of vernacular architecture that seems to align most closely with the context of the Moroccan mountains, particularly Al Haouz and Chichaoua, describes it as "a form of architecture inspired by popular traditions, which has developed and continues to develop unique characteristics in a specific region, often using local materials, traditional methods, and forms" [2].

In Morocco, the architectural and historical authenticity of the Al Haouz and Chichaoua mountains is expressed through earth and stone architecture, showcasing local expertise that has proven its worth over decades. Although considered uncomfortable from the perspective of modern architecture, these constructions—adapted to geographical, cultural, and social circumstances, and "archaic" if one might describe them as such—have demonstrated their effectiveness given the region's climatic conditions.

The challenge posed by these sites represents a major issue in terms of preserving a local vernacular architecture that has shown vulnerability during the 2023 earthquake. This vulnerability places researchers and professionals before a complex challenge, where heritage preservation, the integration of modern protection techniques, and socioeconomic valorization converge.

Thus, the ability of historical buildings to withstand natural disasters while maintaining their architectural integrity is questioned.

1.1 A mode of living in constant reformation

The mountains of Al Haouz and Chichaoua have indeed been inhabited in the form of stone and earthen structures, built on the mountains themselves, blending in with them and merging into their landscapes.

Nevertheless, over time, these regions experienced a transformation of their habitat. The historic homes in the heart of fortified villages were gradually abandoned or kept as simple sheepfolds or stables. New houses were built, closer to the road and schools. Some villages became depopulated, while others developed.

Vernacular architecture is therefore not static; it evolves with the needs of its users and the available resources. As Paul Oliver, a key figure in the study of vernacular architecture, aptly states, "this type of construction is a direct result of a society's response to its climatic, topographical, and social conditions" (Oliver, 2006) [3].

Materials Research Proceedings 47 (2025) 224-231 https://doi.org/10.21741/9781644903391-26

Fig. 1: Province of Chichaoua. Photo source: First author.

Fig. 2 : Commune of Adassil, Douar Tidraguine. Photo source: First author.

1.2 An architecture that withstands the test of time

The earthquake of September 8, 2023, measuring 6.8 on the Richter scale, struck the provinces of Al Haouz and Chichaoua, causing not only material destruction but also revealing structural and organizational shortcomings in the response of authorities, professionals, and civil society. This natural disaster acted as a revealing agent, exposing the vulnerability of vernacular buildings that, while culturally significant, are often not designed to withstand seismic events.

This mode of construction, which is adopted, renewed, or abandoned based on its endurance over time, did not withstand the test of the earthquake. Rethinking reconstruction became essential. As highlighted by Hassan Fathy—an Egyptian architect renowned for his work on vernacular architecture—in one of his most famous quotes, "Vernacular architecture is the pure expression of a community's culture, its relationship to the environment, local materials, and adaptation to climate. It contains valuable lessons for the future of sustainable construction." This emphasizes the importance of rehabilitating traditional building techniques to create sustainable housing that is harmonious with the environment and tailored to the needs of local communities.

2. Post-Earthquake Interventions: Resettlement and Reconstruction Grants

In response to the disaster, a special fund was created by the Moroccan state for reconstruction, called the "Special Fund 126 for the management of the effects of the earthquake that affected the Kingdom of Morocco" (CAS 126). Funded by the state's general budget, local authorities, public enterprises and institutions, the private sector, and international organizations, it is also open to any donor inside or outside of Morocco through an officially mediated bank account.

This account provides grants ranging from 80,000 MAD for the rehabilitation of affected but still standing structures, to 140,000 MAD for completely demolished buildings that need to be reconstructed.

The survivors of the earthquake have found themselves temporarily housed in tents while awaiting the reconstruction of their homes, either relocated away from their original sites or remaining close to their damaged or destroyed dwellings. This temporary situation has already lasted a year, as the reconstruction process now requires official building permits established by an architect—such permits were not previously required in these mountains before the earthquake.

Materials Research Proceedings 47 (2025) 224-231 — https://doi.org/10.21741/9781644903391-26

In addition to this procedure, the allocation of grants follows an administrative process that involves disbursements in stages: 20,000 MAD for the foundations, 20,000 MAD for the walls, and 20,000 MAD for the roof slab. Each payment is contingent upon the completion and acceptance of the preceding phase.

Fig. 4 : Commune Addassil, Douar Tighoula. Photo source: First author.

Fig. 3 : Commune of Adassil, Douar Tidraguine. Photo source: First author.

2.1 A desired vernacular reconstruction.

The on-site intervention is officially framed by a robust set of specifications developed by the state, respecting the local character. This document was delivered to the architects involved in the reconstruction by a judicial officer.

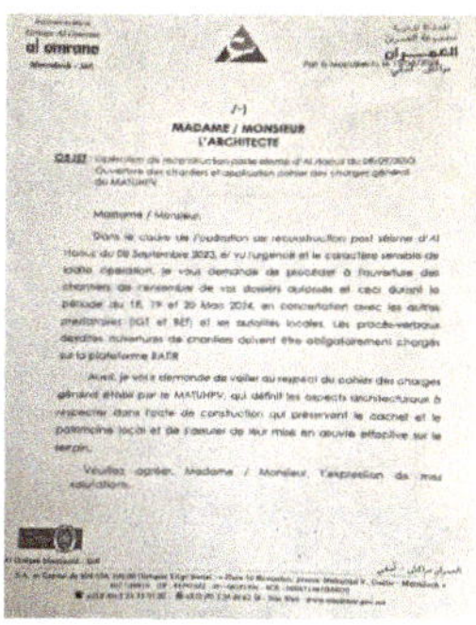

al omrane

MADAME / MONSIEUR
L'ARCHITECTE

Madame / Monsieur,

Fig. 5 : Letter addressed to the architects by the project manager Al Omrane.

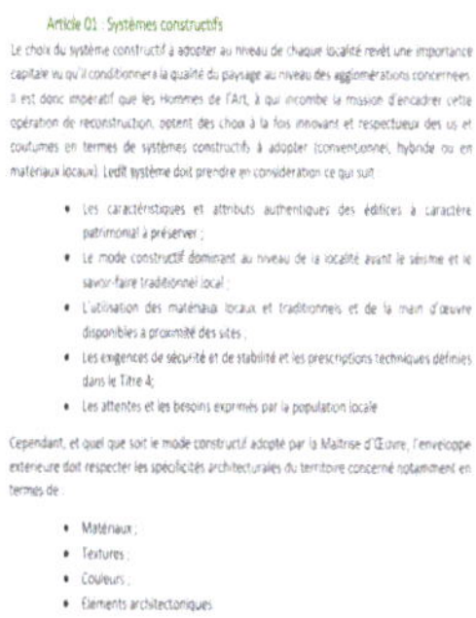

Article 01 : Systèmes constructifs

Le choix du système constructif à adopter au niveau de chaque localité revêt une importance capitale vu qu'il conditionnera la qualité du paysage au niveau des agglomérations concernées. Il est donc impératif que les Hommes de l'Art, à qui incombe la mission d'encadrer cette opération de reconstruction, optent des choix à la fois innovant et respectueux des us et coutumes en termes de systèmes constructifs à adopter (conventionnel, hybride ou en matériaux locaux). Ledit système doit prendre en considération ce qui suit :

- Les caractéristiques et attributs authentiques des édifices à caractère patrimonial à préserver ;
- Le mode constructif dominant au niveau de la localité avant le séisme et le savoir-faire traditionnel local ;
- L'utilisation des matériaux locaux et traditionnels et de la main d'œuvre disponibles à proximité des sites ;
- Les exigences de sécurité et de stabilité et les prescriptions techniques définies dans le Titre 4;
- Les attentes et les besoins exprimés par la population locale

Cependant, et quel que soit le mode constructif adopté par la Maîtrise d'Œuvre, l'enveloppe extérieure doit respecter les spécificités architecturales du territoire concerné notamment en termes de :

- Matériaux ;
- Textures ;
- Couleurs ;
- Éléments architectoniques

Fig. 7 : Excerpt from the specifications provided to the architects.

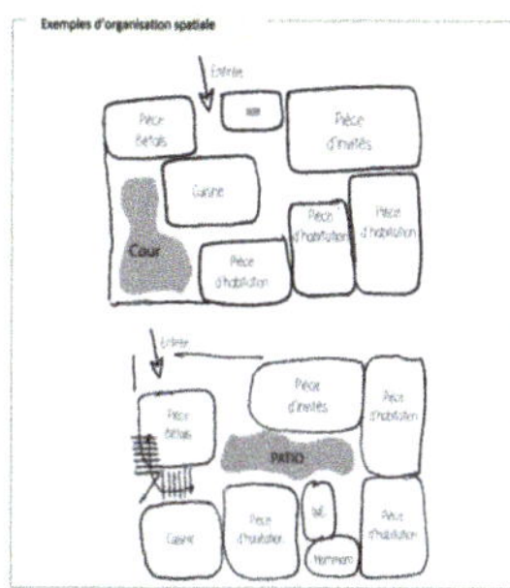

Fig. 6 : Example of plans to be used for the reconstruction, excerpted from the specifications.

Its objectives are to ensure the production of resilient, compliant constructions that adhere to seismic safety standards and provide a quality living environment that meets the expectations of the target population in alignment with their customs and traditions.

Materials Research Proceedings 47 (2025) 224-231 https://doi.org/10.21741/9781644903391-26

It recommends architectural and technical guidelines that respect local and traditional building practices while preserving architectural and landscape heritage. The plans drawn up in the specifications respect local customs, accommodating both livestock and residents, and central instructions are provided to ensure these guidelines are followed.

2.2 Technological solutions for sustainable recovery

The fieldwork explored the post-crisis reconstruction efforts in the region, the challenges and issues encountered, and at times, the innovative solutions implemented to ensure sustainable recovery. It also highlighted a discrepancy between what was planned and what was actually realized on-site.

In its Article 7, the set of specifications recommends the use of local materials with thermal properties that enhance insulation and provide optimal comfort, drawing inspiration from ancestral methods mastered by the local population, which are known for their effectiveness in thermal comfort.

On the other hand, it also recommends integrating renewable energy solutions, particularly the use of photovoltaic panels and solar panels. This technology, adopted on-site, creates a new urban landscape that stands out amidst the mountains.

Fig. 8: The use of photovoltaic panels, Commune of Assif El Mal. Photo source: First author.

2.3 A set of specifications not respected.

Although the specifications are based on an approach that integrates seismic reinforcement and the preservation of the local architectural heritage, they also stipulate the use of local materials or, at the very least, respect for local character. However, the authorized and constructed buildings on-site tell a different story.

The approved plans found at the reconstruction site show single-story houses with three bedrooms, a living room, and a kitchen. This reflects an urban image, excluding livestock from the equation. It raises the question of whether, in addition to the architectural transformation, a societal transformation is also on the horizon.

Fig. 10 : The reconstruction of a house with a post-and-beam structure and concrete blocks. Photo source: First author.

Fig. 9 : Photo of authorized plans found on site. Photo source: First author.

Materials Research Proceedings 47 (2025) 224-231 https://doi.org/10.21741/9781644903391-26

From Slowly Transforming Mountains to Hybrid Architecture

Indeed, the shaken mountains were already transforming the earthquake, especially concerning the buildings lining the roads. However, this mutation, characterized by a very slow pace conditioned by the scarcity of resources and difficulties of accessibility, has been accelerated by post-earthquake interventions and grants. These same mountains are now home to hybrid architecture.

The mountainous landscape is changing along the access roads, with stone and earthen constructions adjacent to those with post-and-beam structures made of reinforced concrete.

Fig. 11: Commune of Assif El Mal, Anabdour. Photo source: First author.

3.1 Differentiated interventions and a divided population.

In an attempt to respect local character, some architects working on-site are trying to use local materials, but the approval of their plans takes longer. Their efforts also include training for local artisans to promote sustainable and resilient reconstruction.

Some even contribute from their own pockets to purchase compressors and equipment necessary for compacting the earth; when it breaks down, the construction halts.

Qualitative interviews with residents have revealed a desire to find harmony between tradition and modernity, as well as an awareness of the need to enhance the safety of their homes. Most are opting for post-and-beam structures made of reinforced concrete with partition walls of concrete blocks. However, this choice does not protect them from the cold of winter or the intense heat of summer, putting it at odds with the established specifications for reconstruction.

Fig. 12 : Assif Lmal, Emi-n Ouassif, Chichaoua. Photo source : First author.

3.2 A guided reconstruction.

Attempts at repair using modern materials have been noted, creating a contrast with traditional techniques. This hybridization underscores the challenges of preserving cultural identity while integrating more resilient solutions. However, the use of concrete blocks predominates the landscape of the post-earthquake mountains. This mode of reconstruction, encouraged by the Sheikhs and Moqqadems, seems to be successful according to interviews conducted with the affected population.

Fig. 14 : A reconstructed house in the commune of Assif Lmal. Photo source: First author.

Fig. 13 : Landscape entirely made of concrete, commune of Assif Lmal. Photo source: First author.

One reality that can no longer be ignored is that change is on the horizon.

Conclusion

The vernacular architecture of the provinces of Al Haouz and Chichaoua reflects a long history of adaptation to the region's climatic and geographical conditions. However, the initial official intention, in response to increasing seismic risks, tends to modernize these constructions to ensure their durability.

Materials Research Proceedings 47 (2025) 224-231 https://doi.org/10.21741/9781644903391-26

By combining traditional techniques with modern solutions, it is possible to preserve this unique heritage while protecting local communities from natural hazards. However, the realization of this intention has manifested quite differently in the affected sites.

In this article, the examined temporalities have demonstrated a shift in the epistemology of local architecture. The pre-earthquake identity of the shaken mountains was characterized by stone and earthen architecture. However, the post-earthquake context has revealed a collective awareness, both local and national, that is double-edged, establishing a new mode not only of construction but also of habitation.

Thus, in the future, it becomes essential to consider the perspectives of an evolving identity in these Moroccan mountains.

References

[1] A. Rapoport, House, Form, and Culture, Prentice-Hall, Englewood Cliffs, NJ, 1969.

[2] P. Oliver, L'architecture vernaculaire: une définition difficile à cerner, in: Encyclopedia of Vernacular Architecture of the World, Vol. 1, Cambridge University Press, Cambridge, 1997, pp. 17-24.

[3] P. Oliver, Built to Meet Needs: Cultural Issues in Vernacular Architecture, Architectural Press, Oxford, 2006. https://doi.org/10.4324/9780080476308

[4] H. Fathy, Natural Energy and Vernacular Architecture: Principles and Examples with Reference to Hot Arid Climates, in: University of Chicago Press, Chicago, 1986.

[5] Décret n°2.23.811 du 25 Safar 1445 (11 septembre 2023) portant création d'un compte spécial intitulé « Fonds spécial pour la gestion des effets du tremblement de terre ayant touché le Royaume du Maroc ».

[6] Ministère de l'Aménagement du Territoire, de l'Urbanisme, de l'Habitat et de la Politique de la ville. (2023). Cahier des charges de l'opération de reconstruction et de réhabilitation dans la préfecture et les cinq provinces touchées par le séisme du 8 septembre 2023. Ministère de l'Aménagement du Territoire, de l'Urbanisme, de l'Habitat et de la Politique de la ville, 19 octobre 2023.

Materials Research Proceedings 47 (2025) 232-241 | https://doi.org/10.21741/9781644903391-27

Vernacular architecture and seismic resilience: Which techniques for which territories?

Nada EL MOADEN

National School of Architecture, Rabat, Morocco

nada.elmoaden@e.enarabat.ac.ma

Keywords: Vernacular Architecture, Seismic Resilience, Seismic Hazard, Traditional Building Techniques, Territory

Abstract. In the context of climate change and the recurrence of natural disasters, vernacular architecture finds itself at the heart of a debate opposing those who find it resilient in facing nature's hazards, and those who question its ability to cope with these new challenges. By focusing on the seismic hazard, it seems even more antagonistic to attribute earthquake resistance to vernacular architecture. However, the earthquakes that have shaken our globe in recent decades have demonstrated that, in many areas, buildings constructed using traditional construction techniques have been more resistant to seismic shocks than those built using modern processes. How can vernacular architectural techniques contribute to a territory's resilience in facing seismic hazards? Beyond the multidisciplinary nature of the resilience concept, our aim in this article is to present and analyze the technical aspect of the resilience of traditional buildings to earthquakes, through a bibliographical and comparative analysis of different traditional processes used in several territories worldwide and which have demonstrated their ability to cope with earthquakes. Despite the divergence between the ancestral techniques used, there is a notable convergence in terms of their ductile nature, enabling better dissipation of the earthquake energy generated in buildings, without causing serious cracks or consequent collapse. It's also worth mentioning that the diversity of forms and techniques used in these vernacular architectures is not the result of a hasty choice, but rather of a careful consideration of the needs of people who live there, while adapting to the various natural hazards, including earthquakes.

Introduction

The technological development that revolutionized the world in the post-industrial era has not only left its physical mark on our territories through implementing buildings that are sophisticated in terms of form and function but has also contributed to rooting out subliminal ideas of apprehension and doubt when it comes to the quality, sustainability, and safety of traditional constructions [1].

While efforts to preserve vernacular architecture as a cultural heritage reflecting the identity of each territory are undeniable, when it comes to the seismic performance of ancestral building techniques, the opinions of construction professionals often intersect to qualify these techniques as non-resilient in facing earthquakes [2].

On the contrary, several researchers have demonstrated that these traditional construction techniques are earthquake resistant, as Randolph Langenbach argued in his 2010 conference paper *"Earthquake Resistant Traditional Construction is Not an Oxymoron"*, where he defended the seismic performance of traditional construction techniques in the Himalayan region, known for its high seismic activity, through examples of vernacular buildings that were able to withstand earthquakes, unlike other buildings constructed using conventional processes [3].

In addition, we would like to point out that the starting point for this research was the 6.8 magnitude earthquake that struck the Moroccan High Atlas region on September 8, 2023 [4], an area previously considered to have negligible or moderate seismicity [5]. In the aftermath of this earthquake, several villages built using vernacular construction techniques were able to withstand this major seismic shock, whereas buildings constructed using concrete collapsed [6]. Considering the unprecedented nature of this event, scientific research into the seismic resilience of vernacular buildings in the High Atlas is lacking. Thus, we have chosen first to examine the leading seismic-resistant vernacular building techniques adapted to different territories worldwide, to be better prepared to explore seismic-resistant traditional building techniques in the Moroccan High Atlas territory.

To this end, this paper aims to identify earthquake-resilient vernacular building techniques in specific territories. Specifically, it seeks to explore how vernacular architecture techniques can enhance a territory's resilience to seismic hazards.

The theoretical framework of this research focuses initially on the analysis of theories related to the concepts of vernacular architecture and seismic resilience. It examines the different theoretical approaches used by researchers in these fields, trying to find the link we can make between these two concepts. Secondly, an analysis of several vernacular building techniques in relation to earthquake resilience is formulated in line with theories linked to vernacular architecture and seismic resilience concepts.

The methodology adopted consists in conducting an in-depth bibliographical analysis to gather historical and contemporary data on construction techniques, materials used, wall structures and the seismic characteristics of these techniques in several territories on an international scale. Subsequently, a comparative analysis of similarities between territories using the same techniques is carried out to identify the territorial determinants impacting the choice of one technique over others, the limitations of these techniques and the possibility of integrating innovation to strengthen the seismic resilience of these vernacular techniques.

Basic concepts

Before discussing the results of this research, it is necessary to clarify the basic concepts related to our subject, namely *"vernacular architecture"* and *"seismic resilience"*.

Vernacular architecture:

The first time the concept of vernacular architecture was addressed to the general public was on the occasion of Bernard Rudofsky's 1964 exhibition *"Architecture without architects: a short introduction to non-pedigreed architecture"* [7]. The concept was subsequently used by several authors. Thus, through analysis of their definitions of this concept, we can highlight the important characteristics of vernacular architecture: It is about constructions made by non-experts and adapted to the local context [7, 8, 9, 10] through the use of local materials adapted to specific climatic and environmental conditions, and allowing for evolution [8, 9, 12, 13, 14] according to the social and economic changes of the local population. These constructions are characterized by their simplicity and functionality [7, 8, 10, 12, 16], providing practical solutions for local residents and expressing their cultural identity [8, 9, 12, 13, 14, 15]. Some authors have focused on the aesthetic aspect of these constructions, describing them as *"Ugly and ordinary"* [16]. While P. Paumelle and A. Picon have emphasized the sustainability of these constructions, insofar as they incorporate sustainable construction techniques [10, 11].

Seismic resilience:

The term resilience was first used by C. Holling in 1973, who defined *ecological resilience* as *the ability of an ecological system to absorb disturbance and reorganize itself while changing, to maintain essentially the same functions, structures, identities, and feedbacks* [17]. He made a distinction between ecological resilience and engineering resilience, which is limited to a system's ability to return to a state of equilibrium after a disturbance. Also, A. Dauphiné & D. Provitolo argue that resilience involves absorbing disturbances and reorganizing while maintaining essential functions, which is similar to Holling's approach except that his definition is applied to socio-ecological systems [18]. On the other hand, B. Quenault's definition of *urban resilience* focuses on anticipation, reorganization and recovery from disturbance, emphasizing a proactive approach [19].

In the case of *seismic resilience*, this refers to the ability of social, economic and environmental systems to resist, absorb, reorganize and recover effectively after an earthquake, while maintaining their essential functions. It also involves preventive measures, such as infrastructure reinforcement, population preparedness and emergency planning [20].

By cross-referencing the characteristics of vernacular architecture with those of seismic resilience, we can define *earthquake-resilient vernacular architecture* as a form of construction that uses *local materials* and *traditional techniques adapted to the specific climatic and environmental conditions* of a region. It expresses *the culture and identity* of local communities, and is often built by *local craftsmen* or the inhabitants themselves. *Functional and simple*, it *evolves and adapts to social, economic and environmental changes*. In addition to these characteristics, it incorporates seismic resilience principles, such as *absorption and recovery capacity*, *adaptation and transformation*, and *infrastructure preparation and reinforcement* to ensure the safety and sustainability of buildings.

Main results

In the following, we will present the main earthquake-resilient vernacular construction techniques analyzed in several territories around the world, highlighting the key elements contributing to the resistance of these techniques to earthquakes.

Table 1: Earthquake-resistant vernacular architecture: leading techniques in Asia

Name of technique	Area concerned	Wall structure and materials	Earthquake resistance
Taq	Jammu and Kashmir region, India.	Taq walls consist of masonry panels (bricks, rubble stones or sun-dried bricks, bonded with mud mortar) between load-bearing wooden pillars [20].	The flexibility between masonry panels and pillars enables the structure to adapt to ground movements without suffering major damage. Also, the timber connections at slab and lintel levels allow the structure to move with the seismic waves (reducing stresses on the walls). The configuration of interior partitions also contributes to the dissipation of seismic energy [20].
Dhajji Dewari	Mountainous regions of Kashmir and Pakistan.	The walls consist of a wooden frame (pine, kail, deodar) with vertical and horizontal posts, filled with masonry (stone or brick) [21].	This method is highly resilient against lateral forces, thanks to the energy dissipation by the masonry infill [21]. Major earthquakes attesting seismic performance of this technique: *Srinagar in 1885, Kashmir in 2005.*
Himis	Black Sea regions, Thrace and other parts of Turkey.	The walls consist of a wooden frame with smaller wooden subdivisions, filled with masonry (brick, adobe or stone [22].	Subdivided wooden frames allow efficient dissipation of seismic energy, reducing structural damage [22]. Major earthquakes attesting seismic performance of this technique: *İzmit in1999 and Van in 2011.*

Materials Research Proceedings 47 (2025) 232-241 https://doi.org/10.21741/9781644903391-27

Bagdadi	Turkey, Lebanon, Syria and parts of Iran and Iraq.	The walls consist of a wooden frame with vertical and horizontal uprights. The spaces between the frames are filled with masonry of brick, stone or cob (a mixture of clay, straw and sand) [2].	Wood can bend without breaking, thus absorbing part of the seismic energy. The wooden frame ensures uniform distribution of seismic forces. Materials such as cob can crack without causing the wall to collapse, thus dissipating energy [2].

Table 2: Earthquake-resistant vernacular architecture: leading techniques in Europe

Name of technique	Area concerned	Wall structure and materials	Earthquake resistance
Colombage (Half-timbered)	France (Alsace, Normandy and Sologne) and other parts of Europe such as Germany and the Netherlands	Half-timbered walls are made up of two main elements: a wooden framework formed by vertical and horizontal beams to form a lattice structure, and a filling (hourdage) of earth, cob, brick, or stone [23].	The systemic interlocking of the beams allows the structure to follow seismic movements, thus absorbing part of the earthquake's energy [23]. Major earthquake attesting seismic performance of this technique: *Basel 1356.*

Table 3: Earthquake-resistant vernacular architecture: leading techniques in Central and South America

Name of technique	Area concerned	Wall structure and materials	Earthquake resistance
Quincha	Coastal regions of Peru.	Quincha walls are made of a wooden frame with cane filling, covered with mud or plaster [24].	This method offers great flexibility and seismic shock absorption capacity, thanks to the flexibility of the wood and the lightness of the filling materials. Quincha walls can deform without collapsing, reducing the risk of serious damage during earthquakes [24]. Major earthquake attesting seismic performance of this technique: *Lima in 1746.*
Bahareque	Costa Rica, Colombia, Nepal, Ecuador, El Salvador and the *Philippines* (the Philippines is	The walls consist of a wooden or bamboo framework, to which an organic matrix of cane or bamboo is attached. This matrix is then covered with mud, clay or plaster, sometimes reinforced with steel	This method creates a light, flexible structure capable of absorbing seismic shocks. Bahareque walls are highly elastic, allowing them to deform without cracking or collapsing, offering remarkable resistance to earthquakes [25].

	in Asia, but is located on the Pacific Ocean like the other above-mentioned countries).	mesh for added strength [25].	Major earthquake attesting seismic performance of this technique: *Manizales, Colombia, 1875.*
Taquezal	Guatemala, El Salvador, Honduras and Nicaragua.	Taquezal walls consist of a wooden or bamboo frame, to which an organic matrix of cane or bamboo is attached. This matrix is then covered with mud, clay, cob, or plaster [26].	Taquezal benefits from the flexibility of wood and the lightness of filling materials, enabling the structures to withstand seismic shocks by absorbing and dissipating earthquake energy [26]. Major earthquake attesting seismic performance of this technique: *Salvador in 2001.*

Discussion of results

Factors influencing the choice of construction techniques used in territories:

According to the world map of seismic risks *(fig 1)*, we note that the territories where seismic activity is preponderant correspond mostly with territories using the vernacular construction techniques we have just analyzed.

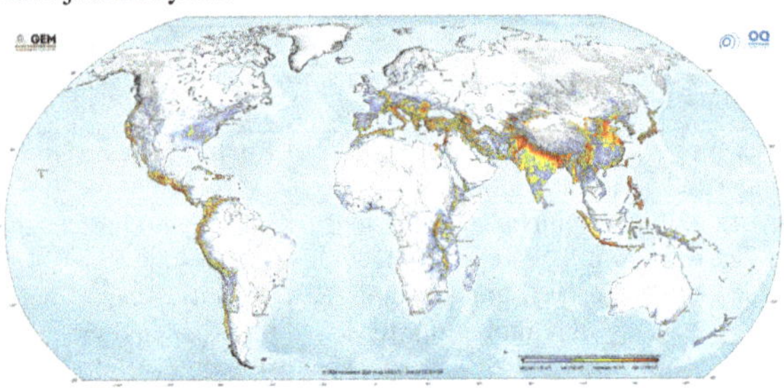

Fig 1: World map of seismic risks 2023
(www.globalquakemodel.org)

Materials Research Proceedings 47 (2025) 232-241 https://doi.org/10.21741/9781644903391-27

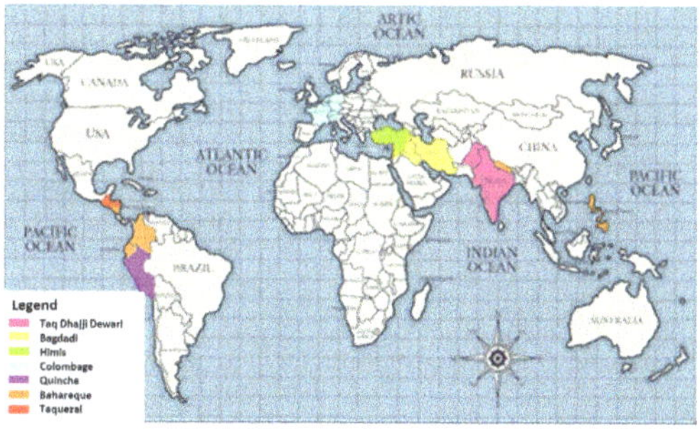

Fig 2: Geographical location of areas using the techniques studied (Author)

We can also see that the territories that use the same vernacular construction techniques are mostly adjacent to each other *(fig 2)*. In fact, even in the case of Bahareque, which is used in Central America and the Philippines in Asia, there are geographical similarities between these two territories, as they border on the Pacific Ocean.

Beyond geographical location, we have identified several factors *(fig 3)* that influence the choice of a specific traditional building technique in a given territory, namely:

Geological conditions: This is a factor that links all the techniques chosen, as the territories where they are used are known for their seismicity (notably Taq and Dhajji Dewari in Kashmir, which is a region of significant seismicity).

Climatic conditions: In arid climates, the Bagdadi technique is preferred. For tropical climates, there is a preference for the Bahareque technique, and for varied climates, we note preponderance for Colombage.

Availability of local materials: This factor plays an important role in the choice of a construction technique depending on the target territory. For example, Colombage is used in Europe, where wood is abundant, while Quincha and Bahareque are common in Latin America, where bamboo and clay are readily available.

Local knowledge and traditions: In each territory, local artisans typically possess specialized expertise in ancestral techniques deeply rooted in their traditions.

Economy and accessibility: We can give the example of Himis, which is used in regions where economic resources are limited, as it offers sustainable structures at lower cost. Quincha and Bahareque also use inexpensive local materials, making construction accessible to a wide population.

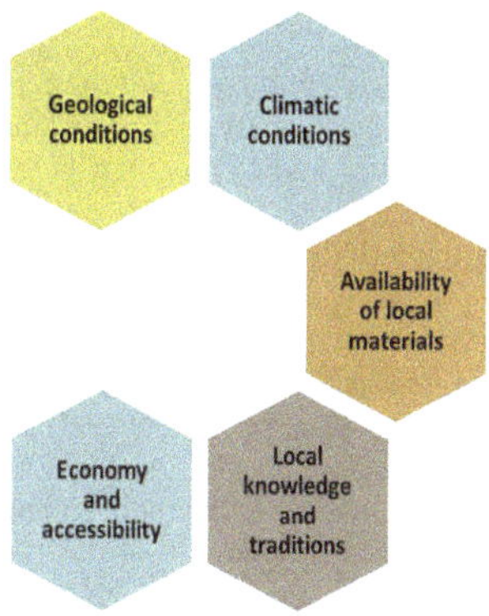

Fig 3: Factors influencing the choice of using construction techniques in the territories (Author)

Similarities between vernacular construction techniques:

We then analyzed the similarities between these vernacular construction techniques, highlighting the materials and similarities that make them resistant to seismic hazards. We were able to pair up the techniques with the most common points (fig 4).

Materials Research Proceedings 47 (2025) 232-241 https://doi.org/10.21741/9781644903391-27

Fig 4: Main similarities between vernacular construction techniques (Author)

Taq and Dhajji Dewari: Both techniques use wooden frames to reinforce masonry walls. Taq combines masonry walls with wooden frames, while Dhajji Dewari uses masonry panels filled with mud or bricks, also reinforced with wooden frames [20, 21].
Himis and Bagdadi: These two techniques use subdivided wooden frames with stone, brick, mud or plaster infills. Himis uses stone or brick walls with a wooden frame for subdivision, while Bagdadi uses mud or plaster walls with a wooden frame [2, 22].
Half-timbered (colombage) and Quincha: These two techniques use wooden frames with infills of natural materials such as cob, brick, stone, mud and straw [23, 24].
Bahareque and Taquezal: These techniques use wooden or bamboo frames with mud or masonry infills. Bahareque uses wooden or bamboo frames with mud infills, while Taquezal combines wooden frames with mud or masonry infills [25, 26].

Seismic characteristics of vernacular building techniques:

Following analysis of the results, we have highlighted the main characteristics of the vernacular techniques studied that contribute to seismic resilience. These are flexibility and ductility, lightness of materials and modular construction, and each technique responds to these characteristics in a differential way (see Figure 4).

With regard to *flexibility and ductility*, wooden frames integrated into masonry walls (Taq and Dhajji Dewari) or stone walls (Himis) offer shock absorption, which is crucial for earthquake resistance [20, 21, 22]. Wooden frames reinforcing walls in cob or brick (Colombage) [23] or with earth or mud filling (Taquezal) also contribute to flexibility and earthquake resistance [26].
Lightness of materials is above all a specific feature of Quincha and Bahareque techniques, insofar as the use of cane, bamboo, and mud creates light yet flexible structures that absorb seismic shocks well [24], [25].
Regarding the *modularity of constructions*, is a feature present in all the techniques on display apart from Quincha, Bahareque, and Taquezal.
The more this feature is present in a technique, the greater its seismic performance.

Construction Technique	Flexibility and Ductility	Lightness of Materials	Modular Construction
Dhajji Dewari	Yes	Average	Yes
Taq	Yes	Average	Yes
Himis	Yes	Average	Yes
Baghdadi	Yes	Average	Yes
Half-timbering	Yes	Average	Yes
Quincha	Yes	Yes	No
Bahareque	Yes	Yes	No
Taquezal	Yes	Average	No

Fig 5: Seismic characteristics of vernacular construction techniques (Author)

Materials Research Proceedings 47 (2025) 232-241 https://doi.org/10.21741/9781644903391-27

Advantages and limitations of using vernacular building techniques in facing seismic hazard:

Indeed, these ancestral techniques offer several *advantages*, including the preservation of cultural identity, the sharing of knowledge, and the strengthening of social ties, as well as greater acceptance and involvement of communities in construction projects, since these techniques are an integral part of the culture and know-how of the local population.

Conversely, these techniques face *limitations* due to their low resistance to significant seismic shocks and modern requirements for building comfort and functionality.

In this respect, it is crucial to strike a balance between the advantages of traditional techniques and today's seismic construction requirements. By combining ancestral knowledge with technological advances, it is possible to create earthquake-resistant structures that meet modern safety standards. This hybrid approach makes it possible to benefit from the advantages of these traditional techniques, while ensuring both sustainability and safety of buildings, through the use of *innovative materials* (e.g. fiber-reinforced polymers or addition of reinforcing metal alloys), the use of *advanced structural systems* (e.g. : seismic isolators, energy dissipation devices, and advanced bracing techniques) and *the integration of technology* to minimize earthquake damage (examples: the use of seismic sensors and the installation of early warning systems capable of instantly informing inhabitants of at-risk areas).

Conclusion and perspectives:

This research emphasizes the significance of incorporating vernacular architecture within the framework of seismic resilience. Our analysis suggests that traditional building techniques present promising potential in this regard. Nevertheless, it is essential to investigate innovations and modern technologies to enhance the resilience of vernacular architecture further.

A comparative analysis of seismic-resistant vernacular construction techniques across different territories has highlighted the impact of different geological, climatic, economic and cultural contexts on seismic resilience, underlining the need to adapt techniques to the specificities of each territory. We can highlight that, despite the varying ancestral techniques used, they all converge on key aspects like flexibility and ductility, which are crucial characteristics that grant them significant seismic performance.

It's important to remember that this research aims to investigate the foremost earthquake-resistant vernacular construction techniques from various parts of the world, to better understand and apply these methods in the Moroccan High Atlas region. This region, is known for its remarkable vernacular architecture, featuring structures built with local materials like stone masonry, adobe, and pisé. Future research should delve into the traditional building techniques unique to this area and assess their performance during the earthquake of September 8, 2023.

Fig 6: Abendour village in Chichaoua with its buildings that withstood the earthquake of September 8, 2023 (Author)

References

[1] R. Langenbach, Don't Tear It Down! Preserving the Earthquake Resistant Vernacular Architecture of Kashmir, UNESCO, New Delhi, 2009.

[2] M. Doğan, Seismic analysis of traditional buildings: Bagdadi and Himis, Anadolu, Univ J. Sci. Technol A Appl. Sci. Eng 11(1) (2010) 35-45.

[3] R. Langenbach, Earthquake Resistant Traditional Construction is Not an Oxymoron, in: International Conference on Disaster Management and Cultural Heritage, The Royal Government of Bhutan (2010).

[4] Information on https://www.universalis.fr/encyclopedie/seisme-au-maroc-2023

[5] Ministry of Housing and Urban Policy, Règlement de Construction Parasismique 2011. Kingdom of Morocco, 2011.

[6] A. Koutous, Diagnostic post-sismique des constructions des régions impactées par le séisme du Haut-Atlas : Focus sur les constructions en matériaux locaux, Technical report, Ministère de l'Habitat et de la Politique de la Ville, 2023.

[7] B. Rudofsky, Architecture Without Architects: A Short Introduction to Non-Pedigreed Architecture, Museum of Modern Art, New York, 1964.

[8] P. Oliver, Built to Meet Needs: Cultural Issues in Vernacular Architecture, Routledge, 2006. https://doi.org/10.4324/9780080476308

[9] R. Brown, D. Maudlin, Concepts of Vernacular Architecture, In C. G. Crysler, S. Cairns, & H. Heynen (Eds.), The SAGE Handbook of Architectural Theory, SAGE Publications Ltd., 2012. https://doi.org/10.4135/9781446201756.n21

[10] S. Moriset, N. Sánchez Muñoz, E. Sevillano Gutiérrez, Versus: lessons from vernacular heritage to sustainable architecture, CRAterre-ENSAG, 2014.

[11] A. Rapoport, House Form and Culture, Prentice-Hall, 1969.

[12] M. De Certeau, The Practice of Everyday Life, University of California Press, 1984.

[13] H. Lefebvre, The Production of Space, Blackwell, 1991.

[14] U. Eco, Travels in Hyperreality, Harcourt Brace Jovanovich, 1986.

[15] R. Venturi, Learning from Las Vegas: The Forgotten Symbolism of Architectural Form, MIT Press, 1977.

[16] C.S. Holling, Resilience and Stability of Ecological Systems, Annu Rev Ecol Syst. 4(1) (1973) 1-23. https://doi.org/10.1146/annurev.es.04.110173.000245

[17] A. Dauphiné, D. Provitolo, Risques et catastrophes : Observer, spatialiser, comprendre, gérer, Armand Colin (2e éd.), 2013.

[18] B. Quenault, La résurgence/convergence du triptyque « catastrophe-résilience-adaptation » pour (re)penser la « fabrique urbaine » face aux risques climatiques, Développement Durable & Territoires, 5(3) (2014). https://doi.org/10.4000/developpementdurable.10683

[19] Information on https://www.futura-sciences.com/planete/definitions/terre-risque-sismique-20683

[20] M. Dar, S. Ahmad, Traditional earthquake resistant systems of Kashmir, International Journal of Civil and Structural Engineering Research, 2(2) (2015) 86-92.

[21] Q. Ali, T. Schacher, M. Ashraf, B. Alam, A. Naeem, N. Ahmad, M. Umar, In-Plane Behavior of the Dhajji-Dewari Structural System: Wooden Braced Frame with Masonry Infill, Earthquake Spectra 28(3) (2012) 835-858. https://doi.org/10.1193/1.4000051

[22] G. Koca, Seismic resistance of traditional wooden buildings in Turkey, in: VI Convegno Internazionale ReUSO, Messina, Italy (2018) 1-10.

https://doi.org/10.47818/DRArch.2023.v4si114

[23] C. Boudaud, Analyse de la vulnérabilité sismique des structures à ossature en bois: essais expérimentaux, modélisation numérique, calculs parasismiques, Thèse de doctorat, Université de Grenoble, 2012.

[24] H. Rodríguez-Camilloni, Quincha architecture: The development of an antiseismic structural system in seventeenth century Lima, in: S. Huerta Fernández (Ed.), Proceedings of the First International Congress on Construction History, Vol. 3, Madrid, Spain, 2003, pp. 1741-1752.

[25] M. López, J. Bommer, P. Méndez, The Seismic Performance of Bahareque Dwellings in El Salvador, in: 13th World Conference on Earthquake Engineering, Vancouver, B.C., Canada, 2004.

[26] L. Holliday, T.H.-K. Kang, K.D. Mish, Taquezal Buildings in Nicaragua and Their Earthquake Performance, Journal of Performance of Constructed Facilities, ASCE 26(5) (2012) 1-10. https://doi.org/10.1061/(ASCE)CF.1943-5509.0000266

Seismic vulnerability of vernacular architectural heritage in the Al Haouz region in Morocco

Khalid EL HARROUNI[1,a *], Hassane KHARMICH[1,b], Khadija KARIBI[1,c], Khadija BABA[2,d] and Satwant RIHAL[3,e]

[1]Ecole Nationale d'Architecture de Rabat, BP 6372, Rabat Instituts, Rabat, Morocco

[2]Ecole Supérieure de Technologie de Salé, Université Mohammed V, Morocco

[3]College of Architecture & Environmental Design, Cal Poly State University, San Luis Obispo, California 93407, U.S.A

[a]k.elharrouni@enarabat.ac.ma, [b]h.kharmich@enarabat.ac.ma, [c]k.karibi@enarabat.ac.ma, [d]khadija.baba@est.um5.ac.ma, [e]srihal@calpoly.edu

Keywords: Vernacular Architectural Heritage, Seismic Vulnerability, Al Haouz Earthquake, Regulations, Traditional Materials

Abstract. Vernacular architectural heritage in the Al Haouz Region of Morocco faces significant seismic vulnerability. This poses a serious threat to preserving these valuable structures made of local materials including earth, wood, and stone. It is therefore crucial to analyze and assess the seismic vulnerability of these buildings to develop effective strategies for their protection and conservation. The seismic vulnerability of vernacular architectural heritage refers to the susceptibility of these structures to damage or collapse during seismic events. Vernacular architecture, characterized by its use of local materials and traditional building techniques, is particularly vulnerable due to its lack of compliance with modern seismic design codes and standards. This paper explores the seismic vulnerability of these vernacular buildings made of local materials in heritage sites and thus proposes adaptation strategies to improve their seismic resilience. The assessment approach is based on a qualitative method including visual inspections and observations to identify visible signs of deterioration or structural weakness following the Al Haouz earthquake disaster of September 2023. Various factors need to be considered including the quality of construction materials, and the geological and geotechnical characteristics of the Al Haouz region, such as soil type and topography, which play a crucial role in determining the seismic vulnerability of these buildings. The approach then uses the seismic regulatory framework for the seismic protection of earthen buildings. Appropriate measures and sustainable and regulatory practices can be implemented to mitigate seismic risks.

Introduction

The seismic vulnerability of vernacular architectural heritage in regions like Al Haouz[1], Morocco, is a critical issue, particularly given the traditional materials and construction techniques used in these areas. Vernacular architecture refers to structures built using locally sourced materials and traditional construction methods, often reflecting the cultural and environmental context of the region. In the Al Haouz region, as in many other rural areas of Morocco, buildings are commonly made from materials like earth including rammed earth (*pisé*) and adobe, stone, and wood, which, while abundant and affordable, can pose significant risks during seismic events.

It is therefore crucial to analyze and assess the seismic vulnerability of these buildings to develop effective strategies for their protection and conservation. The seismic vulnerability of

[1] Al Haouz Province represents a significant portion of the Marrakech-Safi region. Covering an area of 6,231 km², this province is composed of 40 communes and is an integral part of the 75 Moroccan provinces.

vernacular architectural heritage refers to the susceptibility of these structures to damage or collapse during seismic events. Vernacular architecture, characterized by its use of local materials and traditional building techniques, is particularly vulnerable due to its lack of compliance with modern seismic design codes and standards.

This paper explores the seismic vulnerability of these vernacular buildings made of local materials in heritage sites and thus proposes adaptation strategies to improve seismic resilience. The assessment approach is based on a qualitative method including visual inspections and observations to identify visible signs of deterioration or structural weakness following the Al Haouz earthquake disaster of September 2023. Various factors need to be considered including the local seismic activity of the Al Haouz region, the construction materials quality and techniques, the building design, the age and maintenance, and the geological and geotechnical characteristics, such as soil type and topography, which play a crucial role in determining the seismic vulnerability of these buildings. The approach then uses the seismic regulatory framework for the seismic protection of earthen buildings (the Earthquake-Resistant Regulations for Earth Constructions, RPCT2011 [1]).

Appropriate measures and sustainable and regulatory practices can be implemented to mitigate seismic risks.

Seismic Vulnerability of Vernacular Architecture

Earth and stone buildings in the Al Haouz region generally have thick walls and flat roofs. Earth materials, such as rammed earth, provide excellent thermal insulation but remain fragile and susceptible to seismic forces. Stone buildings, although more robust, can also suffer significant damage during earthquakes, especially if they are not properly maintained or braced.

Seismic Risk Assessment

Historical seismic events[2] have caused considerable damage to earth and stone structures, highlighting the need for detailed seismic risk assessments and risk mitigation strategies. Common failure modes include wall cracking, out-of-plane collapse of exterior walls, collapse of unsupported or unbraced sections, and roof detachment.

Our architectural heritage is all the more precious because it often contains the productions of skilled craftsmen, true works of art, and architecture that are difficult and expensive to reproduce. However, this heritage has been affected on several occasions by more or less violent earthquakes.

Morocco's geological position at the meeting point of several interacting tectonic plates is the reason why, historically, Moroccan cities have been destroyed several times by several earthquakes. From 1932, the Cherifian Scientific Institute (now the Scientific Institute) organized macro-seismic surveys using the meteorological observation network.

The seismicity of Morocco is presented in Fig. 1 [2] showing the location of epicenters of earthquakes in Morocco and neighboring regions during the period 1901-2010. On the mainland of Morocco, Fig. 1 indicates that seismic activity is mainly concentrated along the Rif and Atlas Mountains, where the Al Haouz earthquake in 2023 was the most violent, causing the death of more than 2,800 people and the large number of destroyed earthen buildings showed that this

[2] The most emblematic earthquakes:
Before the 20th century: May 11, 1624, November 1, 1755 in Lisbon and November 27, 1755 are the three dates found in history books, which affected the medinas. Other more or less documented earthquakes affected historic cities: Larache in 1276, Tetouan and Fez in 1522, coastal cities and Marrakech in 1722, Agadir in 1731, Salé in 1757, Tangier, Fez and Salé in 1773, … .
After the 20th century: The surroundings of Tetouan 1909; The surroundings of Fez 1929; Agadir 1960; Atlantic 1969; Missour 1985; Moulay Idriss 1987; Essaouira 1988; Rissani 1992; Al Hoceima 1994; Rabat 2001; Al Hoceima 2004; Al Haouz 2023.

Materials Research Forum LLC
https://doi.org/10.21741/9781644903391-28

traditional construction made of local materials is very vulnerable to earthquakes when it does not comply with regulatory provisions.

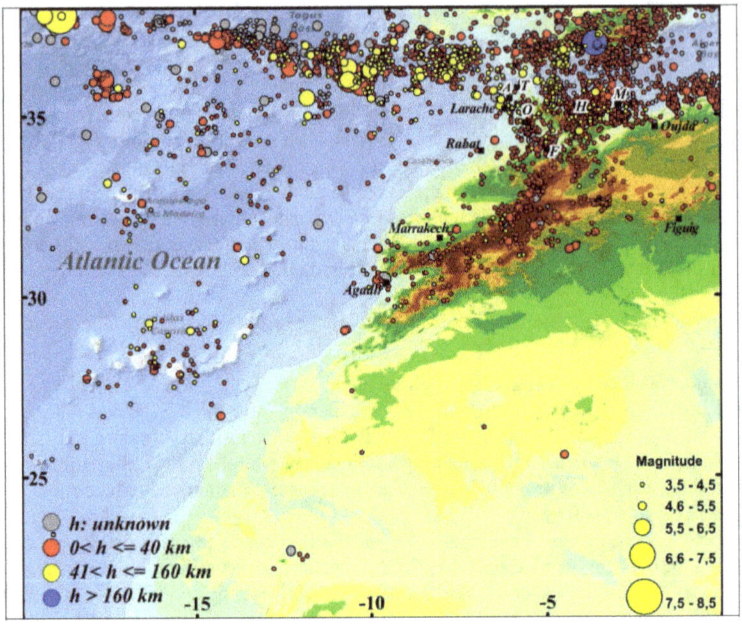

Fig. 1: Seismicity map of Morocco 1901 – 2010.

Al Haouz Earthquake Review

The latest Al Haouz earthquake event of September 8, 2023, provides valuable information on the types of damage suffered by vernacular architecture, particularly earthen and stone buildings. The detailed analysis of these case studies reveals patterns of structural failures and helps identify key areas for improvement in construction practices and renovation techniques. However, the relevant issue is the non-application of seismic regulations and accepted traditional practices

Impact analysis can provide valuable information on the preservation and fortification of historic sites against natural disasters. The analysis considered the following structured approach:

-Documentation and assessment: Comprehensive documentation on the impact of the Al Haouz earthquake on heritage sites and structures was gathered. This includes visual documentation, structural assessments, and historical records of their original designs.

-Analysis and exchange of experts by videoconferencing and participation in face-to-face meetings organized by the Academy of the Kingdom of Morocco [3,4]; the University and higher education institutions; the Ministry of National Land Planning, Urban Planning, Housing, and Urban Policy [5]; the National Order of Architects and Engineers; and some NGOs such as the national committee of ICOMOS (International Council on Monuments and Sites) [6]. We worked with experts in architecture, civil engineering, historic preservation, and seismic resilience. We organized discussions, seminars, and workshops during which these experts shared their views on the specific vulnerabilities exposed by the Al Haouz earthquake and mainly on the impact on heritage sites.

-Structural diagnosis and analysis: We compared the damage suffered by heritage objects or objects of vernacular architecture in the Al Haouz region. We assessed the differences in their structural integrity, the materials used, or the geographical factors that could have influenced the damage.

-Identification of best practices: We identified preservation methods or structural improvements that could have mitigated the damage. This includes renovation techniques, material improvements, or changes in construction methods while respecting the heritage value.

It must be stated that the Al Haouz earthquake was a very violent event, with a magnitude of 6.8 on the Richter scale; the depth of epicenter of 18 Km, located about 70 Km south of Marrakech, in the High Atlas Mountains. Ground movements were recorded at several sites in Morocco, within a radius that reached 500 Km. The maximum ground acceleration (pga) reached more than 50% g (Fig. 2 and Fig. 3) [7], two and a half times higher than that given by the Earthquake Regulations for Earth Construction (Amax = 20% g, Fig. 4) [1].

Key Factors Contributing to Vernacular Architecture Seismic Vulnerability
Earthquake damage suffered by vernacular traditional construction in Al Haouz region was observed in total or almost total collapse of traditional earthen constructions located at the edge of terraces not stabilized or on stepped land. There are also some constructions whose load-bearing walls are simply placed on the rocky ground without any provision for load transfer /load path.

Side effects and their impacts on the extent of damage were also observed on constructions very close to each other.

Some vernacular buildings resisted the earthquake with reinforcement at the corners, surrounding openings, and the use of traditional wooden lintel, or wooden chaining technique (Fig. 5).

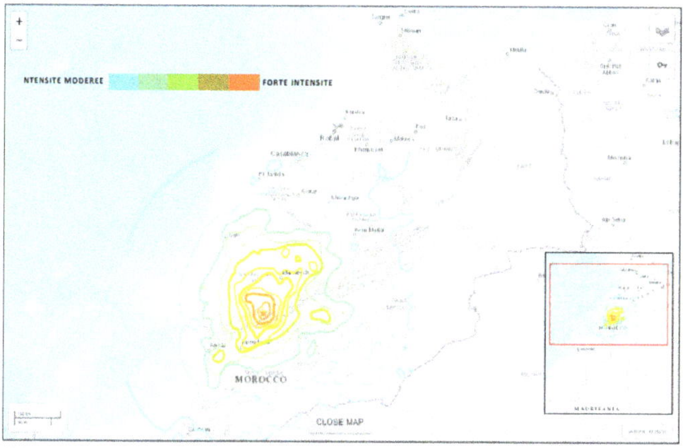

Fig. 2: MMI Intensity Map of the earthquake felt areas M 6.8 Al Haouz, Morocco earthquake of September. 8, 2023 (Source: USGS)

Materials Research Proceedings 47 (2025) 242-250
https://doi.org/10.21741/9781644903391-28

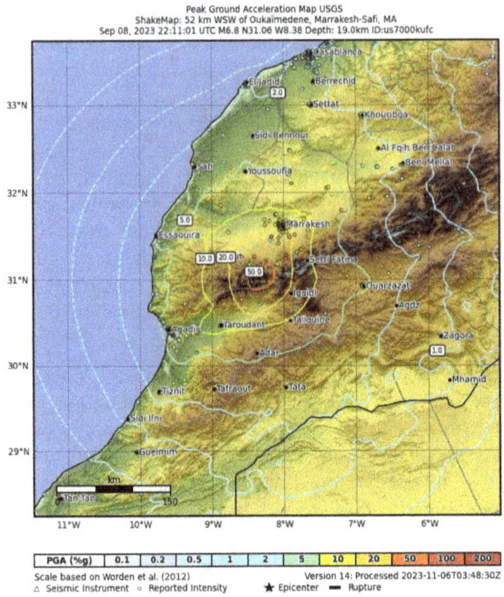

Fig. 3: Peak Ground Acceleration - PGA Map, Al Haouz, Morocco earthquake of September 8, 2023, (Source: USGS)

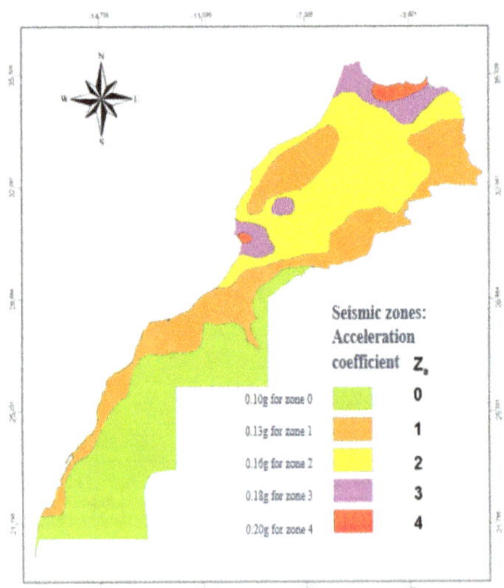

Fig. 4: Map of Seismic zones with the associated acceleration coefficients – maximum of 20% g (Source: RPCT)

Materials Research Proceedings 47 (2025) 242-250 https://doi.org/10.21741/9781644903391-28

Fig. 5: Examples of vernacular buildings that resisted the earthquake

The key factors contributing to vernacular architecture seismic vulnerability include the following:

a) Construction materials and techniques:

-Rammed earth and stone: These materials are heavy and have low tensile strength, making them prone to collapse during an earthquake. Rammed earth and adobe, in particular, can crumble under strong ground motions.

-Lack of reinforcement: Traditional buildings often lack adequate reinforcement, which is necessary to improve structural integrity during seismic activity.

b) Building design:

-Irregular shapes and heights: Vernacular buildings often have irregular shapes and varying heights, which can create stress concentrations and increase the likelihood of collapse during an earthquake.

-Weak connections: The connections between different structural elements, such as walls and roofs, are often weak, leading to partial or total collapse in a seismic event.

c) Age and maintenance:

-Aging structures: Many of these buildings are centuries old and have not been adequately maintained or retrofitted to withstand seismic forces.

-Poor maintenance: The lack of regular maintenance exacerbates vulnerabilities, as deteriorating materials and structural elements further weaken the building.

d) Local seismic activity:

Seismic history: The Al Haouz region is located in a seismically active area, which increases the risk to the vernacular architecture. Past seismic events can provide insight into the types of damage these structures are prone to.

e) Site effects:

Site factors such as topography, geology, and geotechnics, including soil type, greatly influence the vulnerability of vernacular architecture and the traditional built environment to seismic events.

Case study of Tinmel Mosque

The historic Tinmel Mosque [8,9,10], the birthplace of the Moroccan Almohad Empire built in 1153, is included in the indicative list of World Heritage Sites in Morocco by UNESCO. This valuable historical monument was severely impacted by the Al Haouz earthquake of September 8, 2023, while restoration work was underway. During ICOMOS Morocco/Spain visit [6], cleaning work on the collapsed elements has been observed, as sorting of reusable materials such as solid bricks, an inventory, and storage of architectural elements and decorative elements, as well as support sections of walls and arches still standing. This work is carried out by the Ministry of Habous and Islamic Affairs, which plans to quickly launch a new call for tenders for recovery work on this monument.

The Tinmel Mosque is inscribed in a rectangle of approximately 48 x 43 meters on the edge of the palm grove irrigated by the Nfis River in Al Haouz region. Its walls and its exterior minaret –

which incorporates the mihrab – are made of bricks and mortar based on earth, stones, and lime, which gives it this ocher color that integrates it into the landscape.

Fig. 6: Axonometric Perspective of the Tinmel Mosque Before the Earthquake

Fig. 7: The four facade walls localization of the Tinmel Mosque building (Satellite image before earthquake)

Fig. 8: The Tinmel Mosque and its stable eastern facade (E) after the earthquake

Summary Discussion about Tinmel Mosque

The wall of the north facade is the "weakest" from a structural point of view, followed by the two walls of the (S) south and (W) west facades. The most stable wall is that of the eastern facade (E).

Hence the difference in damage observed on the four facades. Indeed, the wall that suffered the most significant damage (almost total collapse) is that of the north facade.

Likewise, the wall of the facade (E), with its three entrances, is the one that most resisted the seismic shock. Furthermore, the minaret, mainly made of stone masonry located in the middle of the south facade, is half destroyed. The wall of the (N) north facade remains the most fragile because it does not have any bracing element and is weakened by a large opening in the middle (north door).

The walls' thickness remains insufficient given the significant height, which is approximately 7 meters (too great a slenderness). Rammed earth is the main construction technique used in the construction of the Tinmel Mosque building. The rammed earth is reinforced with solid wood logs which are applied at the level of the compaction recovery joints.

As for the arcades inside the construction, they are entirely made of earth brick masonry with lime-based jointing mortar.

Conclusion

Even though the Al Haouz earthquake of September 8, 2023, was violent and catastrophic, some buildings with local materials were able to survive, thanks to their implementation of state-of-the-art rules and standards. For the built environment, built with local materials (earth, wood, stone), of heritage or vernacular type, which was severely damaged, the reasons were multiple: non-compliance with the rules and standards, the effects of the site, the quality of the materials and their implementation, construction without foundation or adequate bracing or reinforcement to ensure lateral structural stability.

The intensity of the Al Haouz earthquake has necessitated the adoption of more stringent earthquake-resistant regulations to enhance the safety of earthen constructions in disaster areas. In response to the recommendations from the National Committee for Earthquake Engineering and the National Committee for Earth Construction, the Moroccan Government has adopted a new draft of two decrees in September 2024. These decrees amend the existing earthquake-resistant construction regulations by introducing a new catalog of speed, speed zones, and accelerations at the communal level.

The goal is not only to learn from the damages caused but also to create a roadmap for preserving such heritage sites in the face of future natural disasters e.g. earthquakes etc. Documenting these experiences and deriving best practices from them, can serve as a guide for safeguarding other national architectural heritage in seismic zones.

References

[1] Royaume du Maroc, Bulletin Officiel, Décret N° 2-12-666 Approuvant le Règlement Parasismique pour les Constructions en Terre et Instituant le Comité National des Constructions en Terre (2013).

[2] E. Cherkaoui and A. El Hassani, Bulletin of the Scientific Institute, Rabat, Earth Sciences Section, n°34 (2012), p. 45-55.

[3] Académie du Royaume du Maroc, Journée d'étude, Penser les horizons de dignité après le séisme : Après la catastrophe, ouvrir un nouvel horizon de dignité. Reconstruire, Réparer, Préparer, Rabat (2023).

[4] Académie du Royaume du Maroc, Journée d'étude, Construire avec les matériaux locaux pour faire face au double défi du séisme et du réchauffement climatique. Consolider les acquis après le séisme, Rabat (2023).

[5] ICOMOS Maroc, Rapport ICOMOS Maroc Séisme d'El Haouz et Impact sur le Patrimoine (2023).

[6] Royaume du Maroc, Ministère de l'Aménagement du Territoire National, de l'Urbanisme, de l'Habitat et de la Politique de la Ville – Habitat et Politique de la Ville – Secrétariat Général Direction de la Qualité et des Affaires Techniques, Diagnostic Post-Sismique des Constructions des Régions Impactées par le Séisme du Haut-Atlas. Focus sur les Constructions en Matériaux Locaux, Auteur : Ahmed Koutous (2023)

[7] USGS United States Geological Survey https://earthquake.usgs.gov/earthquakes/eventpage/us7000kufc/executive

[8] Royaume du Maroc, Direction du Patrimoine Culturel, Archives de la Division des Etudes et des Interventions Techniques, Service de la Programmation et des Interventions Techniques, Tinmel ; Mosquée Tinmel ; Plans archives (2016).

[9] Académie du Royaume du Maroc, Journée d'étude, L'archéologie de sauvetage : Tinmel et les sites patrimoniaux de la Région d'Al Haouz, Rabat (2023).

[10] A. S. Ettahiri, A. Fili, F. Gabayet, S. Gaime, Y. Khiara, J.P. Van Staëvel, Le site de Tinmel : après la catastrophe, un premier bilan archéologique (2023).

Modern architecture and the lessons of vernacular shantytown housing in Morocco

Fatimazohra SAAID[1,a], Siham EL RHARBI[2,b]

[1]Ministry of Territorial Planning, Urban Planning, Housing and City Policy, Rabat, Morocco

[2]Practicing architect and architecture studio teacher at SAP+D, UM6P, Benguerir, Morocco

[a]fzsaaid@gmail.com, [b]elrharbi.siham@gmail.com

Keywords: Shantytown, Bidonvilles, Vernacular Housing, Modern Architecture, CIAM, Morocco

Abstract. Vernacular architecture offers responses that are adapted to its inhabitants needs and in harmony with its environment. This observation celebrated in 1964, during the exhibition entitled "Architecture without architect" by Bernard Rudofsky at the MOMA in New York, brought to light the values of *vernacular* architecture and recognized it as an unprecedented source of inspiration for modern architecture. However, as early as the 1940s, even before Rudofsky's exhibition, architectural practices initiated in Morocco began to demonstrate an interest in vernacular architecture. The modern architects of the CIAM Morocco Members developed new architectural approaches focused on the study of the formal diversity of vernacular housing in their connection to the lifestyles of the inhabitants. Far from any stigmatization, the attitude of these modern architects will broaden, from this period, the analysis of the vernacular in architecture to the typology of the shantytown. An achievement of popular inventiveness, the vernacular habitat of the shantytown restores for these architects attributes comparable to those of traditional vernacular environments. These approaches which were presented at the CIAM 9 in 1953, were drawn up as a reformulation of the principles of modern architecture. Although there are trends that focus on shantytowns as vernacular architecture that can solve the problem of housing and tomorrow's cities, the approaches of CIAM-Maroc architects focused on the shantytown are manifested as innovative avant-garde architectural attitudes. Thus, this paper focuses on studying the formulations of these modern architects based on the shantytown, to show the practical values of this vernacular typology and highlight the dialectical relationship between the teachings of vernacular architecture of the shantytown and the renewal of modern architecture.

Introduction

Vernacular architecture was for a long time, considered to be all of the traditional buildings designed with local materials in harmony with their natural environment, through popular building skills and know-how.

Numerous authors such as Amos Rapoport [1] and Sibyl Moholy-Nagy [2] have highlighted the anthropological value of traditional vernacular architecture, distinguishing its primitive, spontaneous, and anonymous character. The architect Bernard Rudofsky (1905-1988) was also very interested in this last characteristic when he introduced vernacular architecture into his exhibition at the MoMA Museum of Modern Art in New York in 1964. The Austrian architect exhibited photographs of examples of rural vernacular architecture from Africa and the Orient, and presented them under the title « Architecture without architects ». The exhibition, which was a celebration of the collective achievement of local community and the innovative architectural solutions they had developed, rooted in basic human needs. The exhibition event was one way to criticize the modern architecture doctrine

Besides this interest in traditional vernacular architecture, many authors, theorists, and architects are now recognizing new solutions for architecture and urban planning in contemporary vernacular architecture. Texts by architects specializing in anthropology and sociology, such as Jean-Pierre Frey [3], Thierry Paquot [4], and Jean-Paul Loubes [5], broaden the perception of vernacular housing to include the phenomenon of shantytowns and invite urban designers and theorists to look at its morphological, social and anthropological qualities. Shantytowns, long seen as a manifestation of social misery and spatial exclusion, are emerging as an expression of contemporary vernacular architecture without architect, a new urban laboratory to discover alternatives in architecture and urban planning.

However, although the interest shown by certain architects in shantytowns appears to be a particularly recent paradigm of thought, the fact remains that shortly after the Second World War, young architects working in Morocco during the colonial period found in the shantytowns of Casablanca a source of inspiration in architecture and urban planning.

Then, the aim of this article is by no means to idealize shantytowns, but to understand how the solutions provided by architects working in Morocco from the end of the Second World War were informed by the discoveries of the vernacular habitat of the shantytown. and how could this urban vernacular habitat be a potential urban laboratory for the future?

1920: The Genesis of Shantytowns in Morocco

From the 1920s onwards, Morocco experienced rapid economic and social development. The development of new towns, accompanied by fast industrialization, triggered a major economic and social boom in several cities, particularly Casablanca.

The establishment of industrial firms in several large cities, such as Casablanca, and the construction of the city's port by the French Protectorate, turned several large cities into real centers attracting a rural workforce in search of work. Soon, the outskirts of the medina and of the new modern cities were taken over by these masses of workers who had found no housing, apart from self-built makeshift dwellings, forming what came to be known as shantytowns. These makeshift dwellings were essentially built from salvaged materials (shanties and sheet metal).

For the Protectorate administration, the shantytown was perceived as a hostile and non-hygienic place. From the 1930s onwards, the colonial administration declared war on these new gathering places for Moroccan workers. This struggle took different forms, ranging from simple control by gathering the occupants of the shacks into supervised perimeters to the introduction of hygienic approaches advocating the sanitation of the shantytown. However, the situation quickly spiraled out of control, when the shantytown was detected by the Hygiene Services as a hotbed of contamination and proliferation of contagious diseases and as a potential danger to public health by the colonial administration.

The colonial administration's housing programs followed one another throughout the 1920s and into the 1940s. With the Moroccan riots, these measures became a health and safety emergency. It was therefore urgent for the colonial administration to resolve this social problem through spatial solutions. [6]

The colonial authorities decided in the 1940s to call on French architect and town planner Michel Écochard. Appointed Head of the Urban Planning Department in Morocco, he and his multi-disciplinary team, representing the CIAM Maroc, the representative group of the Modern Architectural Movement in Morocco, establish a new solution to the housing crisis, based on the analysis of the needs and habits of the shantytown dwellers.

The CIAM-Maroc Method: the Shantytown as a "Research Urban Laboratory"

Écochard advocated a new way of looking at the problem of shantytowns dwellers. He developed a new approach based on genuine fieldwork. With his multidisciplinary teams working with him in the town planning department, carried out in-depth studies of the different types of local housing

Materials Research Proceedings 47 (2025) 251-257 https://doi.org/10.21741/9781644903391-29

in both urban and rural areas. However, the distinctive feature of his approach is that it considers the reality of shantytowns, long considered hostile and stigmatized by the colonial administration of the French Protectorate in Morocco.

These in-depth investigations, drawing on the different disciplinary of the CIAM-Maroc members, have enabled us to understand the different typologies of habitat in Morocco including shantytowns, to the housing patterns and cultural traditions of their inhabitants. Écochard developed a new approach, which takes account of the shantytown environment and overcomes conventional prejudices and reductive acceptance. He focuses on the intrinsic social and spatial qualities of the lifestyles and built forms of the shantytown. In this case, the work of CIAM Maroc group had a similarity between ethnographic tools and the methods of architectural practice of the CIAM Algiers group on the Mahieddine shantytown.

According to Avermaete.T, the approach developed by Écochard, and his team illustrates a new attitude that represents an epistemological shift [7] in conventional practices of modern architects in Europe and the United States. The CIAM Maroc group gave priority to surveys and photographs, ethnography, and urban sociology. They depicted the urban form of the shantytown as considerable material for their proposal "housing for the greatest number", «*...this particular mode of analysis portrayed the Bidonville as a meeting point between a so-called « traditional culture » that was still part of everyday dwelling habits and the modern culture of cities...*» [7].

The CIAM-Maroc members noted the ambivalence of the shantytown's environment and its qualities, which gave off a strong formal symbolism. In this sense, Mas, the leading figure in this group and in the Urban Planning Department highlights the advantages of the shantytown and emphasises that « *si le bidonville est grave parce qu'il reflète des problèmes politiques, économiques et humains considérables, nous avons vu que sa forme n'est pas entièrement à abandonner...* » [8].

It thus appears that to respond to the problem of housing for the greatest number of people on Moroccan colonial land, Écochard, and his team implemented an avant-garde approach, far removed from the ideals and practices of modern architects.

In 1953, at the International Congress of Modern Architects CIAM 9 in Aix-en Provence, the panels exhibited by the CIAM Maroc members focused on the housing projects developed in Morocco and exhibited pictures of residents carrying out various activities in their environment, even in shantytowns.

1948, Discoveries in the Shantytown in Morocco

The urban principle of the extendable grid, "The 8X8 grid"

Under the name of "housing for the greatest number", housing programs in Morocco's aim was to provide an appropriate response to the masses of workers. In 1948, the town planning department began its approach by developing satellite housing estates in Rabat and Casablanca. The development of these housing estates took the form of housing neighborhoods based on a grid known as the "8X8 grid" (Fig. 1). Inspired by the vernacular urban morphology of the shantytown and the medina urban fabric, the grid provides for different types of individual and collective housing, developed according to an urban system defining that allows, for a first stage, the improvement of the shantytown in the 8X8 grid, and the possibility of simultaneously integrating infrastructure, housing and facilities.

Materials Research Proceedings 47 (2025) 251-257 https://doi.org/10.21741/9781644903391-29

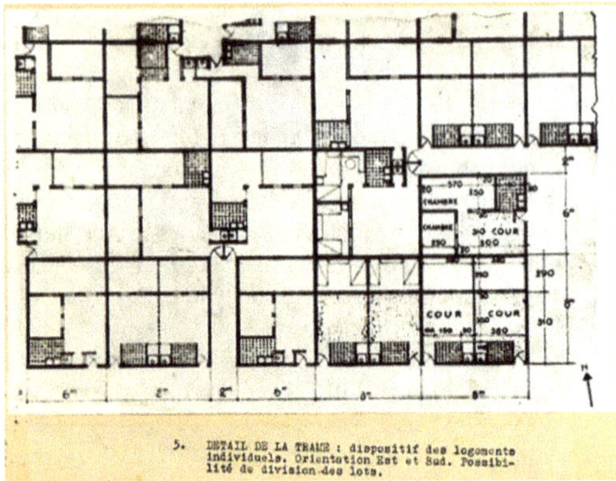

Fig. 1: The Grid 8X8. Source : Archives Ministry of Habitat and Urban planning Morocco.

The Minimum Shantytown Dwelling as A Model

In terms of layout, the minimum cell chosen is a patio house. This choice was made as a result of studies carried out by the CIAM Maroc group and its mobile unit, the "Atelier Ambulant", which based its work on the observation of ways of living in the different forms of urban and rural vernacular housing in the medina, in villages, and in shantytowns.

The work focuses on the interiors of the shacks and how they are organized. Mas explains: « *par un souci d'embellissement et de décoration- peintures murales aux couleurs vives, arbustes entretenus avec amour dans les patios, l'habitant du bidonville témoigne d'une grande ingéniosité, autant que d'un goût de vivre dans la fraicheur surprend dans ce cadre misérable* » [8]. The Urban Planning Department's investigations have brought to light the self-building skills of rural slum dwellers. According to Marion Von Osten, the building skills of the rural population living in shantytowns stem both from the building traditions they have acquired in the villages, and from the building practices they have discovered in the old medina, « *The self-built settlements of the shantytowns were, from the 1930s onward, the locus of the first encounters and negotiations with the modern city for some people moving to the city from rural areas... This building culture is centered not on the architect but on the builder, who plans and organizes the house together with the family who will be living in it. Thus, newcomers to the self-built hut settlements of the bidonvilles applied the strategy of building a growing house,..* » [9].

The minimum size of the house is 8 meters on both sides « *une trame d'habitations répartie sur des parcelles de 8m X 8m* » [10]. which can be oriented to the south-east in any combination. Écochard chose as a dominant architectural typology a minimum of L-shaped ground-floor patio house (Fig.2). Curiously, this echoes the typological L-shaped patio shantytown dwelling, a plan drawn up by Pierre Mas in the Casablanca shantytown (Fig.3).

Yet, the housing solution with the greatest number is based on the grid principle, the basic element of which is a minimum dwelling, with local public facilities and services from urban areas. These urban areas have been created to foster a convivial community living environment. In shantytowns in Casablanca, Écochard noted that, such as the medina, the shantytown has a variety of facilities such as the Koranic school, the mosque, the mill, and the souk, services that favorably facilitate community life for the people who live there.

In its proposal for housing for the greatest number, Écochard adopts these facilities based on the principle of Anglo-Saxon principle related to urban sociology: the neighborhood unit.

Écochard and his team set this principle for a population of 9.000 inhabitants and considered that each neighborhood unit should be organized around schools and facilities.

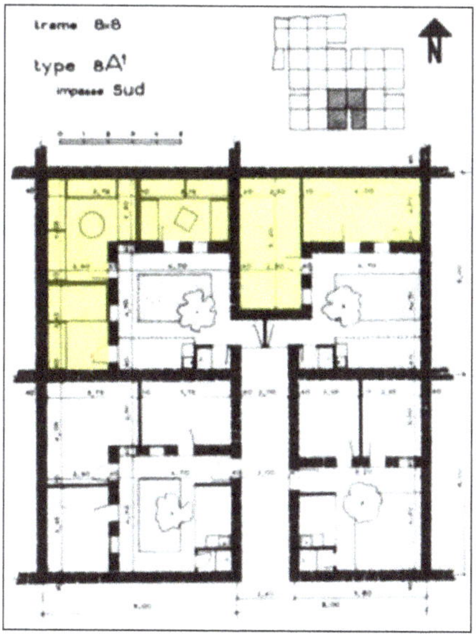

Fig.2: The L-shape of the two minimum housing types used in the 8x8 grid. Grid CIAM Morocco, 1953.

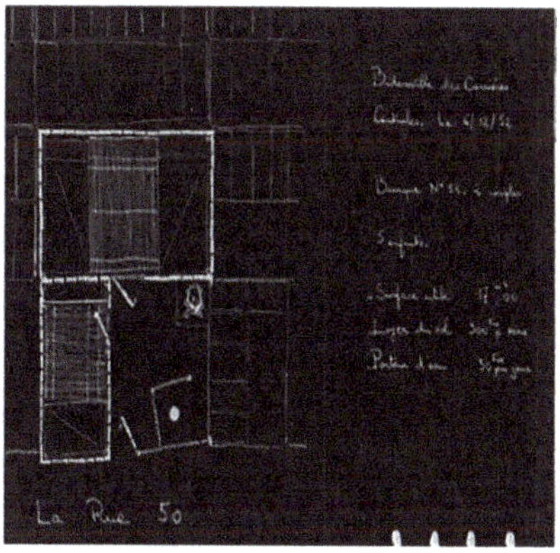

***Fig. 3**: L-shaped accommodation in a Shantytown Dwelling according to a survey carried out in 1952 at slum in Casablanca.*

Community living spaces

Yet, the housing solution the greatest number is based on the grid principle, the basic element of which is a minimum dwelling, with local public facilities and services from urban areas. These urban areas have been created to foster a convivial community living environment. In shantytowns in Casablanca, Écochard noted that, such as the medina, the shantytown has a variety of facilities such as the Koranic school, the mosque, the mill, and the souk, services that favorably facilitate community life for the people who live there.

In its proposal for housing for the greatest number, Écochard adopts these facilities based on the principle of Anglo-Saxon principle related to urban sociology: the neighborhood unit. Écochard and his team set this principle for a population of 9.000 inhabitants and considered that each neighborhood unit should be organized around schools and facilities.

Housing and intermediate spaces

Within each neighborhood unit, priority is given to pedestrian traffic, while fast-moving mechanical traffic is relegated to the periphery and continues along the city's main thoroughfares and roads.

These provisions have been implemented using intermediate spaces. In this respect, the approach is reminiscent of the discoveries made in the shantytowns. Michel Écochard points out that, despite the ambivalence of the shantytown environment, it constitutes a real urban center

« *…Un véritable centre urbain* » [11], which is well organized and connected« *en blocs réguliers avec ses rues centrales, ses ruelles, ses places, …* » [11].

The houses are arranged around enclosed public spaces of different sizes, an arrangement that first and foremost allows for the rational layout of the roads in the neighborhoods and an urban organization regulated by passageways, squares, and small squares running directly alongside the dwellings (Fig. 4). The cul-de-sacs, squares, and tree-lined squares create public spaces for community life.

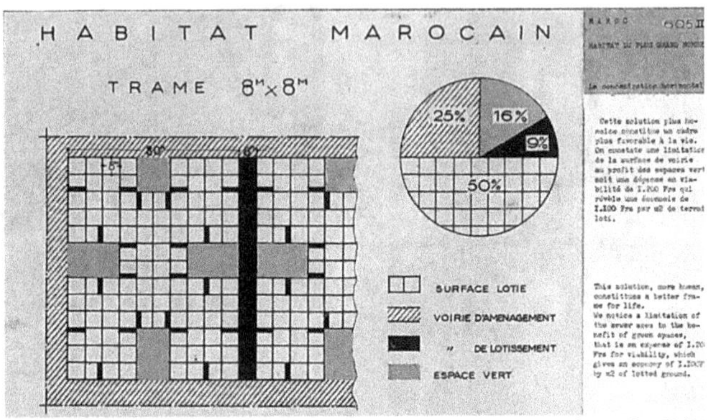

Fig 4 : Open spaces and areas planted with trees in the grid. Extract from the GAMMA grids exhibited in 1953 at CIAM 9. In The Making: Traversing the project exhibition In the Desert of Modern City. Colonial Planning and After. (2018).

The Vernacular Habitat of the Shantytown, *a New Promise for Today's Architecture...*

Although it was contested at the International Congress of Modern Architects in Aix-en-Provence in 1953, the approach of the CIAM Maroc group in the Moroccan colonial context, enlightened by the teaching value of the shantytown and its aesthetic, morphological, and anthropological dimensions, paved the way to new alternatives in modern architecture and urban planning.

These architects can be credited with broadening the concept of vernacular architecture to include the shantytown as a new source of architectural inspiration. Based on their findings in the shantytowns, the CIAM Maroc Group has identified recurring architectural and urban models and arrangements in urban and rural vernacular habitats, this has led to the development of an adapted and evolving architectural and urban model. This model has been transposed to several cities in Morocco. Gradually, the principles of this model were used to design other urban and architectural models until 1964, which evolved into the urban form of low-cost housing that we know today. Then, it is interesting nowadays, to see how we could recognize the vernacular architecture of shantytowns and think to their great potential for solutions to current urban problems, as promise to imagine models of convivial cities, with adaptable and multifunctional housing.

Reference

[1] A. Rapoport, House, Form and Culture, Prentice-Hall, 1969.

[2] S. Moholy-Nagy, Native Genius in Anonymous Architecture, Horizon Press, 1957.

[3] P. Frey, Learning From Vernacular : pour une nouvelle architecture vernaculaire, Arles : Actes Sud, 2010.

[4] T. Paquot, Les Bidonvilles, La découverte, Repères, 2022.

[5] J.P. Loubes, Traité d'Architecture Sauvage, Manifeste pour une Architecture Située, Collection La roue de la bicyclette, 2010.

[6] J.L. Cohen, M. Eleb, Casablanca: mythes et figures d'une aventure urbaine, Hazan, (2004).

[7] T. Avermaete, Between Dogon and Bidonville: CIAM, Team 10 and Perspectives on African Vernacular, in Jean-Francois Lejeune and Michelangelo Sabatino (eds.), North-South: The Vernacular, The Modern and The Mediterranean, London: Routledge, 2009, pp. 263–275.

[8] P. Mas, Problèmes d'habitat musulman au Maroc. In : Bulletin économique et social du Maroc (BESM), XVII, n° 60, 1954, pp.201-218. https://cinumed.mmsh.univ-aix.fr/idurl/1/83897

[9] M. Von Osten, In The Making: Traversing the project exhibition In the Desert of Modernity: Colonial Planning and After. [Doctoral Thesis artistic]. Malmö Faculty of Fine and Performing Arts, Lund University, 2018.

[10] M. Écochard, Urbanisme et construction pour le plus Grand Nombre. [Conférence], la Chambre de Commerce et d'Industrie de Casablanca à l'occasion de l'inauguration de l'Institut Technique Français du Bâtiment et des Travaux Publics du Maroc, Les Annales de l'Institut Technique Français du Bâtiment et des Travaux Publics, (oct. 1950).

[11] M. Écochard, Casablanca, le Roman d'une Ville. Éditions de Paris, 1955.

Materials Research Proceedings 47 (2025) 258-267 https://doi.org/10.21741/9781644903391-30

The transformations of Andean vernacular housing in the arid north of Chile

Natalia JORQUERA SILVA[1,a*] and Tomás SEPÚLVEDA SCHWEMBER[2,b]

[1]Department of Architecture, Universidad de La Serena, Amunátegui s/n, La Serena, Chile

[2] Universidad Bernardo O'Higgins, Centro de Estudios Históricos y Humanidades/ Director Ñaupa Estudios, *Avenida Viel 1497, Santiago*, Chile

[a]natalia.jorquera@userena.cl, [b]naupa.estudios@gmail.com

Keywords: Andean Architecture, Vernacular Building Culture, Seismic Risk, Climate Resilience

Abstract. This article presents the transformations undergone in recent decades by the vernacular Andean housing of the Atacama Desert in northern Chile. This architectural typology is a product of the Andean worldview, where architecture and materials (stone or adobe walls and vegetation-based roofs) respond both to the adverse desert climate of this part of the Andes Mountain and the intense Chilean seismic activity. The characteristics of this housing, which had been preserved for around five centuries, have rapidly transformed in just a few decades due to global economic, political, and social processes that rebound in local transformations in the desert, especially based on intense mining activity and mobility of the labor force. These changes have encouraged the use of industrialized materials from urban areas, which have gradually replaced the natural materials traditionally used in vernacular housing. This shift has altered not only the appearance and harmonious relationship with the surrounding landscape and impacted community-building practices but also the thermal-acoustic performance and structural capacity to withstand the frequent earthquakes that affect the area. Based on years of architectural and ethnographic research on fieldwork, a comparative analysis will be presented between the characteristics of vernacular Andean housing and new houses built with industrial materials, with an emphasis on seismic and climatic performance, as both are major conditions in the Atacama territory.

Introduction

The Andean area is a vast geographic region that extends along the Andes Mountains, the longest mountain system in the world, located in the western part of South America. It covers territories of several countries: Venezuela, Colombia, Ecuador, Peru, Bolivia, Chile, and Argentina, and is characterized by having great ecological diversity, due to its mountainous configuration, which generates different altitudinal levels or ecological zones (Fig. 1).

In this region, different communities that have shared the geographic space dominated by the presence of the Andes Mountains have developed, since pre-Hispanic times, a complex system of knowledge and symbolic construction of the cosmos, territory, and life [1], allowing them to adapt to the challenging environmental conditions of the Andes. Some societies reached high degrees of complexity (such as Tiwanaku and the Incas), evident in the extensive agricultural terraces on the mountains, irrigation canals, transportation systems, and some monumental constructions. Today, numerous Indigenous communities exist as heirs of these cultures, primarily Quechua, Aymara, Atacameños in the north of the Atacama Desert. All of them are the inhabitants of the study area presented.

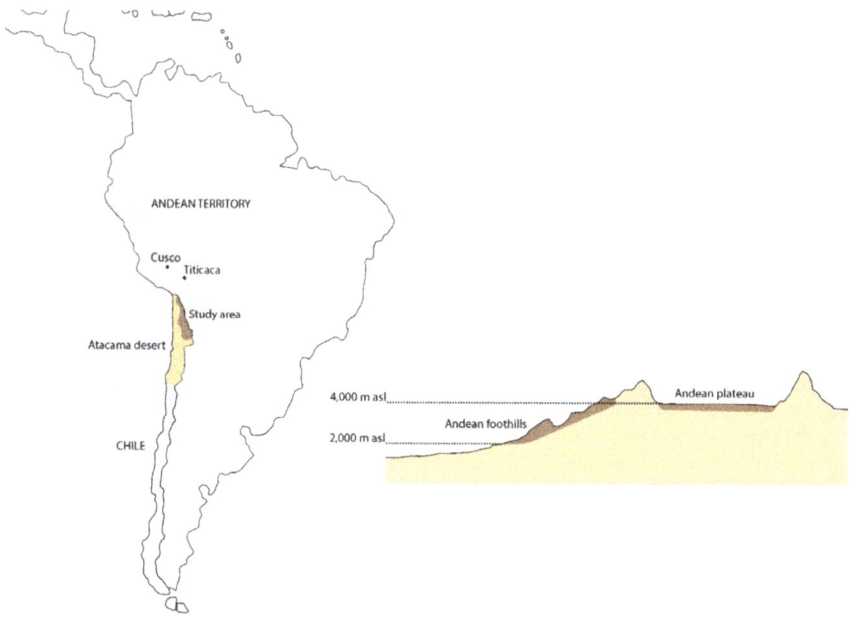

Figure 1. Map and section of the Andean territory and the study area.

In this geographic and cultural context lies the Atacama Desert in northern Chile, considered the driest desert in the world, with human settlements located at altitudes between 2,000 and 4,500 meters above sea level, thus occupying two geographic bands: the foothills, characterized by watercourses that give rise to cultivated oases, and the high plateau (Fig. 2). The study of the vernacular Andean housing located in this area provides a rich opportunity to explore how communities have developed adaptive and resilient architectural solutions in response to this geographic and material environment.

Adopting Marvin Harris's cultural materialism [2, 3] as a key theoretical framework for analyzing the spatiality, infrastructure, and materials used in these dwellings, it reveals how ecological, economic, and social conditions shape construction practices and settlements.

Cultural materialism focuses on how material needs and the availability of resources determine the cultural and social forms of a community. By applying this approach to Andean housing, it can be understood that the choice of materials such as earth and stone, as well as construction techniques, are strategic responses to the extreme conditions of the desert and the highlands, where water scarcity, the intense seismic activity that characterizes Chile and drastic temperature changes demand efficient and durable structures. Furthermore, the knowledge, beliefs and rituals that accompany the construction and organization of settlements can be interpreted as reflections of this relationship between communities and their material environment.

Figure 2. Comparison between the Andean foothill and plateau.

This material basis and the relationship with the environment also explain the animistic ontology that characterizes traditional Andean societies and religions, where mountains, springs, the earth, and the ancestors who inhabited the desert symbolically engage with local communities. This interaction occurs through collective and individual ritual practices that foster harmony between humans, environmental elements or phenomena, and the archaeological remains of their ancestors.

Following cultural materialism, John Murra's theory of "Vertical Archipelagos" [4] explains how Andean societies have organized their settlements across different ecological floors for accessing complementary resources since ancient times. The notion of an archipelago is used here to describe a settlement system that integrates various ecological floors and locations that concentrate subsistence resources, characterized as "islands," separated by vast stretches of desert and complex circuits of trans desert mobility in the case of Atacama. This concept is fundamental for analyzing the location and distribution of vernacular housing in the Atacama landscape, where settlements are determined not only by proximity to immediate resources but also by their integration into a broader network of circulation, exchange, and reciprocity among different ecosystems. The human traces of this way of inhabiting the desert have been recorded in thousands of archaeological sites and multiple heritage, architectural, ethnographic, and ethnohistorical records.

Vernacular houses in this context are part of a settlement system characterized by the use of a wide variety of places, through small towns and pastoral ranches, shelters, rocky eaves, caravan camps, and ancient human and animal trafficking routes, considering all ecological floors of territory and its distinctive qualities.

In the case of the vernacular housing of Atacama, the theory of Vertical Archipelagos helps to understand how Andean communities planned their settlements based on the availability of resources, facilitating access to building materials, water, and land suitable for grazing or agriculture, according to the altitude and climate of the desert. This, in turn, determines how settlements are organized and interconnected.

In this relationship of interdependence between dwelling and environment, the traditional life of local communities and how they organize and construct their settlements and homes have been progressively affected and frequently abandoned in recent decades due to transformations inherent to modernization processes, urbanization, migratory phenomena, and water scarcity, especially related with the exponential growth of large-scale mining in the Atacama Desert.

These changes have led to an increasing incorporation of industrialized materials from the cities, primarily concrete blocks and zinc sheet roofs, which have progressively replaced the natural materials used in traditional housing. This transition has significantly altered local architecture, weakening the harmonious relationship these structures maintained with the landscape and affecting both living conditions and traditional community-building practices. Additionally, the thermal-acoustic performance of the homes and their structural capacity to respond to the frequent earthquakes that impact Chile have also been affected.

Characteristics of Andean vernacular housing

As a result of the cultural contact produced in the Southern Andes between the traditions introduced by the Spanish invaders (circa 1500)—which, in turn, brings together Mediterranean legacies from Europe, Africa, and Asia—and the indigenous construction traditions of the Americas, the Andean housing emerged. This type of housing is present, with local particularities, in numerous towns across the macroregion that encompasses southern Peru, western Bolivia, northwestern Argentina, and northern Chile—including the Atacama Desert and Andean plateau—up to the present day.

From the perspective of cultural materialism [2], these houses are not mere functional responses to environmental conditions but rather complex expressions of sophisticated cultural adaptation that articulate the relationship between society, territory, and economy. Furthermore, housing is part of a larger domestic space, consisting of the house, and, according to Villagrán [5], includes three other local economic-social units: the garden plot (*'chacra'*), or family cultivation space, relative to each house; the field (*'campo'*), or collective space dedicated to gathering and grazing, and the hill. Each one of these units, has as well, animist religious connotations as the place where reside together non-human spirits of mountains, water outcrops, earth, and the souls of the ancestors.

Figure 3. The four units of housing space.

Architecturally, Andean housing possesses common characteristics and repetitive typological patterns throughout the mentioned macroregion: it is a hermetic rectangular parallelepiped with a slightly pyramidal shape, featuring a sharply sloped gabled roof, a central entrance door, and sometimes one or two small window openings—all characteristics that reveal its function as a refuge against the harsh climate of the Andes [6]. The dimensions are rather small, with average measurements of 3 meters wide by 6 meters long in plan; the height generally reaches 3 meters at the ridge [7]. These small dimensions reflect the fact that enclosed living space is limited to sleeping and cooking, while social, sacred, and productive activities occur outside, in the surrounding territory, where there is a deep connection with the elements of nature [8]. Simultaneously, the small dimensions of housing are due to the scarcity of long wood suitable for constructing roofing structures.

The basic housing module described can be duplicated or tripled as the family grows. Additionally, it is common for the housing to be surrounded by small structures, such as chicken coops, other domestic animal enclosures, and an oven, forming a small complex.

Building in the Andean territory is a sacred act, where 'Mother Earth', or *Pachamama*, provides all the necessary materials. Thus, in an environment that might appear hostile and resource-scarce,

Materials Research Proceedings 47 (2025) 258-267
https://doi.org/10.21741/9781644903391-30

Andean communities have effectively managed water, found stone quarries, and clay soils to erect walls, as well as gathered local woods and grasses to create roofing structures, demonstrating a deep knowledge of an extensive territory. Consequently, Andean housing is built with adobe or stone with clay mortar, featuring roofing structures made from the scarce local woods (cactus, Chañar *Geoffroea decorticans* and Algarrobo *Neltuma chilensis*), bound with llama leather or wire, and topped with a layer of vegetable fiber, a stratum of clay for insulation, and a finish of wild straw (Fig. 4).

Figure 4. Exterior, interior and roof details of an Andean vernacular house in the town of Huatacondo.

All variations and local construction particularities adhere to what the local natural environment offers, according to the different ecological floors and its own characteristics in different areas. For example, looking for describe botanically salt flat basin of *Salar de Atacama*, based on local communities' categories of environment, Villagrán et al. [5] subdivide the geographic grounds of foothills and the Andean plateau in four ecological floors, associated to altitude, vegetation characteristics and shared cultural significances of landscape. The authors call these floors: *Pampa, Tolar, Pajonal y Panizo* (Table 1):

Table 1. Prepared by the authors based on Villagrán et al. [5].

Floor local name	Other denominations	Description
Pampa	***Piso Desértico o Prepuneño***	Term used to refer to desert plains or those with very sparse vegetation, characterized by shrubs with very little coverage.
Tolar	***Piso Puneño***	Semi-desert environment, characterized by the predominance of shrubs designated by the generic name of *tolas*.
Pajonal	***Piso Altoandino***	Environment with a steppe physiognomy and dominated by grasses in clumps and plants in cushions, located immediately above the *tolar* and characterized as a colder area with a predominance of *ichu* (straws).
Panizo	***Piso Subnival***	Environment located on the summits of hills and defined by the absence of vegetation.

Therefore, as in all vernacular architecture, Andean housing is materialized using the natural resources of the immediate environment as construction materials [9, 10], blending harmoniously with the surrounding landscape.

Transformations of Andean vernacular housing

The socio-economic changes that have occurred in recent decades, particularly the intense mining activity in the Atacama Desert—mainly associated with copper and lithium—and its subsequent

Materials Research Proceedings 47 (2025) 258-267 | https://doi.org/10.21741/9781644903391-30

water usage, have led to ecological, social, and economic impacts in the region, affecting ancestral construction practices. This has resulted in a transformation of settlements and traditional housing. Additionally, the pressure from international tourism, centered in San Pedro de Atacama (the communal capital), has led to modifications of many vernacular homes, converting them into hotels and restaurants.

Other very specific phenomena have directly affected vernacular construction techniques: since 1978, the use of artisanal explosives, which were used to extract stones from quarries in the hills, has been prohibited. A similar situation happens with the Cardón cactus used for beams and other carpentry, whose felling has been prohibited by the National Forestry Corporation since 2009. Finally, a significant factor hindering the preservation of traditional techniques is the generational continuity of builders. One of the grandmothers from the town of Camar recounts having built her stone house with her deceased husband; however, due to her lack of physical strength for necessary maintenance, she has moved into a new home made of concrete blocks.

Nowadays, most Andean houses keep some traditional elements and are in an intermediate state between the vernacular and the industrialized. One of the most common modifications is the replacement of doors, windows, and roofing, where metal carpentry has replaced traditional cactus-based carpentry and zinc sheets have replaced wild straw. The stone or adobe walls are often preserved, but in many cases, they are coated with cement. Therefore, many Andean houses are in a state of architectural hybridization, where it is still possible to recognize Andean patterns—such as the general dimensions of the houses and their morphology —mixed with contemporary construction elements (Fig. 5).

However, there is another percentage of housing that is entirely new and made from industrialized materials, which does not respect any—or hardly any—Andean construction patterns and is instead conceived as housing inspired by the architecture of large cities. As such, it features a broad repertoire of exogenous architectural elements, such as large windows, balconies with balustrades, and bow windows (Fig. 6). This new housing is often two stories high and is mostly constructed with unreinforced concrete block masonry, resulting in wall thicknesses of 20cm.

Figure 5 and 6. Andean houses in a state of constructive hybridization in the town of Caspana (left) and contemporary houses integrated into the Andean town of Huatacondo (right).

Comparative analysis of seismic vulnerability between traditional and contemporary housing

Seismic vulnerability in unreinforced masonry structures—such as the adobe and stone masonry that make up Andean houses and the concrete blocks used in contemporary homes—is determined by two fundamental factors: the geometry of the building and the connections between the various constructive elements [11].

Regarding geometry, there are several parameters that contribute to better seismic performance in unreinforced masonry, both in terms of global capacity of the buildings and the local in-plane and out-of-plane capacity of walls: symmetry, simple and regular shape, the ratio between the length and width of the building in plan (less than or equal to 2), the ratio between the height and width of the building, the density of the structure, the vertical slenderness of the walls, the free length of the walls, and the location and size of openings relative to the mass of the walls [12, 13, 14, 15].

These characteristics are fully met by Andean housing, which has a symmetrical plan shape with a ratio of 1:2, a high structural density due to thick walls and a small space, a pyramidal shape in height (ensuring a low center of gravity), thick walls (approximately 60 cm in thickness and 2 m in height, giving a slenderness ratio of 3.33), and small and centered openings that do not weaken the simple masonry walls.

Regarding contemporary homes, although if there is a lack of geometric pattern that allows generic analysis, there are certain common aspects that make them more seismically vulnerable, mainly the fact that despite being constructed with modern concrete blocks, they remain unreinforced masonry structures. Contemporary houses do not feature pyramidal morphology, they exhibit significantly higher wall slenderness, and they have wide window openings to the wall mass. In many cases, there is a second level that is larger than the first one (Table 2).

Table 2. Comparison of seismic performance between the Andean vernacular housing and the industrialized housing.

COMPARISON OF SEISMIC PERFORMANCE	
Andean vernacular housing	**Industrialized housing**
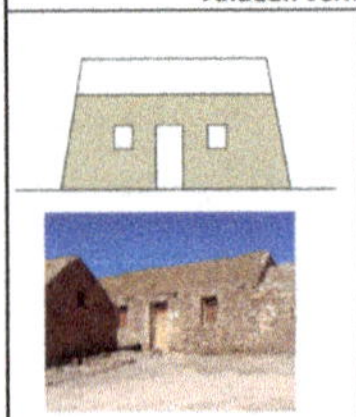 	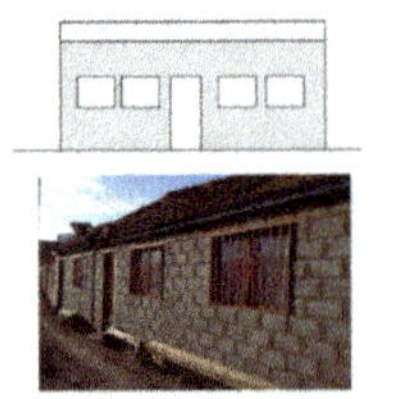 
One-store house	One, two, or three-story houses
Symmetrical plan	Symmetrical plan
High density of the structure	Low density of the structure
Pyramidal shape that ensures a low center of gravity	Parallelepiped volumes, where sometimes the upper floor is larger than the lower one
Walls with low slenderness	Walls with medium-high slenderness
Small and well-placed openings	Large openings that are not always well-placed

Comparative analysis of thermal comfort between traditional and contemporary Housing

The thermal performance of a building is one of the most important variables for assessing environmental sustainability, as it directly influences energy consumption and the ability to create comfortable interior spaces with a low impact. This performance fundamentally depends on two factors: Thermal Transmittance and Surface Mass [16].

The Thermal Transmittance value U (W/m^2K) is an expression of the thermal behavior of the wall: the lower it is, the better is the thermal insulation of the building envelope elements. The Surface Mass of the envelope elements is directly related to the density of the materials (kg/m^3) and affects the damping and delay of heat waves; therefore, the surface mass value should be sufficiently high, ideally above 200 kg/m^2 [17]. Considering that traditional Andean housing is built with adobe or stone walls, while contemporary housing typically uses industrialized hollowed concrete blocks (which are low in density), the following is a comparative table with values for Thermal Transmittance and Surface Mass for the three materials:

Materials Research Proceedings 47 (2025) 258-267 https://doi.org/10.21741/9781644903391-30

Table 3. Comparison between Transmittance and Surface Mass

Material	Thermal Transmittance U (W/m²K)	Surface Mass (kg/m²)
Adobe	0.5 - 0.7 W/m²K	425 kg/m²
Stone	1.0 - 1.5 W/m²K	810 kg/m²
Concrete blocks	2.0 - 3.0 W/m²K	96 kg/m²

The previous values indicate that adobe is the best in terms of thermal insulation, having the ability to retain heat and maintain stable interior temperatures due to its low transmittance; this is followed by stone, and lastly by the hollowed concrete block, which, with a higher thermal transmittance, loses more heat compared to adobe and stone, resulting in lower thermal performance.

Regarding the evaluation of Surface Mass, which indicates the ability to efficiently store and release heat, the stone is the most efficient, possessing a high surface mass. This is followed by adobe, which also performs well, while the hollowed concrete block has the lowest surface mass and, therefore, a limited capacity for heat storage, which can lead to more pronounced fluctuations in interior temperature.

In addition to the values mentioned, there are also differences in architectural design that make traditional housing more suitable for the local climate. Specifically, the design of the sloped roof ensures effective drainage of rainwater and snow in winter, while the small openings guarantee temperature control. These characteristics are often not respected in many contemporary homes, where roofs are frequently flat or nearly flat, and window openings are large.

Conclusions

As in all vernacular architecture, there is a deep interrelationship between the environment and the ways of living and building; if one of the factors changes, the entire system is altered. This is the case of Andean housing, where it is evident how socio-economic, productive, and environmental changes have rapidly transformed the characteristics of vernacular housing and its harmonious integration with the surroundings.

In just a few decades, local and collective construction methods, based on the natural materials of the environment, have been replaced by industrialized materials with standardized and global characteristics, which, among other things, undermine traditional construction practices. Nevertheless, many homes retain certain traditional elements, resulting in a state of architectural "hybridity" rather than a total replacement of Andean architecture.

The comparative analysis between vernacular housing and industrialized housing reveals that the former, due to its geometry, has a better seismic performance. Instead, new concrete block housing, although it has allowed many architectural liberties, is considered a seismically vulnerable example, as the technique corresponds to unreinforced masonry, putting local inhabitants at risk.

Table 4. Comparison of thermal performance between the Andean vernacular housing and the industrialized housing.

COMPARISON OF THERMAL PERFORMANCE	
Andean vernacular housing	**Industrialized housing**
 	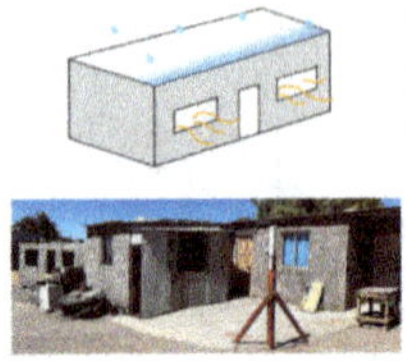 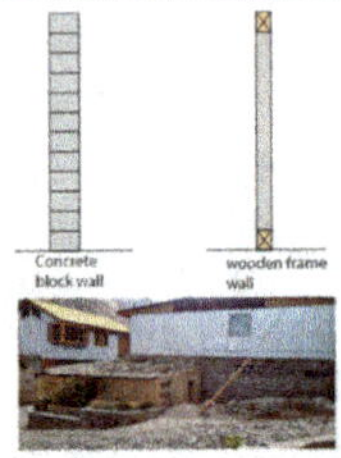
Steep roof pitch to evacuate rain and snow	Flat or nearly flat roof that does not ensure proper drainage of rain and snow
Few openings to ensure thermal control	Big openings that don't ensure thermal control
Thick walls that ensure thermal inertia	Thin walls that do not ensure thermal inertia or insulation
Adobe and stone have low thermal transmitance, which ensures good thermal insulation	Concrete block with high thermal transmitance is unable to provide insulation
Adobe and stone have a high surface mass, which ensures their capacity to efficiently store and release heat	Concrete block with low surface mass has a limited capacity for heat storage

Regarding thermal performance, it is evident that industrialized housing does not provide the same thermal benefits needed to respond to the climate of the foothills and the Andean plateau of the Atacama Desert. In summary, contemporary housing is considered less suitable and resilient compared to the vernacular constructions that existed to face the geographical challenges of the region.

It should be noted, however, that certain patterns present in contemporary homes, such as large window openings or smooth surfaces covered with ceramics, reveal that traditional Andean housing does not entirely meet the current requirements of Andean communities. Good lighting, which allows residents to spend more time indoors, as well as the ease of cleaning the spaces, are legitimate needs for inhabitants who now spend an important part of their time in urban centers and mines, who face every day the demands of work and modern life, that have internet access, and therefore choice other ways to build and create architecture. Combining the favorable aspects of traditional Andean housing in terms of seismic performance, thermal comfort, and harmony with the landscape with the current needs of Andean communities can ensure relevant architecture and development with identity.

Acknowledgments

The project that gave rise to this article has been financed by the National Fund for Cultural Development and the Arts-FONDART, a national funding area, Convocatoria 2024, of the Chilean Ministry de las Culturas, Las Artes y el Patrimonio, folio nº 750219.
To Vanessa Torres, for her support in the preparation of the graphic elements.

References

[1] B. H. Zenteno, Acercamiento a la visión cósmica del mundo Andino. Punto Cero, 14(2009), 83–89, http://www.scielo.org.bo/scielo.php?script=sci_arttext&pid=S1815-02762009000100010

[2] M. Harris, El materialismo cultural, first ed., Alianza Universidad, Madrid, 1982.

[3] M. Harris, *El desarrollo de la teoría antropológica. Historia de las teorías de la cultura.* Siglo XXI editores, Mexico, 1996

[4] J. Murra, *El Mundo Andino. Población, medio ambiente y economía,* Instituto de Estudios Peruanos, Pontificia Universidad Católica del Perú, 2002. https://doi.org/10.18800/9972510727

[5] C. Villagrán, V. Castro, G. Sánchez and M. Romo, La tradición surandina del desierto: Etnobotánica del área del Salar de Atacama (Provincia de El Loa, Región de Antofagasta, Chile), *Estudios Atacameños. Arqueología y antropología surandinas*, 16 (1998), 7–105. https://doi.org/10.22199/S07181043.1998.0016.00002

[6] N. Jorquera, S. Cornibert, S. and Y. Díaz, Estado actual y transformaciones de la arquitectura de la vivienda tradicional Likan-antai: Los casos de Ayquina, Caspana y Toconce, Chile, *Estudios Atacameños* vol 67 (2021) e3747, http://dx.doi.org/10.22199/issn.0718-1043-2021-0016.

[7] V. Šolc, Casa Aymara en Enquelga, Chungará 43((2011) 89-111. https://doi.org/10.4067/S0717-73562011000100006

[8] H. Van den Berg, Capítulo 5: Cristianización del mundo aymara y aymarización del cristianismo. *Ciencia Y Cultura* 21 (2008), 181–244, http://www.scielo.org.bo/scielo.php?script=sci_arttext&pid=S2077-33232005000100006&lng=es&nrm=iso>

[9] J. Tillería, La arquitectura sin arquitectos, algunas reflexiones sobre arquitectura vernácula. Revista AUS, 8 (2010), 12-15, https://www.redalyc.org/pdf/2817/281722857004.pdf.

[10] N. Jorquera, El patrimonio vernacular, fuente de saberes tecnológicos y de sostenibilidad, in: Cordero, (Eds.), Taller [Sur], Universidad Austral de Chile, Valdivia, Chile, 2012, 175-185.

[11] P. Lourenço, N. Mendes, L. Ramos, and D. Oliveira.Analysis of masonry structures without box behaviour. International Journal of Architectural Heritage 5 (2011):369–82. https://doi.org/10.1080/15583058.2010.528824

[12] Jorquera, N., Vargas, J., Lobos, M. & Cortez, D. (2016). Revealing Earthquake-Resistant Geometrical Features in Heritage Masonry Architecture in Santiago, Chile. En *revista International Journal of Architectural Heritage* 11(4): 519-538. https://doi.org/10.1080/15583058.2016.1266414

[13] C. Arnold and R. Reithermann, Influencia de la configuración sobre el comportamiento sísmico, Limusa, México, 1991.

[14] E. Cruz, E. Importancia de la configuración estructural. In IX Congreso Nacional de Ingeniería Estructural, Instituto Tecnológico y de Estudios Superiores de Monterrey, 280–87. Mexico, 1995.

[15] R. Meli, Structural engineering of historical buildings, ICA Foundation, Mexico, 1998.

[16] L. Dipasquale, N. Jorquera, G. Giuffrida, R. Pulselli, L. Mannucci, R. Caponetto Evaluación de los beneficios ambientales de una cabaña construida en tierra en el valle del Elqui, Chile. Actas del 22° Congreso SIACOT, La Serena, 2024.

[17] G. Giuffrida, L. Dipasquale, R. Pulselli, R. Caponetto, R. Compared Environmental Lifecycle Performances of Earth-Based Walls to Drive Building Envelope Design. Sustainability 16 (2024), 1367. https://doi.org/10.3390/su16041367

Biosourced materials and vernacular architecture: A bibliometric analysis of properties and recent developments

GHARBI Islam[1,a] *, BENAICHA Mouhcine[1,b] and MOULOUDI Hicham[1,c]

[1]National School of Architecture, Rabat, Morocco

[a]islamgharbi.archi@gmail.com, [b]m.benaicha@enarabat.ac.ma, [c]h.mouloudi@enarabat.ac.ma

Keywords: Bio-sourced Materials, Vernacular Architecture, Bibliometric Analysis, Seismic Resistance, Building Resilience, Etymology

Abstract. Vernacular architecture, which employs traditional materials and techniques tailored to local contexts, is vital for managing natural risks and enhancing building resilience. This study explores the use of bio-sourced materials as a sustainable solution in vernacular architecture. By valorizing plant waste, these materials offer alternatives to conventional resources and reduce the environmental footprint of the construction sector. A bibliometric analysis provides insights into the evolution of research in vernacular architecture, highlighting key trends, influential contributors, and collaborations within the field. This work aims to define the etymological and epistemological significance of vernacular architecture, investigate its recognition, and identify its lessons. Additionally, the study includes a survey of vernacular seismic resistance techniques and an inventory of available bio-sourced materials in Morocco that can bolster the seismic resilience of vernacular constructions.

1. Introduction

The construction industry is recognized as a significant contributor to global environmental pollution, being a major consumer of non-renewable resources and a substantial source of waste. In 2020, the construction and operation of buildings were responsible for 36% of the world's final energy consumption and 37% of carbon dioxide emissions linked to energy use [1]. To combat climate change effectively, the building sector must adopt strategies that significantly lower its energy and resource consumption. Recent efforts to lessen the environmental footprint of buildings have largely concentrated on operational impacts. While these initiatives have been effective, they only address energy use during the building's operational phase—such as heating, cooling, lighting, and rainwater management—without considering impacts from other phases of the life cycle, including construction and end-of-life disposal.

The life cycle assessment (LCA) methodology categorizes environmental impacts into operational impacts and embodied impacts. Operational impacts pertain to the energy and water usage required for maintaining a comfortable indoor environment, while embodied impacts arise from all other life cycle stages. These include the extraction and production of materials, transportation, processing at the end of the building's life, as well as maintenance and replacement activities [2]. There is a pressing need to explore alternatives that not only minimize the embodied impacts of buildings but also address the anticipated shortages of raw materials and the depletion of fossil fuels in the upcoming years (with projections of zinc depletion by 2025, lead by 2030, and copper by 2040) [3]. This challenge also extends to concrete, which is the second most used construction material globally, following water, and involves sand sourced from seabed extraction specifically for concrete production, as well as oil, which is essential for modern civilization.

One effective approach to mitigate the embodied impacts of buildings is to replace traditional construction materials, which are primarily derived from fossil fuels, with bio-sourced alternatives [1]. Today, it is feasible to create construction materials from biomass that possess mechanical,

Materials Research Proceedings 47 (2025) 268-278 https://doi.org/10.21741/9781644903391-31

thermal, acoustic, and other properties comparable to those of conventional materials, while also providing enhanced environmental performance throughout their life cycle [4–7].

Additionally, there is a growing discourse around de-territorialized architecture, which no longer represents a specific geographical context, unlike pre-industrial architecture which reflected the values, culture, and lifestyles of human societies. However, it is noteworthy that over 80% of the global building stock is classified as vernacular architecture, indicating that a significant portion of the world's population resides in vernacular homes [8].

This research focuses on vernacular architecture, which is closely linked to the geographical location where it is built. The objective is to understand the challenges of vernacular architecture and its significance in academic research, with an evaluation of vernacular seismic resistance techniques and bio-sourced materials available in Morocco. Our main hypothesis is that bio-sourced materials available in Morocco could provide sustainable solutions well adapted to local specificities, particularly in high-risk areas where traditional construction techniques play a crucial role in the resilience of buildings.

In this contribution, the first challenge is to attempt to define the etymological and epistemological meanings of the term. What is vernacular architecture? Why is this architecture less recognized? What can it teach us?

The second stage will focus on bibliometric analysis. This research method examines and summarizes the structure of knowledge in a specific field, identifies emerging trends, and highlights influential research and publications. This approach relies on quantitative techniques that provide a clear overview of the current state of knowledge and guide future research.

The final stage will involve an inventory of vernacular seismic resistance techniques, accompanied by a survey of bio-sourced materials available in Morocco that could enhance the seismic resistance of vernacular constructions.

2. Reflections on Vernacular Architecture

2.1 Etymology and Epistemological Significance

According to research by Marie-France Bisson, the term "vernacular" was first used in the context of architecture in English architectural history [9]. In literature, vernacular originally refers to the native language of a place or country [10]. It was first applied to architecture in 1857 by architect George Gilbert Scott, who used it to describe domestic, ordinary, functional architecture specific to a place or country—architecture that is not fundamentally monumental [9].

A later definition by Amos Rapoport refines this notion of vernacular architecture. The term specifically refers to buildings unique to a certain geographical context, constructed in direct response to the local physical and cultural environment. Vernacular architecture relies on local construction techniques and materials sourced from the immediate environment. It often results from an informal local tradition passed down empirically through generations and adapts to the immediate daily needs of the inhabitants [11].

Vernacular architecture remains closely tied to natural and local resources, typically avoiding materials and techniques associated with industrialization, favoring raw materials available in the immediate surroundings. This approach generates specific typological models and distinctive elements characteristic of the region, which can sometimes be quite limited. Often described as "architecture without architects," it is a product of culture and reflects that culture through its aesthetics, form, and construction methods. More than that, it is dictated largely by the climate (wind, rainfall duration and intensity, seasons, humidity, temperature, etc.) and the nature of its environment, relying on passive adaptation and keen environmental observation.

As a result, vernacular architecture not only reacts and adapts to natural disasters but also aims to enhance community resilience without relying on costly technologies or complex infrastructures, reducing dependency on external resources. This type of architecture offers critical

Materials Research Proceedings 47 (2025) 268-278
https://doi.org/10.21741/9781644903391-31

lessons in environmental, climatic, and social integration, as well as risk mitigation. Passed down from generation to generation, it embodies a deep understanding of the specific natural risks of each region.

Currently, there is a growing number of studies on vernacular architecture. In an era when buildings are often viewed as consumer goods with a lifespan equivalent to that of a human (or 1.5 times at best), the concept of programmed obsolescence in construction seems all too real as we knowingly allow the building of low-quality, short-term structures [12].

It is time to rethink our construction approach, and vernacular architecture can provide the essential ingredients for change. Pierre Frey, in his work Learning from Vernacular, writes, "Vernacular architecture offers itself to the world and architects as an enchanting catalog of suggestions and starting points for solutions open to their inventiveness" [13].

2.2 Bibliometric Analysis

Bibliometric analysis is a key tool for understanding the evolution of research on vernacular architecture over time. This section focuses on the collection and interpretation of data from scientific publications, enabling the identification of key trends, influential actors, and collaborations that shape this field. The statistical analysis was conducted using the SCOPUS database.

The examination of published documents shows a consistent increase in academic interest in vernacular architecture. This trend is illustrated by the rising number of publications over the years, reflecting a growing recognition of the importance of this subject within academic circles (Fig. 1).

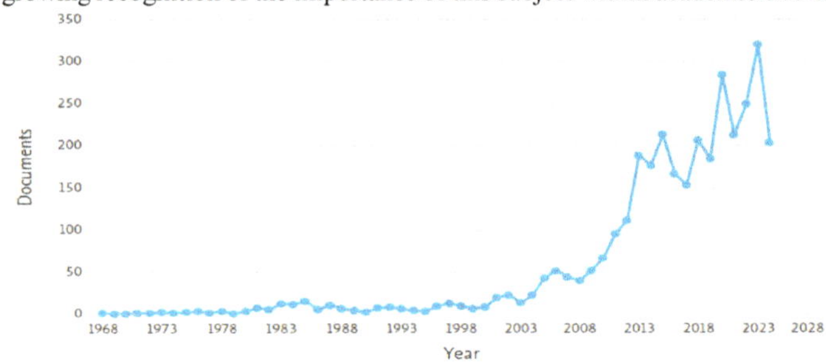

Fig. 1. Trends in Academic Publications on Vernacular Architecture Over Time.

The analysis of the types of documents published, including journal articles, conference papers, and book chapters, reveals a diversity in research formats, indicating that vernacular architecture is approached from various angles and in different contexts (Fig. 2).

Materials Research Proceedings 47 (2025) 268-278
https://doi.org/10.21741/9781644903391-31

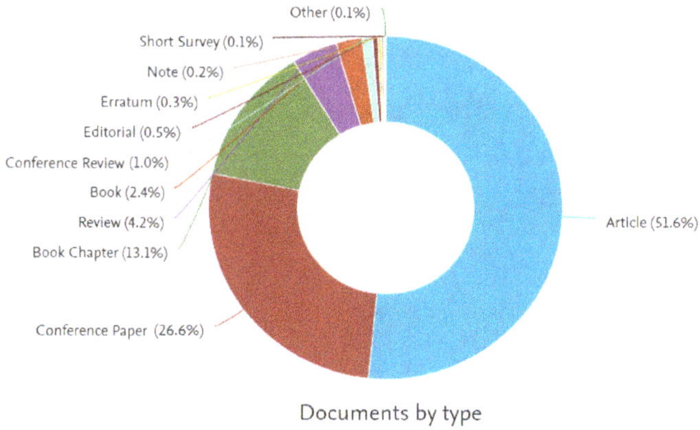

Fig. 2. Distribution of Document Types in Vernacular Architecture Research.

The most prolific countries in terms of publications on vernacular architecture include regions with a strong cultural and historical tradition. Researchers from these countries tend to explore local architectural practices while highlighting specific case studies. The analysis of international collaborations highlights a growing dynamic between academic institutions in these countries and other regions of the world, suggesting an exchange of knowledge and expertise around vernacular practices (Fig. 3).

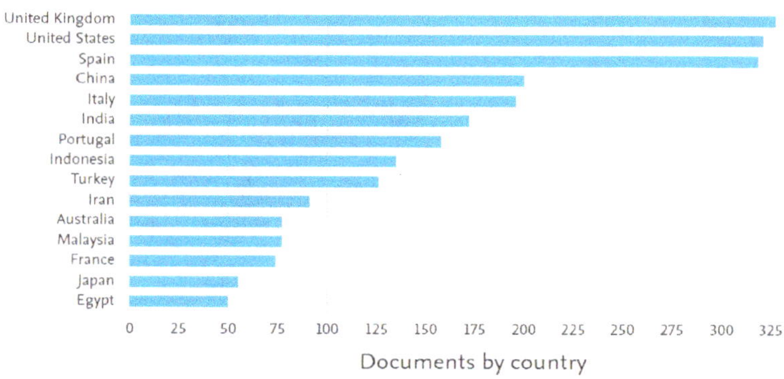

Fig. 3. Number of Vernacular Architecture Publications by Country.

The analysis of specialties sheds light on the multidisciplinarity of this field, with contributions from architecture, anthropology, geography, and social sciences, among others. This diversity indicates that vernacular architecture is not just a technical research subject but is deeply intertwined with cultural, social, and environmental issues (Fig. 4).

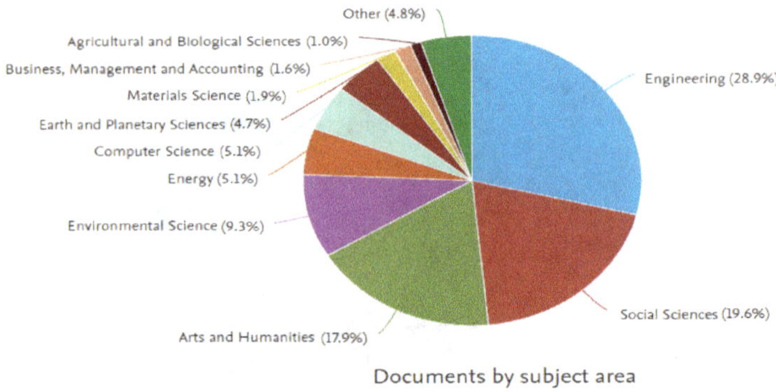

Fig.4. Multidisciplinary Contributions to Vernacular Architecture Research.

The word cloud of the most frequently used keywords in scientific publications highlights the key concepts around which research is organized. In addition to terms such as "climate," "construction," "heritage," "identity," "landscape," and "community," the analysis has revealed two dominant concepts in current literature: "bio-sourced materials" and "seismic resistance." These two notions reflect the growing importance of these themes in recent studies (Fig. 5).

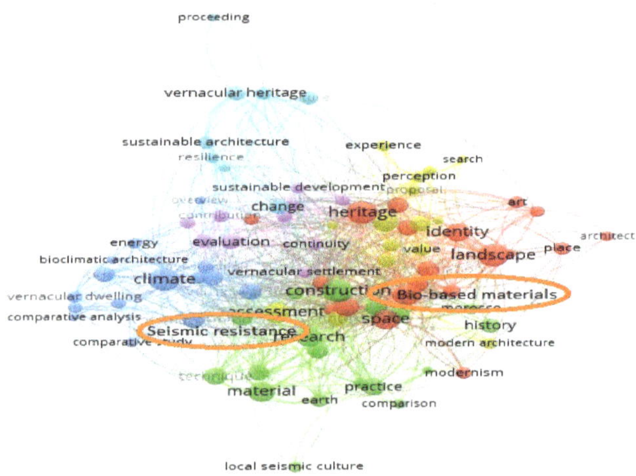

Fig. 5. Key Concepts in Vernacular Architecture Research.

Bio-sourced materials, often associated with vernacular architecture, are studied as sustainable solutions well-suited to local specifics. Meanwhile, seismic resistance is a crucial issue, especially in at-risk areas where traditional construction techniques play a determining role in the resilience of buildings.

This bibliometric analysis provides an overview of the main research trends in vernacular architecture while revealing existing gaps and opportunities for future research.

Materials Research Proceedings 47 (2025) 268-278 https://doi.org/10.21741/9781644903391-31

3. Vernacular Techniques for Seismic Resistance

3.1 Definition of the Study Area: Geographic Scope

This study is a review of the vernacular seismic resistance techniques used by indigenous Mediterranean populations. Indeed, the Mediterranean and its surroundings experience intense tectonic activity (Fig. 6). Responsible for 15% of global seismic activity, Mediterranean communities have historically faced recurring earthquakes, forcing them to make decisions to adapt, safeguard, repair, and/or restore their built environment.

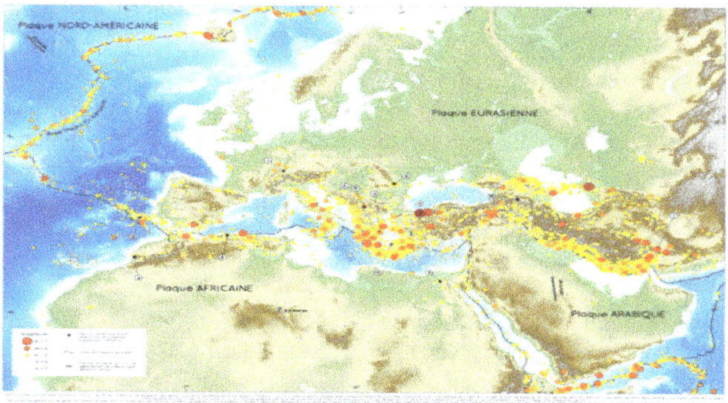

Fig. 6. Mediterranean Seismic Hazard Map [14].

These efforts by indigenous populations have led to the development of a local seismic culture through the implementation of preventive measures or traditional construction techniques that demonstrate resilience. Although these techniques were not initially designed as such, their widespread use following seismic events has recognized them as evidence of a local seismic culture [15].

3.2 Inventory of Techniques

It is essential to keep in mind that the primary objective is not to avoid deformations or damage but to prevent the collapse of the structure, which is the main cause of human losses. Indeed, it is not earthquakes that kill, but the collapse of buildings. The earthquake that struck the city of Agadir in 1960 serves as a historical example demonstrating the consequences of vulnerable construction, as it resulted in the death of 12,000 people despite having a relatively moderate magnitude of "6."

In this section, we will provide an overview of the common seismic measures easily identifiable in Mediterranean vernacular architecture, which have persisted over the centuries, proving their effectiveness against seismic events [16].

The most recurrent practices concern the geometry of buildings:

- Constructions must be simple to ensure seismic stability and symmetrical in terms of mass and rigidity, whether in plan or elevation, to counteract torsion.
- The height-to-base ratio must be low to avoid overturning, and the plan should be compact to resist forces from all directions.
- To resist tensile forces and improve the deformability of the structure, load-bearing elements must be well-coupled, forming closed contours both horizontally and vertically, and should be redundant so that the failure of some can be tolerated.

Materials Research Proceedings 47 (2025) 268-278
https://doi.org/10.21741/9781644903391-31

- The number of openings should be minimized and symmetrically arranged, with relieving arches to better distribute the load path and resist bending.
- Strengthening work following earthquakes is of utmost importance; the better preserved the building is, the less vulnerable it is.
- The ductility of materials is crucial for absorbing and dissipating energy; they deform plastically without breaking, allowing evacuation and reducing the risk of instant collapse.

Beyond these generalities, we can distinguish two types of seismic reinforcement devices in Mediterranean vernacular architecture: one aimed at counteracting horizontal forces and the other using wooden reinforcements to absorb shear movements, promoting deformation [17].

3.2.1 Strategies for Resisting Lateral Forces

- The center of gravity of the construction must be low to concentrate more mass downward: thus, retaining walls are used (Fig. 7a), and in cases of stacked levels, the walls of the lower floors must be thicker than those of the upper floors.
- Buttresses are pillars that lean against the walls to counterbalance the rotation of the facade (Fig. 7b). They can serve an additional function, such as loggias, arcades, or external stairs (Fig. 7c).
- Neighboring buildings can collaborate to reinforce each other (Fig. 7d).
- The use of relieving arches placed between two opposing buildings helps transmit stresses and prevents the buildings from isolating (Fig. 7e).

3.2.2 Strategies for Shear Force Resistance

The incorporation of flexible, lightweight, and deformable materials can effectively resist horizontal loads without reaching their breaking point while also facilitating substantial energy dissipation. Wood, in particular, possesses excellent elastic properties that enable it to adjust to daily and seasonal fluctuations in humidity and temperature, allowing for free movements that respond effectively to seismic activity. Furthermore, wooden components can segment the structure, helping to inhibit the spread of cracks in masonry sections. By establishing both horizontal and vertical connections, wooden elements integrated into structures with strong compressive properties—such as stone, adobe, or brick masonry—can enhance resistance to shear, bending, and torsional forces. Consequently, during an earthquake, while stone walls may fail, the ductility of wood can prevent complete structural collapse, thereby preserving lives.

Materials Research Proceedings 47 (2025) 268-278 https://doi.org/10.21741/9781644903391-31

Fig. 7. a) Retaining walls, Ait Othmane Village, ElHaouz; b) Buttresses, Mechouar de BavDekaken, Fez; c) External stairs serving as buttresses, Ait Othmane Village, ElHaouz; d) Party walls that remained intact after the ElHaouz earthquake, Ait Othmane Village, ElHaouz; e) Relieving arches, Old Medina of Rabat. (Credits: Islam Gharbi, 2024).

Based on the research cited above, we have identified several categories of seismic reinforcement systems that use wooden elements:

- The frame system relies on a grid of wooden pillars and beams that form the main structure and is filled according to locally available materials (Fig. 8a). This is the very principle of the Ottoman house structure, for which several reports demonstrate its validity as a seismic-resistant system, even more so than reinforced concrete or masonry [18].
- The use of wooden inserts to reinforce walls (Fig. 8b).
- Under load, wood deforms without breaking, allowing it to absorb and dissipate seismic forces. These elements thus serve as sliding planes, acting as dampers. They also connect structural elements to facilitate the box behavior of the construction. Finally, they ensure a continuous physical connection between the internal and external layers of the wall, preventing delamination (Fig. 8c).
- The use of wooden elbows to strengthen connections between walls and floors and maintain the geometric integrity of the structure. The absence of elbows is fatal. It prevents the box behavior of the construction and inevitably leads to partial or total collapse (Fig. 8d).
- Perpendicular walls are connected to ensure lateral bracing. Parallel walls are linked to constrain the floors and prevent their out-of-plane collapse. This is the role of the joists (Fig. 8e).
- The extension of window lintels in masonry beyond the edges (Fig. 8f).

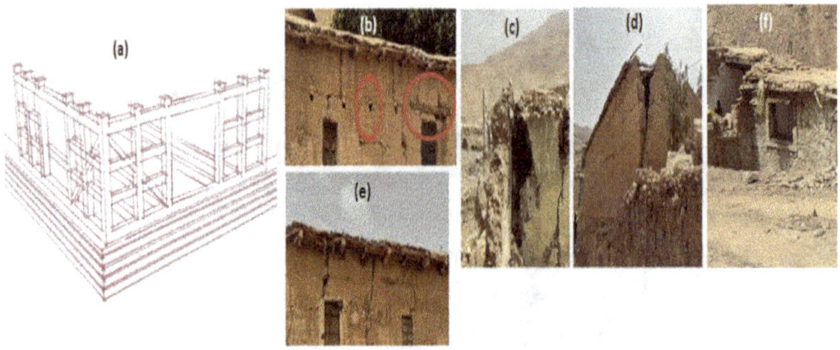

Fig. 8. a) Ottoman house system; b) Wooden inserts for wall reinforcement, Ait Othmane Village, ElHaouz; c) Wall delamination, Ait Othmane Village, ElHaouz; d) Absence of elbows to tighten the corners of the construction, partial collapse of the wall, Ait Othmane Village, ElHaouz; e) Joists ensuring the connection of parallel walls, Ait Othmane Village, ElHaouz; f) Extension of lintels in masonry, Ait Othmane Village, ElHaouz. (Credits: Islam Gharbi, 2024).

Furthermore, other materials, particularly bio-sourced ones, can be added to these structures to increase the ductility of the building and provide a remarkable response to multidirectional stresses.

3.3 Inventory of Bio-sourced Materials in Morocco

Bio-sourced materials, derived from plant or animal biomass, play an increasing role in sustainable construction due to their ability to reduce the carbon footprint of buildings while offering various functional performances. By absorbing and storing CO_2, these materials contribute positively to the carbon balance. Their thermal, acoustic, and hygrothermal insulation properties, as well as their fire resistance and energy performance, make them attractive for use in a variety of products such as insulators, mortars, concretes, panels, and partitions. However, their use in concrete structures is often limited to alternatives such as wood and geo-sourced materials. Recent research explores the possibilities offered by "green" or plant-based concretes as partial replacements for conventional materials. The inventory of available or promising bio-sourced materials in Morocco reveals their potential to enhance the seismic resistance of vernacular constructions, often found in areas with high seismic activity. The materials are classified according to their production sectors: agriculture, forestry, paludiculture, and livestock farming.

Moroccan agriculture, with its 8.7 million hectares of arable land, produces several bio-sourced materials. Straw, for example, used in rammed earth constructions, offers great ductility that effectively absorbs seismic energy. Hemp, authorized in 2022 for medicinal and industrial uses, enables the production of lightweight concrete without cement. Flax, although not widely cultivated in Morocco, improves the durability of concrete by reducing CO_2 emissions, while plants like miscanthus and mycelium represent promising but underutilized options. Agricultural waste such as olive pits or corn and sunflower stalks can also be repurposed as aggregates in concrete. In the forestry sector, forests cover 9 million hectares, and reforestation strategies aim to encourage the use of local woods such as cedar and cork oak, despite regulatory constraints in seismic construction. Paludiculture, though less widespread, offers materials like Typha and reeds, traditionally used in Moroccan architecture for their hygrothermal properties and resistance to cracking. Finally, the valorization of livestock by-products, such as sheep and camel wool or ground eggshells, provides innovative solutions to enhance building resilience against earthquakes. This inventory thus highlights a diversity of bio-sourced materials in Morocco, opening interesting prospects for sustainable architecture adapted to seismic challenges.

Conclusion

This study highlights the vital role of vernacular architecture in addressing natural risks and enhancing building resilience through the use of bio-sourced materials. By integrating traditional techniques and local resources, vernacular architecture not only preserves cultural heritage but also offers sustainable solutions to contemporary construction challenges. The bibliometric analysis provides insights into the evolving research landscape, identifying key trends and influential contributors in the field. Furthermore, the inventory of seismic-resistant vernacular techniques and available bio-sourced materials in Morocco underscores the potential for innovation and adaptation in the face of seismic threats. Ultimately, promoting the use of bio-sourced materials in vernacular architecture can lead to a more sustainable and resilient built environment, benefiting both communities and the planet.

References

[1] United Nations Environment Programme (2021), 2021 Global Status Report for Buildings and Construction: Towards a Zero-emission, Efficient and Resilient Buildings and Construction Sector, Nairobi, 2021.

[2] S. Pauliuk, N. Heeren, P. Berrill, T. Fishman, A. Nistad, Q. Tu, P. Wolfram, E.G. Hertwich, Global scenarios of resource and emission savings from material efficiency in residential buildings and cars, Nat. Commun. 12 (2021) 5097. https://doi.org/10.1038/s41467-021-25300-4

[3] J. (viaf)314846734 (role)edt Choppin, N. (viaf)305911349 (role)edt Delon, Matière grise : Matériaux/Réemploi/Architecture, (2014). http://lib.ugent.be/catalog/rug01:002930005

[4] G. Bumanis, L. Vitola, I. Pundiene, M. Sinka, D. Bajare, Gypsum, Geopolymers, and Starch—Alternative Binders for Bio-Based Building Materials: A Review and Life-Cycle Assessment, Sustainability. 12 (2020) 5666. https://doi.org/10.3390/su12145666

[5] M. Sinka, A. Korjakins, D. Bajare, Z. Zimele, G. Sahmenko, Bio-based construction panels for low carbon development, Energy Procedia. 147 (2018) 220–226. https://doi.org/10.1016/j.egypro.2018.07.063

[6] A.D. La Rosa, G. Recca, J. Summerscales, A. Latteri, G. Cozzo, G. Cicala, Bio-based versus traditional polymer composites. A life cycle assessment perspective, J. Clean. Prod. 74 (2014) 135–144. https://doi.org/10.1016/j.jclepro.2014.03.017

[7] Y. Florentin, D. Pearlmutter, B. Givoni, E. Gal, A life-cycle energy and carbon analysis of hemp-lime bio-composite building materials, Energy Build. 156 (2017) 293–305. https://doi.org/10.1016/j.enbuild.2017.09.097

[8] M. Vellinga, P. Oliver, A. Bridge, Atlas of Vernacular Architecture of the World, Routledge, London, 2024. https://doi.org/10.4324/9781003572459

[9] Bisson Marie-France, Vernaculaire moderne? : vers une compréhension de la notion d'architecture vernaculaire et de ses liens avec la modernité architecturale, Université du Québec à Montréal, 2007. https://archipel.uqam.ca/849/

[10] C.O. Nnaji, Monastic Philosophy of the Origins of University Education, Open J. Philos. 05 (2015) 228–233. https://doi.org/10.4236/ojpp.2015.54028

[11] H. Kamalipour, M. Zaroudi, Sociocultural Context and Vernacular Housing Morphology: A Case Study, Curr. Urban Stud. 02 (2014) 220–232. https://doi.org/10.4236/cus.2014.23022

[12] M. Méquignon, Durée de vie et développement durable, Les Cah. La Rech. Archit. Urbaine. (2012) 225–232. https://doi.org/10.4000/crau.587

[13] V. Didelon, Pierre Frey. Learning from Vernacular : pour une nouvelle architecture vernaculaire, Crit. d'art. (2011). https://doi.org/10.4000/critiquedart.1321

[14] G.D. Danciu L., Nandan S., Reyes C., Basili R., Weatherill G., Beauval C., Rovida A., Vilanova S., Sesetyan K., Bard P-Y., Cotton F., Wiemer S., The 2020 update of the European Seismic Hazard Model - ESHM20: Model Overview, 2021. https://doi.org/https://doi.org/10.12686/a15

[15] J. Ortega, G. Vasconcelos, M.R. Correia, Seismic-resistant building practices resulting from Local Seismic Culture, in: Seism. Retrofit. Learn. from Vernac. Archit., CRC Press, 2015: pp. 17–22. https://doi.org/10.1201/b18856-3

[16] H.V. Mariana R. Correia, Paulo B. Lourenco, Seismic Retrofitting: Learning from Vernacular Architecture, 1st Editio, CRC Press, London, 2015. https://doi.org/10.1201/b18856

[17] F. Ferrigni, C. universitario europeo per i beni culturali di Ravello, E.-O.M.H. Agreement, Ancient Buildings and Earthquakes: Reducing the Vulnerability of Historical Built-up Environment by Recovering the Local Seismic Culture : Principles, Methods, Potentialities, Edipuglia, 2005. https://books.google.co.ma/books?id=T0s9DeosdA0C

[18] A. Dutu, J. Gomes Ferreira, L. Guerreiro, F. Branco, A.M. Gonçalves, Timbered masonry for earthquake resistance in Europe, Mater. Construcción. 62 (2012) 615–628. https://doi.org/10.3989/mc.2012.01811

Materials Research Proceedings 47 (2025) 279-284 | https://doi.org/10.21741/9781644903391-32

VerSus+, heritage for people: Transmit vernacular knowledge to foster a sustainable architecture

Letizia DIPASQUALE[1,a *], Saverio MECCA[1,b], Camilla MILETO[2,c], Fernando VEGAS[2,d], Maddalena ACHENZA[3,e], Bakonirina RAKOTOMAMONJY[4,f], Sebastien MORISET[4,g], Mariana CORREIA[5,h]

[1]Department of Architecture, University of Florence, Florence, Italy

[2]Research Centre PEGASO, Universitat Politècnica de València, Valencia,Spain

[3]Department of Architecture, Università degli Studi di Cagliari, Cagliari, Italy

[4]CRAterre, Grenoble, France

[5]Portucalense University (UPT), Porto, Portugal

[a]letizia.dipasquale@unifi.it, [b]saverio.mecca@unifi.it, [c]cami2@cpa.upv.es, [d]fvegas@cpa.upv.es, [e]mariam.achenza@unica.it, [f]b.rakotomamonjy@craterre.org, [g]sebastien.moriset@grenoble.archi.fr, [h]marianacorreia@upt.pt

Keywords: Vernacular Architecture, Sustainable Architecture, Intangible Heritage

Abstract. This paper presents the results of the project 'Versus-Heritage for People', funded by the EU's Creative Europe programme. The main objective of the project was to actively involve and raise awareness among a wide audience of the value of vernacular knowledge in shaping more sustainable and resilient development models. The material and immaterial elements that make up a vernacular heritage (including local materials, construction methods, models of living and social interaction, technical and ecological knowledge for the management of territories, natural resources, settlements, etc.) have enormous potential for generating forward-looking models that can improve the environmental and social quality of our habitats, foster a sense of identity and belonging, and relate in a balanced way to the capacities of our planet. The paper explores strategies and tools for managing and transmitting the knowledge and values associated with vernacular heritage. The result is the collaboration of the five academic institutions involved in the project - Universitat Politècnica de València as coordinator (ES), University of Florence (IT), University of Cagliari (IT), CRAterre-ENSAG (FR) and Universidade Portucalense (PT).

Introduction

Vernacular architecture represents a symbiotic relationship between built environments and local knowledge. It is a form of architecture shaped not only by the physical environment—climate, topography, and available materials—but also by the knowledge systems that have evolved within specific cultural and ecological contexts. This knowledge is multifaceted, encompassing technical skills, ecological awareness, social practices, and cultural meanings that have been passed down through generations. One of the key aspects of vernacular architecture is its reliance on local knowledge to solve practical challenges, particularly those related to climate and resource management. Vernacular solutions arise from generations of experience in managing local environmental conditions, which have been tested and refined over time. Beyond technical know-how, vernacular architecture also embodies deeper layers of cultural and symbolic knowledge. Buildings often reflect the social structures, belief systems, and rituals of the community. As the ICOMOS Charter on Built Vernacular Heritage (1999) states: "Vernacular building is the traditional and natural way by which communities house themselves. It is a continuing process including necessary changes and continuous adaptation as a response to social and environmental constraints (…) It is the fundamental expression of the culture of a community, of its relationship with its territory, and, at the same time, the expression of the world's cultural diversity" [1].

The combination of material and immaterial elements that constitute vernacular heritage—such as local materials and resources, traditional building and craft techniques, living models, and social interaction patterns, as well as technical and environmental knowledge for managing territories, natural resources, and settlements—holds great potential for enhancing the environmental and social quality of our habitats. It can foster a deeper sense of identity and belonging while promoting a more balanced and harmonious relationship with the planet's natural capacities. This architecture is not only cost-effective but also has a low environmental impact, so plays a vital role in contemporary society by imparting essential principles for sustainable and respectful design.

The lessons from vernacular heritage have been thoroughly explored in the VerSus project: *Lessons from Vernacular Heritage in Sustainable Architecture* (grant 2012-2792/001-001 CU7 COOP7), co-funded by the European Union from 2012 to 2014 [2]. The VerSus project, relevant to both scientific and technical communities, as well as the general public, had a far-reaching impact. It garnered over 10,000 document downloads from its website, earned international recognition, and received the 2017 European Award for Architectural Heritage Intervention in the Dissemination category. The European Union also hailed it as a "success story" for its significant influence, policy contributions, and innovative approach, serving as an inspiration for future initiatives.

Despite its success within European and international communities, the project's knowledge and insights must permeate deeper into society to raise awareness across all sectors. The intangible heritage of knowledge and skills—embedded in vernacular heritage must be made explicit to foster education and creative approaches for protecting it. This requires a dynamic understanding and awareness of its evolving nature. This is crucial for the preservation of both tangible and intangible heritage and to promote more sustainable architecture for the future.

To tackle this challenge, the *VerSus+ | Heritage for PEOPLE* project was launched. It focused on disseminating this knowledge to all levels of society, with particular emphasis on the next generation (children and youth), as well as local, regional, and national heritage authorities. Additionally, the project engages experts in architecture (architects, engineers, cultural managers, historians, ethnographers, and students), alongside craftspeople, construction companies, tourism sectors, cultural associations, and educational institutions. The *VerSus+ | Heritage for PEOPLE* project was co-funded by the European Union between 2019 and 2023 [3].

Objective of the project and methodology

The VerSus+ | Heritage for PEOPLE project focused on three core aims:

- *Promote international cooperation*: encourage collaboration between experts across Europe in sustainable and vernacular heritage.
- Raise awareness of vernacular habitats: highlight the sustainable and cultural values of vernacular architecture, focusing on its lessons for contemporary and future societies.
- *Disseminate knowledge*: spread the techniques and principles of vernacular architecture across European society to foster cultural and environmental sustainability.

Building on these general goals, four specific objectives were established:

- To foster a multidisciplinary methodology to analyze vernacular heritage, considering both tangible and intangible aspects and the relevance of heritage values for architecture and local society.
- To engage a diverse audience (general public, children, migrants, administrators, craftspeople, etc.) through various strategies (participatory, creative) and tools (traditional, digital, visual, interactive) to transmit the principles and lessons of vernacular heritage

- To empower local artists and craftspeople, promoting traditional materials and techniques in contemporary society, making local artisans key agents in revitalizing and innovating local architecture through promotion, appreciation, and intercultural exchange.
- To share and disseminate the project's methodology, strategies, activities, and results at local, regional, and European levels through various media (presentations, talks, social networks, publications).

The objectives of the VerSus+ Heritage for PEOPLE project guided the development of a multidisciplinary methodology aimed at transferring the principles and lessons of vernacular heritage to society. The activities were specifically designed to promote citizen participation, education, and the dissemination of knowledge, all in close collaboration with local administrations and community stakeholders. Four main strategies guided the project's actions. Based on these four strategies, a variety of actions were developed, each addressing one or more strategies and targeting different groups.

- *Research and Documentation:* The first step in knowledge transfer is a deep understanding of what needs to be communicated.
- *Community Engagement:* Involving local communities is critical when working with vernacular heritage, as they are both its users and guardians. Establishing communication channels with the local population, authorities, heritage specialists, and craftspeople is essential for ensuring meaningful participation and engagement.
- *Educational Strategies*: Education, both formal and informal, is key to raising awareness among young people, the future pillars of society. Educational initiatives are crucial for passing down the knowledge of tangible and intangible heritage to future generations.
- *Dissemination Strategies:* Dissemination, whether on a local, regional, national, or international scale, allows the project's concepts, content, and activities to reach diverse audiences, including the scientific community and the general public. Digital platforms and networks provide valuable tools for extending this reach.

Case studies

The project selected small Mediterranean islands as pilot locations there apply the identified methodology. These geographically constrained territories, while limited in resources, offer significant potential for rethinking development models rooted in self-sufficiency. By fostering sustainable interactions with local environmental, cultural, and economic resources, these regions can serve as models for sustainable growth. However, they are also highly vulnerable, with traditional heritage under constant pressure from contemporary transformations, particularly the impact of mass tourism.

The pilot experiences acted as a proving ground for initiatives focused on social participation, education, communication, and dissemination, utilizing diverse contexts and media. These actions were intended to create a ripple effect, influencing not only the region and country but also offering potential for broader application in other European and international settings. Two Mediterranean islands were chosen for their size, accessibility, and notable vernacular heritage: Formentera in Spain's Balearic Islands, and Sant'Antioco in Italy's Sulcis archipelago. Both islands, though similar in size and population (12,124 and 11,811 residents, respectively), present distinct cultural, architectural, and environmental characteristics, offering varied insights into the challenges and opportunities of preserving local heritage.

	KNOWLEDGE TRANSFER STRATEGIE				TARGET GROUP										
	1. Research and Documentation	2. Engagement	3. Education	4. Dissemination	uni. students	children	young people	authorities	professionals	researchers	craftspeople	artists	migrants	tourists	general pub.
CONFERENCE AND SEMINARS															
International Conference	•		•	•	•					•					
Conference proceedings	•			•						•					
Conference exhibitions	•			•	•					•					
Scientific Seminar	•		•	•	•					•					
Scientific workshop	•			•						•					
Seminar proceedings	•			•						•					
FESTIVALS															
Architecture documentation	•				•					•					
Craftspeople interviews	•	•								•	•				
Artists interviews	•	•								•		•			
Seminar Formentera (Talks)	•	•	•	•	•			•	•	•	•	•		•	•
Didactic Workshop Calasetta	•	•	•	•	•					•					
Didactic Workshop Formentera	•	•	•	•	•					•					
Art and craft call		•		•								•			
Art and craft exhibition		•		•	•	•	•	•	•	•	•	•	•	•	•
Art and craft award				•							•	•			
Dry Stone workshop		•	•		•				•	•	•	•	•	•	
Recycled Materials workshop		•	•		•	•	•				•	•	•	•	•
VerSus workshop	•		•		•										
TOOLS															
Web App	•	•	•	•	•			•	•	•	•	•			•
Website				•	•		•	•	•	•	•	•			•
Instagram/Facebook				•	•		•	•	•	•	•	•	•	•	•
Youtube Channel	•			•	•		•	•	•	•	•	•	•	•	•
Digital interactive games		•	•	•	•	•	•								•
7 Families game		•	•	•		•	•						•	•	•
Booklet				•	•			•	•	•					
Book	•			•	•			•		•					
Videos				•	•	•	•	•	•	•	•	•	•	•	•
Digital exhibition				•	•			•	•	•					•
TOTAL	16	12	11	22	20	5	8	10	10	22	11	12	7	8	11

Fig. 2. Table showing the 4 strategies, 28 actions, and 11 target groups established in the project (credits: authors)

Project results

The VerSus+PEOPLE project carried out nearly thirty activities aimed at diverse audiences, addressing research, community engagement, education, and public dissemination. These initiatives included scientific studies, fieldwork, social participation, training for craftspeople and artists, and public outreach through festivals and digital tools (fig.1).

The scientific component of the project encompassed research, training, and dissemination efforts within a structured academic framework, producing significant scholarly results.

A free, open-access platform, *Heritage for People (https://heritageforpeople.unifi.it/) (Fig.2)*, allows users to upload and analyze cases of tangible and intangible vernacular heritage using VerSus principles, creating a growing collaborative database [4].

Four interactive games were developed to engage users with Calasetta's tangible and intangible heritage on-site, and a card game based on Formentera's landscape units helped the public learn about the relationship between the island's landscape, materials, and architecture. A booklet was also created, offering 20 examples of educational and community participation activities to inspire teachers and cultural managers in heritage-based learning.

The book *Heritage for People. Sharing Vernacular Knowledge to Build the Future* analyzes four strategies for knowledge transfer related to vernacular heritage and offers best practice examples. Videos documenting festival activities, including interviews with artisans, and a digital exhibition that reinterpreted heritage from a contemporary digital art perspective, were also created [5].

Materials Research Proceedings 47 (2025) 279-284 https://doi.org/10.21741/9781644903391-32

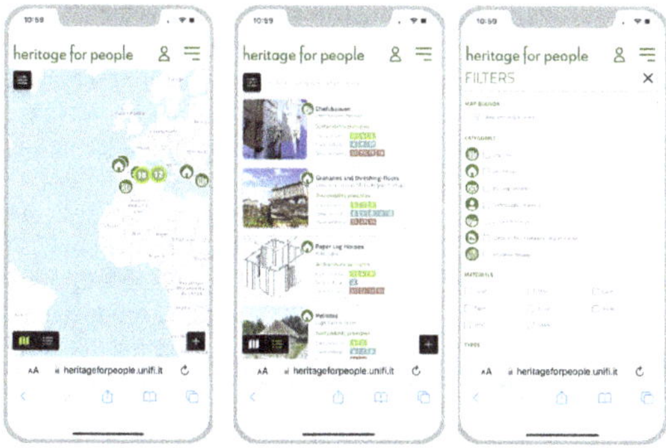

Fig. 2. Collaborative platform Heritage for People (credits: authors)

Fig.3. Educational and practical workshops with students in Formentera, Spain

To support education, social participation, and outreach, the project developed various tools. The project website and social media platforms (Facebook, Instagram, YouTube) highlighted key achievements and featured interviews with artists and craftspeople, showcasing their work.

Two festivals, held in Sant'Antioco (Italy, April 2022) and Formentera (Spain, March 2023), presented a range of activities aligned with the project's core strategies and aimed at engaging a broad audience. The festivals digitally documented vernacular architecture using advanced 3D laser scanning techniques, enabling detailed analysis of materials, construction methods, and their environmental context.

In Formentera, interviews with local artisans and artists shed light on traditional techniques and their sustainability links. A series of ten talks facilitated dialogue between local residents and experts on heritage preservation and the use of traditional materials in modern architecture. Educational VerSus workshops, held in both locations with students from the Universities of Florence, Cagliari, and Valencia, focused on documenting and enhancing vernacular heritage. Students analyzed traditional architecture and explored sustainable principles to inspire and design new uses (Fig.3).

An international arts and crafts exhibition in Formentera displayed 28 selected works, with prizes awarded to those that best reflected the project's themes. Additionally, hands-on workshops offered participants practical learning in traditional techniques such as dry stone construction and traditional practices like fishing and vine cultivation. The VerSus methodology was central to these activities, fostering interactive community engagement through games and learning sessions focused on local heritage

Key events included the *HERITAGE 2023 – International Conference on Vernacular Heritage* (Valencia, September 2022), with 254 authors from 25 countries, 134 peer-reviewed papers, and four exhibitions, and the seminar *Which Future for Vernacular & Earthen Architecture?* (Porto, June 2023), which gathered over 60 experts from 20 countries, produced 31 papers, and featured an international workshop along with student work showcasing the relevance of vernacular architecture in sustainable design.

Conclusion

The VerSus+ / Heritage for People project successfully aimed to transfer sustainability principles from vernacular heritage to contemporary life and architecture. The project engaged a wide range of audiences, including youth, specialists, residents, tourists, and migrants. By focusing on knowledge sharing, education, participation, and dissemination, the project effectively reached its target groups in the pilot locations. The methodologies and strategies developed in Sant'Antioco and Formentera have demonstrated their adaptability to other contexts, providing valuable insights into the challenges and opportunities encountered in each unique setting.

References

[1] ICOMOS 1999, Charter on the Built Vernacular Heritage, available on: https://www.icomos.org/images/DOCUMENTS/Charters/vernacular_e.pdf

[2] M. Correia, L. Dipasquale, S. Mecca (eds.) 2014, Versus: Heritage for Tomorrow. Vernacular Knowledge for Sustainable Architecture, Firenze University Press, Florence, 2014. https://doi.org.10.36253/978-88-6655-742-5

[3] C. Mileto, F. Vegas, M. Correia, G. Carlos, L. Dipasquale, S. Mecca, M.Achenza, B. Rakotomamonjy and N. Sánchez. "The European Project "Versus+ / Heritage For People". Objectives And Methodology." submitted to ISPRS - Int. Arch. Photogramm. Remote Sens. Spatial Inf. Sci. (2020), XLIV-M-1-2020, 645–649, https://doi.org/10.5194/isprs-archives-XLIV-M-1-2020-645-2020.

[4] L. Dipasquale, J. Ammendola, L. Montoni, E.P. Ferrari, M. Zambelli, Harnessing Vernacular Knowledge for Contemporary Sustainable Design through a Collaborative Digital Platform: submitted to Heritage (2024), 7, 5251-5267. https://doi.org/10.3390/heritage7090247

[5] L. Dipasquale, S. Mecca, L. Montoni, Heritage for people Sharing vernacular knowledge to build the future, DIDA Press, 2023

Materials Research Proceedings 47 (2025) 285-293 https://doi.org/10.21741/9781644903391-33

Geotechnical and geophysical characterization of traditional construction materials in the Tafilalet Region: Towards sustainable building

Ayoub SOUILEH[1,a] *, Mohammed CHRACHMY[2,b], Latifa OUADIF[1,c], Mohamed AZROUR[2,d], and Achraf MABROUK[3,e]

[1]L3GIE, Mohammadia Engineering School, Mohammed V University in Rabat, Morocco

[2]Laboratory of Materials Engineering for the Environment and Natural Resources, Faculty of Sciences and Technologies, Moulay Ismail University of Meknes, 52000, Errachidia, Morocco

[3]LAFH, Faculty of Sciences and Techniques, Hassan 1st University, Settat, Morocco

[a]ayoub.souileh@research.emi.ac.ma, [b]mo.chrachmy@edu.umi.ac.ma, [c]ouadif@emi.ac.ma, [d]m.azrour@fste.umi.ac.ma, [e]achraf.mab@gmail.com

Keywords: Traditional Construction Materials, Sustainable Building, Eco-Friendly Construction, Geotechnical, And Geophysical

Abstract. This study focuses on the geotechnical and geophysical characterization of traditional construction materials in the Tafilalet region, Morocco. The objective is to assess the physical properties of soils used in traditional construction to optimize their use in sustainable building projects. The materials, primarily composed of raw earth, were analyzed through granulometric and compaction tests, along with geophysical methods such as electrical resistivity tomography (ERT) and ground-penetrating radar (GPR). These analyses evaluate soil composition, compaction potential, and stability, particularly to moisture content, providing recommendations for the optimal use of local materials in eco-friendly construction

Introduction

Globally, vernacular architecture is increasingly being recognized for its contributions to sustainable development. Regions such as North Africa, Latin America, and arid zones in the Middle East have long utilized local materials and climate-adapted construction methods, offering valuable lessons for modern building practices. In this context, the Tafilalet region's traditional construction methods stand as an important case study for how indigenous knowledge can be harnessed to address contemporary environmental challenges.

Vernacular architecture refers to traditional, local constructions rooted in ancestral knowledge, which have been adapted to specific environmental and cultural contexts. In today's global context marked by major environmental, social, and economic challenges this form of architecture is increasingly recognized as a key solution for promoting sustainable territorial development[1,2]. The recent International Colloquium on Vernacular Architecture on October 4-5, 2024, organized by the School Nationale for Architecture in collaboration with its research laboratories, bears witness to the growing scientific interest in this subject.

In the framework of sustainable development, traditional construction materials are gaining renewed attention due to their low environmental impact and historical significance, particularly in regions facing harsh climatic conditions. For example, the Tafilalet region in southeastern Morocco, characterized by an arid climate and scarce natural resources, has a long history of using locally sourced materials like raw earth, clay, and adobe bricks. These materials are valued for their thermal inertia and adaptability to the local environment, contributing to the construction of buildings well-suited to hot, dry climates[3,4].

Despite their historical significance, traditional materials encounter challenges in modern construction projects due to the increasing demand for enhanced mechanical properties and durability, particularly in light of evolving safety and performance standards. Characterizing these materials without compromising their ecological benefits presents a significant challenge, necessitating a deeper understanding of their inherent properties and potential improvements.

Recent studies have demonstrated the value of integrating geotechnical and geophysical analyses to evaluate the suitability of these traditional materials for contemporary building practices. Techniques like granulometric analysis and Electrical Resistivity Tomography (ERT) offer valuable insights into subsurface stability, which is crucial for ensuring the long-term durability of structures in regions with challenging soil and moisture conditions.

This research aims to assess the potential of traditional earthen materials in sustainable modern construction in the Tafilalet region. The goal is to demonstrate their performance while preserving their ecological and cultural value, contributing to the development of resilient infrastructure in arid regions.

Materials and methods

The study began by identifying traditional construction materials in the Tafilalet region. These materials, primarily clay, raw earth, and earthen bricks, were chosen for their local availability and historical significance (Figures 1 and 2). Although environmentally friendly and suited to local climate conditions, these materials need to be characterized to enhance their mechanical strength and thermal insulation properties. Samples were collected from various representative sites across the region to ensure geological diversity and enable a comprehensive analysis of soil characteristics.

Geotechnical tests were conducted to characterize the physical and mechanical properties of the identified materials. These tests were performed according to ASTM and ISO standards to ensure reliable and reproducible measurements. The Atterberg limits test was used to evaluate the plasticity properties of the soils by measuring the liquid limit and plastic limit, following ASTM D4318 [5]and ISO 17892-12.[6] These properties are essential for understanding the behavior of the materials when exposed to varying moisture conditions.

The Proctor compaction test was carried out to determine the maximum dry density and optimal moisture content of the materials, ensuring effective compaction and stable structures. This test was conducted according to ASTM D698 [7]and ISO 10372.[8] Additionally, a granulometric analysis was performed to determine the distribution of soil particles, including the proportions of sand, silt, and clay, which directly influence the strength and stability of the materials. The granulometry test followed ASTM D422 [9]and ISO 178 92-4.[10]

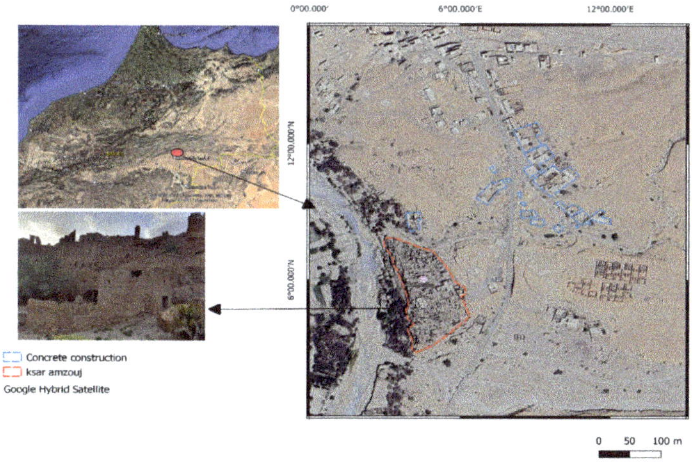

Figure 1: Location Map of Ksar Amjouj: Highlighting Traditional Earthen Structures (in Red) Surrounded by Modern Concrete Buildings (in Blue), Depicted via Satellite Imagery, Morocco.

Figure 2: Rammed earth construction type in the oasis (Kssar Amezouj, Tafilalet region).

Geophysical tests were also conducted to assess the subsurface conditions of potential construction sites in the region. The first method used was electrical resistivity tomography (ERT), which mapped the electrical resistivity of soils to provide information on the composition and geological structure of the subsurface at various depths. This method followed ASTM G57 . [11]Ground-penetrating radar (GPR) was also employed to generate detailed images of underground layers, allowing the detection of anomalies or obstacles, as well as areas where the soil structure is sufficiently homogeneous and stable for construction. The GPR tests were conducted by ASTM D6432[12]. Finally, the results of all analyses (geotechnical tests, and geophysical tests) were subjected to in-depth analysis.

Results

Grain Size Analysis

The granulometric analysis was conducted using the standard sieve test to determine the particle size distribution of the collected soil samples. This method is crucial for understanding the compaction potential, permeability, and load-bearing capacity of the soils, as the relative proportions of clay, silt, and sand significantly affect their behavior under mechanical stress. The particle size distribution is visualized through the granulometric curves shown in Figure 3, which illustrate the cumulative percentage of particles passing through each sieve. According to the literature, the raw material particle size distribution has a direct effect on the properties of the bricks that are created in the end[13].

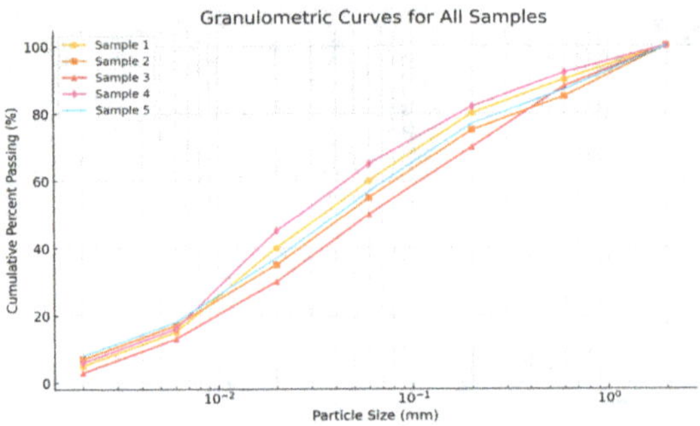

Figure 3 : Granulometric curves for all soil samples.

When compared to similar studies conducted in other arid regions, such as the Saharan and Middle Eastern deserts, the materials from the Tafilalet region exhibit relatively higher thermal conductivity due to the predominance of clay in the composition. However, the compaction properties are on par with those of other raw earth materials found in arid climates. This suggests that while Tafilalet soils are well-suited for maintaining stable structures, additional insulation measures might be necessary when these materials are used in regions with higher temperature fluctuations.

The thermal conductivity

The thermal conductivity tests were carried out to assess the heat transfer properties of the soils, which is particularly relevant for understanding their insulation potential in construction applications. Soils with higher clay content tend to have better thermal conductivity due to the fine particle structure, which allows for more points of contact between particles, enhancing heat transfer. As shown in Figure 3, Sample 1 exhibits the highest thermal conductivity at 1.45 W/m·K, making it suitable for insulation purposes in buildings, particularly in hot climates. In contrast,

Materials Research Proceedings 47 (2025) 285-293 https://doi.org/10.21741/9781644903391-33

Sample 4, with lower clay content and higher sand proportion, has a thermal conductivity of 1.28 W/m·K, making it less efficient in retaining heat.
This sample might require additional insulation materials if used in construction where thermal performance is a concern.

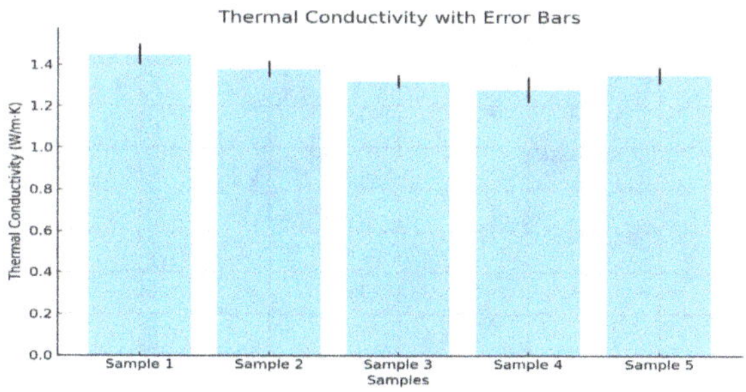

Figure 4: Thermal Conductivity Measurements of Soil Samples with Associated Error Bars

Proctor tests

The results of Proctor tests conducted on five soil samples show significant variations in terms of maximum dry density and optimal moisture content, both key factors for assessing soil compaction and stability. Samples 1, 3, 4, and 5 exhibit relatively similar maximum dry densities, all exceeding 1.9 g/cm³. In contrast, Sample 2 achieves effective compaction at a lower moisture content (just below 11.6%), though this reduced water content may influence its mechanical properties. These results highlight the importance of adjusting moisture levels to optimize soil compaction according to specific soil compositions, thereby improving the stability and durability of construction materials, particularly in the context of projects focused on promoting sustainability and infrastructure resilience.

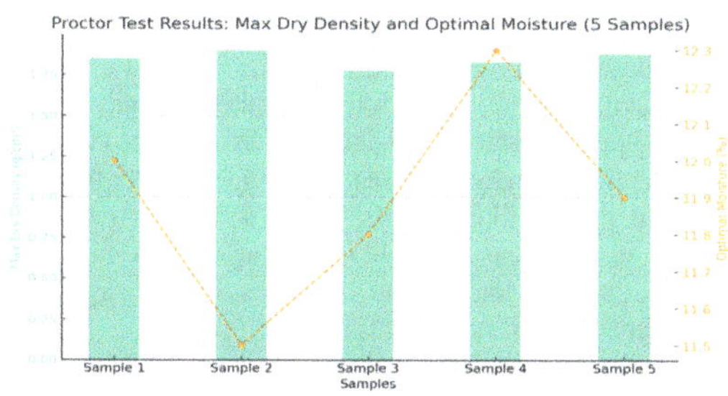

Figure 5: Proctor test results for maximum dry density and optimal moisture content.

The Electrical Resistivity Tomography (ERT)

The Electrical Resistivity Tomography (ERT) technique was used to map the subsurface resistivity of the soils, providing insights into their moisture content and stability. Higher resistivity values typically indicate drier and more stable soils, while lower resistivity is associated with higher moisture content, which can lead to instability[14]. As shown in Figure 6, Sample 3 and Sample 4 exhibit high resistivity values, suggesting that these soils are suitable for construction, particularly in regions where moisture control is a concern. In contrast, areas with lower resistivity may require

Materials Research Proceedings 47 (2025) 285-293 https://doi.org/10.21741/9781644903391-33

drainage solutions or additional stabilization to prevent potential settlement or subsidence due to moisture infiltration.

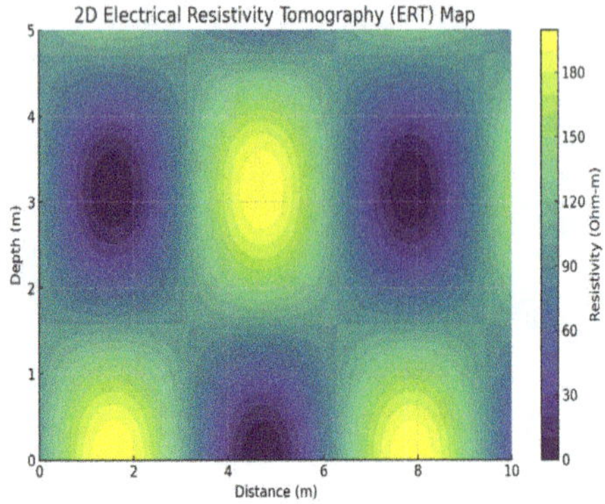

Figure 6: ERT map showing subsurface resistivity variations in the soil.

The Electrical Resistivity Tomography (ERT) results provide a crucial insight into the subsurface conditions of the soils by measuring the resistivity of the soil layers. Resistivity is highly dependent on the moisture content of the soil: soils with higher resistivity are generally drier and, as a result, more stable, whereas lower resistivity suggests higher moisture content, which can weaken the soil's structure and lead to instability.

In this study, the areas with high resistivity, notably those corresponding to Samples 3 and 4, are interpreted as being relatively dry. This suggests that these areas have low moisture content and, therefore, possess better load-bearing capacity and stability. These soils are less prone to settlement or subsidence, making them ideal candidates for construction projects, particularly for foundational structures. In regions where maintaining stability is crucial, such as in the development of roads, buildings, or other infrastructure, these areas can be confidently selected for use without the need for significant soil modification.

In contrast, regions displaying low resistivity are likely to have higher moisture content, indicating potential challenges. Moisture-saturated soils are more prone to issues like compaction, settlement, or even liquefaction in extreme cases, particularly under the influence of heavy loads or dynamic environmental factors like rainfall. These soils may require drainage solutions to reduce moisture content and increase stability, or they might necessitate soil stabilization techniques (such as the addition of cement, lime, or other stabilizing agents) to enhance their load-bearing capacity and prevent deformation over time.

The interpretation of the ERT results helps identify which areas need pre-construction interventions and which can be used as-is for building projects. Soils with high resistivity, like those in Samples 3 and 4, will likely need minimal treatment, while low-resistivity soils will demand more careful management to avoid construction issues related to excessive moisture. This approach not only improves the safety and durability of infrastructure but also optimizes resource allocation, as areas requiring intervention can be identified early in the planning process, saving time and costs related to construction delays or soil failure.

Variation in apparent density due to fiber percentage.

Figure 8 illustrates a general trend of decreasing apparent density as the fiber content increases across all types of fibers[15]. The most significant reduction is observed in the Palm cluster, where the apparent density drops from 1586.3 kg/m^3 (0% fiber content) to 1173.2 kg/m^3 (6% fiber

Materials Research Proceedings 47 (2025) 285-293 https://doi.org/10.21741/9781644903391-33

content). Other fiber types, such as Palm fiber mesh and Petiole, show a similar pattern but with less pronounced reductions.

This analysis suggests that increasing fiber content reduces the density of the materials, potentially making them lighter and improving their thermal insulation properties. This is crucial for optimizing the use of traditional construction materials in sustainable building practices, as lighter materials with better insulation capabilities may enhance energy efficiency and reduce environmental impact.

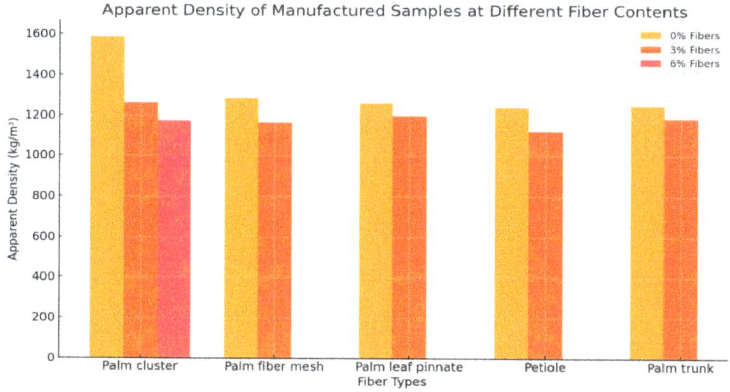

Figure 7 : Apparent density of manufactured samples (kg/m^3) as a function of fiber content (0%, 3%, 6%) for different fiber types: Palm cluster, Palm fiber mesh, Palm leaf pinnate, Petiole, and Palm trunk. The bars represent the apparent density of each material.

Discussion

To illustrate the practical application of this study's findings, a hypothetical project could be considered in the oasis of Ksar Amzjouj, Using the materials from Samples 3 and 4, which showed high resistivity and compaction stability, these soils could be utilized for constructing residential buildings with minimal need for soil modification.

However, for areas where low resistivity soils were identified, such as the location corresponding to Sample 2, drainage systems and additional stabilization agents (e.g., lime or cement) would be required to ensure long-term structural integrity.[16]

The integration of geotechnical and geophysical testing further strengthens the conclusions of our study research, such as the work of Ouakarrouch et al. (2019),[17] which focused solely on the geotechnical aspects of traditional materials. Our addition of geophysical methods (ERT and GPR) provides a more comprehensive understanding of subsurface conditions that optimized materials are used in stable and moisture-controlled environments. This interdisciplinary approach, combining geotechnics, and geophysics, ensures that sustainable building practices can be achieved without compromising on structural integrity.[18]

Moreover, our results highlight the importance of soil resistivity in construction planning. In comparison to studies that focus on soil compaction and strength, such as Ouakrouch et al. (2020), our study emphasizes moisture control as a critical factor in determining site suitability for construction. This geophysical insight is essential for the development of long-lasting structures in arid regions, where moisture fluctuations significant risks to stability.[17]

Conclusion

The study highlights the importance of geotechnical and geophysical characterization of traditional construction materials to ensure durable and stable structures in the Tafilalet region. Compaction tests and geophysical analyses revealed that certain soils have better stability due to lower moisture content, making them suitable for construction without major interventions. Other areas, with higher moisture content, may require additional measures such as drainage or stabilization to ensure a stable foundation. This integrated approach maximizes the use of local materials while ensuring the long-term durability of structures built in an arid environment.

Future research should explore how these traditional materials can be integrated into modern construction techniques without losing their ecological and cultural significance. Collaboration between local governments, architects, and international organizations could promote the wider adoption of these materials in rural and urban settings, fostering resilient infrastructure development. Additionally, the potential for combining traditional materials with modern, eco-friendly innovations such as phase-change materials or natural fiber composites holds promise for further enhancing earthen structures' thermal and mechanical performance in arid regions.

References

[1] Jamil AM, Shahzad AK, Mustafa A. Paradigm Shift of Construction Techniques from Vernacular to Neo-- Vernacular in Villages to Sustain Climate Change. 8.

[2] Pardo JMF. Challenges and Current Research Trends for Vernacular Architecture in a Global World: A Literature Review. Buildings [Internet] 2023 [cited 2024 Oct 14];13:162. https://doi.org/10.3390/buildings13010162

[3] Santos MM, Ferreira AV, Lanzinha JCG. Sustainable Vernacular Architecture to Improve Thermal Comfort in African Countries [Internet]. In: Lanzinha JCG, Qualharini EL, editors. Proceedings of CIRMARE 2023. Cham: Springer Nature Switzerland; 2024 [cited 2024 Oct 14]. page 564–75. https://doi.org/10.1007/978-3-031-48461-2_48

[4] Beraaouz M, Abioui M, Hssaisoune M, Martínez-Frías J. Khettaras in the Tafilalet oasis (Morocco): contribution to the promotion of tourism and sustainable development. Built Heritage [Internet] 2022 [cited 2024 Oct 13];6:24. https://doi.org/10.1186/s43238-022-00073-x

[5] ASTM. (2005). ASTM D4318, standard test method for liquid limit, plastic limit, and plasticity index of soils.

[6] SO 17892-12: Geotechnical investigation and testing – Laboratory testing of soil – Part 12: Determination of Atterberg limits. International Organization for Standardization.

[7] ASTM D698: Standard Test Methods for Laboratory Compaction Characteristics of Soil Using Standard Effort. ASTM International.

[8] ISO 10372: Geotechnical investigation and testing – Determination of soil compaction properties using the Proctor test. International Organization for Standardization.

[9] ASTM D422: Standard Test Method for Particle-Size Analysis of Soils. ASTM International.

[10] SO 17892-4: Geotechnical investigation and testing – Laboratory testing of soil – Part 4: Determination of particle size distribution. International Organization for Standardization.

[11] ASTM G57: Standard Test Method for Field Measurement of Soil Resistivity Using the Wenner Four-Electrode Method. ASTM International.

[12] ASTM D6432: Standard Guide for Using the Surface Ground Penetrating Radar Method for Subsurface Investigation. ASTM International.

[13] Lechheb M, Chrachmy M, Ben Baaziz M, G. El Boukili, Ouallal H, Azdouz M, et al. Mechanical and Thermal Properties of Goulmima's Earth Bricks. Materials Research Proceedings [Internet] 2024 [cited 2024 Mar 6];40:198–206. https://doi.org/10.21741/9781644903117-21

[14] Perrone A, Lapenna V, Piscitelli S. Electrical resistivity tomography technique for landslide investigation: A review. Earth-Science Reviews [Internet] 2014 [cited 2024 Oct 14];135:65–82. https://doi.org/10.1016/j.earscirev.2014.04.002

[15] Chrachmy M, Ouallal H. Improving Thermal Insulation and Mechanical Properties of Building Bricks made from Moroccan Clay. Materials Research Proceedings [Internet] 2024 [cited 2024 Mar 6];40:260–72. https://doi.org/10.21741/9781644903117-28

[16] Lindqvist, S., Dahl, A., Smith, M., Blouin, A., Dillström, P., Szavai, S., & Nicak, T. (2022, July). Advanced structural integrity assessment tools for safe long term operation (ATLAS+). In Pressure Vessels and Piping Conference (Vol. 86175, p. V04AT06A033). American Society of Mechanical Engineers. https://doi.org/10.1115/PVP2022-80693

[17] Ouakarrouch, M., El Azhary, K., Mansour, M., Laaroussi, N., & Garoum, M. (2020). Thermal study of clay bricks reinforced by sisal-fibers used in construction in South of Morocco. Energy Reports, 6, 81-88. https://doi.org/10.1016/j.egyr.2019.11.045

[18] Zaid, Z. B., Tilioua, A., Lamaamar, I., Ansari, O., Souli, H., & Alaoui, M. A. H. (2020). An experimental study of the efficacy of integrating a phase change material into a clay-straw wall in the Drâa-Tafilalet Region (Errachidia Province), Morocco. Journal of Building Engineering, 32, 101670. https://doi.org/10.1016/j.jobe.2020.101670

Materials Research Proceedings 47 (2025) 294-301 | https://doi.org/10.21741/9781644903391-34

From disintegration to valorization: What perspectives and multi-scale approaches towards a sustainable territorial design for the architecture of Kasbahs and Ksour in southern Morocco?

Zakaria ABYAA[1,a*], Khalid EL HARROUNI[2,b], Robin DEGRON[3,c] and Sanae ALJEM[2,d]

[1]Ecole Nationale d'Architecture, Rabat, Morocco & Paris 1 Panthéon Sorbonne University, France

[2]Ecole Nationale d'Architecture, Rabat, Morocco

[3]Paris 1 Panthéon Sorbonne University, France

[a]zakaria.abyaa@etu.univ-paris1.fr, [b]k.elharrouni@enarabat.ac.ma, [c]robin.degron@univ-paris1.fr, [d]s.aljem@enarabat.ac.ma

Keywords: Kasbah, Ksour, Sustainability, Territorial Design, Morocco

Abstract. Despite their historical and cultural importance, Kasbahs and Ksour in southern Morocco, as emblematic vernacular architectures that demonstrate an ingenious adaptation to the oasis environment, are facing progressive degradation, threatening their survival and sustainability. This research explores the disintegration and valorization of these traditional architectural structures known as Kasbahs and Ksour in southern Morocco. These fortified structures, constructed from local materials like rammed earth, represent the cultural and historical identity of the region. The research identifies the multi-faceted challenges these structures face, including natural degradation, socio-economic changes, rural exodus, and inadequate maintenance. It emphasizes the need for a holistic approach to preserve these architectural heritages while adapting them for contemporary use. The study advocates for a sustainable territorial design that integrates community participation, local practices, and tourism development, showcasing successful examples, such as Ksar Aït Ben Haddou. The focus is on balancing heritage conservation with modern requirements, aiming to foster economic development and community involvement. By developing collaborative initiatives, this research calls for an inclusive methodology that respects the historical significance of these sites while promoting resilient local economies and ensuring cultural continuity for future generations. Ultimately, the research underscores the potential of these architectures as catalysts for sustainable architectural, urban and territorial development in Morocco.

Introduction

Over the past decades and since the 80s, addressing the disintegration of the traditional architecture of Medinas, Kasbahs and Ksour in Morocco has garnered significant attention, particularly in light of increasing global and local efforts toward heritage preservation [1]. By definition, Kasbahs and Ksour are native architectural forms to southern Morocco characterized by their fortified structures built with locally sourced materials such as rammed earth (pisé). These traditional constructions are emblematic of the region's historical and cultural identity, embodying a unique genius loci [2].

As a heritage in decadence, these architectural structures face daily use and natural wear, which introduces multiple challenges to their preservation. The Moroccan government, along with various international organizations, has initiated numerous measures to protect and valorize these structures. These measures include heritage conservation programs, sustainable rehabilitation projects, and policies geared towards integrating modern needs without compromising historical integrity [2].

Materials Research Proceedings 47 (2025) 294-301 https://doi.org/10.21741/9781644903391-34

Kasbahs and Ksour in southern Morocco are witnesses to the history and culture of the oasian regions of the Kingdom. These structures, often built of earth or stone, are not only examples of traditional architecture, but they also represent the socio-cultural identity of the populations who built them and who continue to inhabit them. Indeed, the Kasbahs, which historically served as fortresses or residences for notables, as well as the Ksour, which functioned as fortified villages for the protection of oasis communities, have an inestimable heritage value. Highlighting their historical and cultural importance is essential to understand not only their role in the past but also their relevance in the present and future of sustainable development in Morocco.

This research aims to explore the perspectives and multi-scale approaches necessary for transitioning from the disintegration to the valorization of Kasbahs and Ksour in southern Morocco. It focuses on building upon research on local practices leading to interventions that are efficient and harmonious. By promoting the valorization of the physical and historical identity of these architectural forms, the research underscores the cultural heritage's potential as a driver of sustainable territorial development.

Pertinent examples are found in the Draa Valley (Case of Ksar Aït Ben Haddou), where local dynamics around heritage preservation have demonstrated the successful integration of community initiatives. The methods employed in this research will include qualitative data collection and analysis, encompassing official documents, reports, interviews with diverse stakeholders, and participatory observations. Such a comprehensive approach aims to provide a holistic understanding of the collective perception of heritage and the role of community involvement in mitigating various deterioration risks.

Ultimately, this research seeks to shed light on how a sustainable territorial design can be achieved for the architecture of Kasbahs and Ksour, balancing the intricacies of tradition and what's called "modernity" through inclusive and participatory methodologies.

The disintegration of Kasbahs and Ksour in southern Morocco

Moroccan Kasbahs and Ksour face significant threats from a multitude of forces, both natural and anthropogenic, leading to their progressive disintegration. The objective of this research is to analyze these threats, ranging from natural degradation to socio-economic transformations, rural exodus, and lack of maintenance, that endanger this invaluable heritage.

The first major threat to Kasbahs and Ksour is natural degradation, exacerbated by the harsh climatic conditions of the pre-Saharan desert. The fragile nature of the earthen architecture makes these structures highly susceptible to erosion, particularly during sporadic but intense rainfall, which can lead to the rapid deterioration of walls and foundations [3]. Coupled with this, human activities such as modern construction projects and infrastructural developments often lead to disturbances in the landscape that indirectly impact these ancient structures. The lack of a clear and integrated regulatory framework further complicates conservation efforts [3].

The socio-economic changes sweeping through Morocco have had profound impacts on the traditional lifestyles and economies of the regions where Ksour and Kasbahs are located. With the advent of modernization and shifts in economic activities, the traditional roles that these structures played in community life and local economies have diminished. Tourism, while a potential savior, has often led to adverse effects such as the commodification of cultural practices [4]. As inhabitants adapt to new economic opportunities, the communities that once maintained and revitalized the Kasbahs and Ksour are moving away, leading to their neglect and eventual ruin [4].

The rural exodus, driven by the search for better economic opportunities and living conditions in urban areas, constitutes another significant threat. As younger generations migrate to cities, the population of the regions declines, eroding the human resources necessary to maintain these structures. This shift results in the abandonment of many Ksour and kasbahs, accelerating their decay [2]. The departure of local skilled labor and traditional craftsmen, who possess the necessary expertise to repair and conserve these edifices, further exacerbates the situation [5].

Finally, the lack of consistent maintenance plays a critical role in the deterioration of Ksour and Kasbahs. Given their construction from perishable materials, these structures require regular upkeep to combat the effects of weathering and erosion [6]. Unfortunately, inadequate funding, insufficient legal protections, and the absence of coordinated conservation efforts have left many of these vulnerable structures. While there have been numerous initiatives aimed at preserving this heritage, an integrated vision that involves collaboration among various stakeholders remains largely absent [6].

The disintegration of Kasbahs and Ksour in southern Morocco is a multifaceted issue involving environmental factors, socio-economic shifts, demographic changes, and systemic neglect. Addressing these challenges requires a holistic approach that combines local, national, and international efforts to ensure the preservation of this cultural heritage.

Table 1: Causes and root sources for the disintegration of Moroccan Ksour

Category	Causes	Root Sources
Natural and Anthropogenic Degradation	- Natural degradation due to harsh climate (erosion, intense rainfall) and fragile earthen architecture. - Human activities, modern construction, and lack of integrated regulatory framework.	- Weathering - Human Factors
Socio-Economic Transformations	- Modernization diminishes traditional roles of ksour and kasbahs. - Adverse tourism effects/commodification of cultural practices. - Communities move away, leading to neglect.	- Community Shifts
Rural Exodus	- Search for better opportunities in cities. - Decline in regional population. - Abandonment of ksour and kasbahs. - Loss of skilled labor and traditional craftsmen.	- Migration
Lack of Maintenance and Conservation Efforts	- Need for regular upkeep due to perishable construction materials. - Inadequate funding, legal protections, and coordinated conservation efforts. - Vulnerable structures despite numerous preservation initiatives.	- Funding and Conservation

The valorization of Kasbahs and Ksour : a potential for sustainable territorial design

In terms of territorial development, the valorization of Ksour and Kasbahs offers promising prospects. On the one hand, these structures can serve as essential levers for the revitalization of local economies. The integration of Ksour and Kasbahs into urban planning and tourism development initiatives would promote a sustainable oasis economy and generate income-generating activities [7]. In addition, declaring these sites as public utility assets could guarantee adequate funding and facilitate the intervention of public authorities [6]. The involvement of local communities in safeguarding this heritage is also crucial. Considering residents as key players in rehabilitation projects and promoting consultation between all stakeholders involved helps ensure an inclusive and sustainable approach [6]. Successful initiatives, such as those carried out under the national sustainable development program, should be generalized to maximize their positive impact on the living conditions of local populations [8].

The heritage process also contributes to enhancing the value of the Ksour and Kasbahs. For example, the inscription of the Ksar Ait Ben Haddou on the UNESCO World Heritage List in 1987 has helped preserve the authenticity of this site while limiting unauthorized modifications [7]. However, it remains essential to reconcile the conservation of this heritage with the contemporary needs of the inhabitants.

Sustainable territorial design is emerging as a favorable approach at the crossroads between the conservation of architectural heritage, economic and social development, and citizen participation. This integrated and multi-scale methodology is based on the active participation of stakeholders, the hybridization of professional disciplines, and the implementation of innovative and inclusive

Materials Research Proceedings 47 (2025) 294-301 https://doi.org/10.21741/9781644903391-34

practices. Territorial design, as defined by Jégou (2018), is an approach to design applied to the territory that involves listening to stakeholders, co-design, and experimentation. It is a multidisciplinary method that mobilizes tools from the social and human sciences and service design to improve the habitability of territories while giving citizens back the power to act [9]. Among the main objectives of sustainable territorial design is the conservation of local architectural heritage. This involves not only the preservation of historical structures, but also their adaptation for contemporary use. For example, in southern Morocco, the heritage designation of the earthen Kasbah demonstrates the delicate balance between preserving architectural authenticity and adapting to modern uses. Territorial design helps create dynamic economic ecosystems by fostering collaborative practices and strengthening local networks [9].

Sustainable territorial design operates at various scales, from the neighborhood to the region, involving different strata of society. One of the essential aspects of territorial design is the use of innovative tools and techniques to encourage citizen participation and autonomy. Spatial visualizations and representations are key elements to involve local communities in the design process [9].

Figure 1: Ruins of a Ksar, Zagora Province (Author)

The deployment of sustainable territorial design is not without challenges. Managing the different action logics and coordinating the many actors involved requires a thoughtful approach and continuous adjustments. However, these challenges also open opportunities for innovative and inclusive solutions, transforming the local context through a sustainable development approach. Indeed, the notion of sustainable territorial design represents an innovative way forward for the conservation of architectural heritage and the economic and social development of local communities. By integrating various disciplines and emphasizing the active participation of citizens, this multi-scale approach aims to create habitable and dynamic environments. It proposes not only to preserve local history and culture but also to encourage resilient and prosperous development for the future, thus laying the foundations for a broader and deeper commitment to truly integrated and sustainable development practices at the architectural, urban, and territorial levels as much as at the local, regional and global levels.

Perspectives on multi-scale approaches for sustainable territorial design: The case of the architectural heritage conservation of Ksar Aït Ben Haddou

Figure 2: Conservation Master Plan of Ksar Ait Ben Haddou, 2015 (Decree No. 2.15.187 dated 30-03-2015 - BO No.: 6350 of 09-04-2015)

Integrating architectural heritage conservation into territorial development requires a holistic and multi-scale approach. The Kasbahs and Ksour of Morocco, emblematic elements of the country's architectural and cultural landscape, provide a perfect example of the complexity and richness of this task. It is crucial to take into account a plurality of economic, social, environmental, and cultural dimensions to ensure their effective preservation and integration into contemporary territorial development dynamics.

The conservation of Kasbahs and Ksour can be a driver of economic development. By transforming these historic structures into tourist attractions, local communities can generate substantial income. Architectural and cultural tourism can thus play a key role in financing restoration and maintenance projects for these sites. For example, Henri Terrasse, for almost a century, has been emphasizing the importance of protecting and promoting Kasbahs to improve the tourist attractiveness of the regions concerned [10, 11] In addition, integrating local economic activities, such as crafts and local products, into these sites can diversify sources of income and create jobs for the local population. Since the French protectorate, such reflections have already emerged. Salima Naji (2018) confirms that "Terrace had or was going to finance research on archaeological sites in southern Morocco, the Dades or Draa valleys, release funds to redo the facades of the tighremts of the "Valley of Roses", between Ouarzazate and Kelaa M'gouna, which he had classified as early as 1943".

The social dimension of the conservation of Kasbahs and Ksour is based on the involvement of the local community in the restoration and management processes. According to Naji (2018), restoration projects must include training programs for local craftsmen to perpetuate ancestral construction and renovation techniques. Furthermore, community participation should not be limited to mere involvement in the work, but should also integrate the perceptions and needs of the local population so that conservation projects enjoy broad support and meet the expectations of the inhabitants [11].

Kasbahs and Ksour are intrinsically linked to their natural environment, using local materials such as raw earth and stone, methodologies that ensure harmonious integration with the landscape [12]. Restoration and conservation should therefore encourage the use of traditional techniques that respect the environment and minimize the ecological footprint of interventions. For example, the use of local materials and bioclimatic techniques, such as those observed in the Errachidia craft center, show how Moroccan vernacular architecture can serve as a model for sustainable development [11].

Table 2: Ksar Ait Ben Haddou Building Materials

Building Material	**Nature**	**Use**
Adobe (earth blocks)	Traditional, durable if maintained	Constructing walls and structures
Stone and earth mortar	Traditional	Building wall bases
Earth and reeds	Traditional	Extended "hat" over parapet wall
Earth and dess	Traditional, local	Covering of terraces
Lime plaster (Taddelakt)	Traditional	Coating water channels
Wood	Traditional, compatible with earth structures	Constructing roofs, incorporating taddelakt
Gypsum and lime	Modern but compatible with traditional materials	Stabilizing exterior wall coatings while allowing breathability

The Ksar of Aït Ben Haddou stands as a testament to the enduring strength of earth architecture, strategically utilizing materials that are abundantly sourced from its surroundings. Dominantly, the raw earth is employed in the form of adobe bricks, which are a mixture of local clay, straw, and water. These bricks are known for their excellent thermal properties, keeping the interiors cool during the summer months and warm throughout the winter. In addition to adobe, rammed earth (pisé) techniques are also prominent; this involves compacting layers of damp earth within wooden frameworks to create solid and durable walls. Stone bases often complement these earth structures by providing much-needed stability and protection from ground moisture, thereby extending the longevity of the buildings.

Furthermore, the traditional construction techniques at Aït Ben Haddou emphasize not just the functionality but also the aesthetic harmony with the landscape. The use of natural pigments in the plastering processes adds to the visual coherence, mirroring the hues of the surrounding terrain. Roof coverings often incorporate reeds and earth, creating extended "hats" that shield the walls from rain and provide additional insulation. These architectural elements are not merely decorative but have practical implications: they manage moisture, optimize sunlight exposure, and enable effective water drainage. Taddelakt, a traditional waterproof plaster made from lime and natural soap, is also commonly used to coat water channels and walls, providing both resilience to erosion and a polished finish.

The preservation of Kasbahs and Ksour is not limited to the conservation of buildings but also includes the transmission and valorization of the associated intangible heritage, such as know-how, social practices, and cultural rituals [11]. Conservation projects are called upon to include actions to document, preserve, and promote these intangible elements. In this sense, it is vital to develop awareness-raising programs to strengthen the appropriation of this heritage by residents and visitors.

The integrated and multi-scale approach to the conservation of Kasbahs and Ksour requires close coordination between the different stakeholders, a rigorous assessment of socio-economic and environmental impacts, and continuous adaptation to local and global challenges. By combining the contributions of various disciplines and actively involving local communities, it is possible to develop conservation strategies that not only preserve these historic structures, but also invigorate them and fully integrate them into the contemporary socio-economic fabric.

Figure 3: Overview of Ksar AIt Ben Haddou (Author, 2024)

Conclusion

Contemporary challenges such as climate change and growing sustainability needs require a reorientation of research and action priorities. Integrating territorial design into planning and adaptation to emerging realities is a promising way to address these challenges comprehensively and inclusively.

Future research could focus on developing integrated participatory methodologies and real-time experimentation. As each territory has unique characteristics, it is crucial to generalize good practices while adapting solutions to local contexts. The evolution of architectural conservation and restoration systems, for example, can benefit from special attention to human and ecological well-being indicators as well as natural resource management and bioclimatic design.

Finally, to sustainably anchor these innovations, public decision-makers must engage in profound transformations of territorial policies. Territorial design can thus become a new frontier of public action, where action research and collective participation play a central role. This will require enhanced education and training of local stakeholders, as well as the promotion of a culture of open innovation and collaboration at all levels of society.

Thus, the increasing integration of territorial design in research and action provides a robust roadmap to address the complex challenges related to the valorization of Kasbahs and Ksour in southern Morocco.

References

[1] K. El Harrouni, Le patrimoine bâti au Maroc: vulnérabilité, sauvegarde et requalification. Editions universitaires européennes, 2019.

[2] M.T. Gil-Piqueras, P. Rodríguez-Navarro. Tradition and sustainability in vernacular architecture of Southeast Morocco. *Sustainability*, 2021, vol. 13, no 2, p. 684. https://doi.org/10.3390/su13020684

[3] M.T. Gil-Piqueras, P. Rodríguez-Navarro, The ksour of the mdagra oasis (Er-Rachidia, Morocco): An inventory. *The International Archives of the Photogrammetry, Remote Sensing*

and Spatial Information Sciences, 2020, vol. 44, p. 961-968. https://doi.org/10.5194/isprs-archives-XLIV-M-1-2020-961-2020. https://doi.org/10.5194/isprs-archives-XLIV-M-1-2020-961-2020

[4] A. Dłużewska, M. Dłużewski, Tourism versus the transformation of ksours-Southern Morocco case study. *Bulletin of Geography. Socio-economic Series*, 2017, no 36, p. 77-86. https://doi.org/10.1515/bog-2017-0015

[5] H. Ilahaine, Common property, ethnicity, and social exploitation in the Ziz villey, southeast Morocco. In : *IASCP conference*. 1995.

[6] N.D. Salimi, Les ksour et les kasbahs au Maroc, contraintes de conservation et perspectives de valorisation durable. *Espace Géographique et Société Marocaine*, 2022, vol. 1, no 57.

[7] N. Oussoulous, Le sud marocain un nouveau processus de patrimonialisation importée. *Annales du patrimoine*, 2019, no 19, p. 117-132. https://doi.org/10.35268/0186-000-019-014

[8] K.S. Alaoui, M. Daoud, A. Bouaouinate, Les ksour du Tafilalet au cœur du développement du pays présaharien. *Espace Géographique et Société Marocaine*, 2022, vol. 1, no 61.

[9] M. Jolivet-Duval, S. Safin, S. Huron, Territorial design, spatial representations, and citizen participation: Case review and tool analysis. *Sciences du Design*, 2021, vol. 14, no 2, p. 55-75. https://doi.org/10.3917/sdd.014.0055

[10] H. Terrasse, S. Naji. *Kasbas BerbèRes de l'Atlas et Des Oasis : Les Grandes Architectures Du Sud Marocain.* , Illustrated by Théophile-Jean Delaye, [Reproduction en fac-similé], Actes Sud; Centre Jacques-Berque, 2011-1938.

[11] S. Naji, « La kasbah berbère, ou comment un particularisme architectural devint l'un des principaux archétypes touristiques chérifiens ». *Fabrique du tourisme et expériences patrimoniales au Maghreb, XIXe-XXIe siècles*, édité par Cyril Isnart et al., Centre Jacques-Berque, 2018. https://doi.org/10.4000/books.cjb.1556

[12] M.A Odghiri, M. Mahdane, Le Patrimoine Millinaire des Oasis du Maroc, Quelle inspiration pour l'architecture contemporaine? *African and Mediterranean Journal of Architecture and Urbanism*, 2023, vol. 5, no 1.

A macroelement formulation for modeling soil-structure interaction in seismic conditions

Youssef ZAKARIA[1,a *], Younes SALAMI[1,2,b], Reda JAAFRI[1,c] and Mounia FARAH[1,d]

[1]Hassania School of Public Works, LaGCHEC Laboratory, BP 8108, Oasis, Casablanca, Morocco

[2]National School of Architecture of Fez, Fez-Shore Park, Sidi Hrazem Road, Fez, Morocco

[a]youssef.zakaria.civilengineer@gmail.com, [b]younes.salami@enafes.ma, [c]jaafri.reda@gmail.com, [d]farah.mounia@ehtp.ac.ma

Keywords: Macroelement, Soil-Structure Interaction, Hypoplasticity, Seismic Conditions, Cyclic Loading, Bearing Capacity

Abstract. The objective of this study is to present a new macroelement modeling, based on the principles of the hypoplasticity theory, and aimed at better representing the soil-structure interaction (SSI) during seismic events. Its target is to overcome the limitations of the three-dimensional model by Grange (2008), particularly when it comes to significant instabilities or material fatigue the structure reaches during cyclic loading, which leads to a larger gap between reality and the model. A physico-mathematical analysis of the soil-structure interaction is therefore proposed, and the model is numerically integrated into a computation code to verify its performance against the experimental results. This approach could contribute to improving seismic standards in terms of performance-based design.

Introduction

Soil-structure interaction is a physical phenomenon that was long overlooked by civil engineers. It occurs when vibratory waves, such as seismic tremors or shock waves, propagate through the soil and reach the embedded foundations of a structure. Indeed, it is common practice to establish a close material connection between the soil and the structure through foundations to withstand the effect of dead or live loads applied to the structure. These loads are diffused through the soil owing to the implanted foundations, provided that their intensity remains within reasonable limits and does not exceed the threshold values set by design standards.

However, when dynamic forces—defined as mechanical loads with varying intensity over time and uneven spatial distribution—disturb the soil, as in the classic case of earthquakes, the presence of the structure spontaneously alters the dynamic response of the propagated wave. This phenomenon particularly occurs near the foundations, where the soil's grain size distribution has been most disrupted by the insertion of the foundations, and where the medium's resistance to loading has changed significantly due to the stiffness of the structural material, which differs from that of the soil.

Moreover, seismic vibrations generated within the soil cause the structure to tremble, the latter picking the tremors up indirectly since the soil acts as an intercessor imparting the dynamic stresses and their effects to the structure. Soil-structure interaction, therefore, refers to this mutual relationship between the dynamic responses of the "soil" and the "structure" entities and is exceedingly cumbersome to quantify or model. However, given its tremendous impact during violent earthquakes and its detrimental outcomes on human lives and civil engineering structures' durability, researchers in the construction field have developed advanced techniques to better

Materials Research Proceedings 47 (2025) 302-308 https://doi.org/10.21741/9781644903391-35

apprehend this occult interaction and to ensure safer structures, among them is the macroelement method, the focus of our research.

Definition of the macroelement concept

As the name suggests, understanding the notion of macroelement requires a mindset capable of reconciling both the small and the large, with the prefix "macro" referring to a large-scale geometric dimension, and the term "element" conveying an idea of unity and indivisibility. It aims to shed light upon the soil-structure interaction (SSI) and to predict from the design phase to what extent the structure can hold out against external forces before failure, thus ensuring performance-based sizing of buildings to protect life and property from irreversible loss.

On the philosophy of soil-structure interaction

To characterize the resilience of the structure against seismic loads, the macroelement—conceived as a mesoscopic linking element between the soil and the structure—strives to define a threshold of physical quantities that, once exceeded, would mark the onset of structural disintegration and loss of serviceability. These quantities usually consist of forces per unit area (or mechanical stresses) and displacements (or deformations). They can be calculated using various computational methods available in the literature, such as the finite element method (FEM). This technique virtually decomposes a loaded structure into multiple sub-portions, most often triangular or quadrangular in shape, and characterized by degrees of freedom (translations and rotations) at their vertices. By integrating their values over the material domain of the structure and accounting for boundary conditions, FEM helps determine the overall effects of the applied loads and work out the stiffness of the system, i.e., its resistance to the applied forces.

However, estimating the stresses and deformations brought about by the vibrative effect of the soil-structure system at the foundations and their vicinity requires a particularly demanding level of detail. Consequently, the usually exploited FEM is in this context almost impractical, as it is a direct method of dynamic calculation that, applied to the entirety of an undivided studied system, would require an enormous amount of time and a massive memory space. Since it would compile the numerical code that will handle the SSI effect on specialized software, and would translate the fundamental laws of seismic calculation into a programming language, this method would compel designers to respect the thorough precision level of FEM to reflect the great variability of the soil and of the structure physical parameters under dynamic conditions, which is a cumbersome task. Furthermore, the complexity of the algorithms governing FEM makes it unfeasible to model SSI in three dimensions using this method, which imposes to wholly ignore the z-axis —a drawback to overcome.

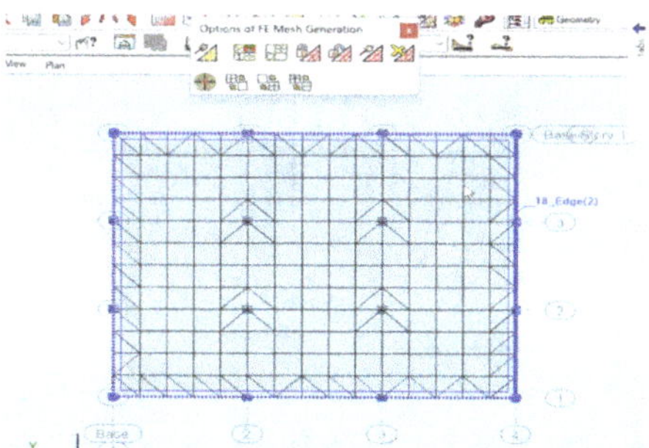

Fig. 1: Finite element method applied to a structure.

Materials Research Proceedings 47 (2025) 302-308
https://doi.org/10.21741/9781644903391-35

Faced with this frustration, engineers initially opted to turn a blind eye on SSI to streamline their designs. Such an approach, while flawed, is admissible for very rigid soils such as rocky terrain, since the stiffness gradient between the structure and the soil is negligible in such cases. Ergo, the structure can be considered a natural extension of the soil, which justifies merging the two systems and ruling out any potential interaction between them. However, this scenario is scarce, as soils are predominantly clayey, meaning they are soft and have low rigidity. In comparable situations, SSI becomes critical, as soil stiffness is significantly lower than that of the structure. Moreover, SSI affects seismic waves by attenuating their magnitude through a damping coefficient. The latter is included in the calculations and better assays the seismic load which was overestimated by classic dynamic methods. Furthermore, accounting for SSI increases the natural oscillation period of the soil-structure system—that is, the amount of time the structure takes to complete one cycle of vibration around its equilibrium position; thus, the soil is perceived as more flexible, and the spatial range over which the structure oscillates is broadened.

Therefore, soil-structure interaction proves not only compulsory but also behoove for designers to consider, as it optimizes the quantity of the materials used and, thereupon, the sizing of the structures and of their foundations. Yet, no procedure established by researchers over the last decades, since the 1930s when SSI was first uncovered, has easified the integration of this interaction into calculation codes. This is partly because understanding it relies on the two major pillars of civil engineering, too hard to master conjointly and expertly: structural mechanics (or structural design calculations and analysis) and geotechnics (or soil and rocks mechanics). Consequently, engineers have circumvented SSI by using a workaround that aligns better with their field of expertise. Structural engineers, for example, would typically place a linear spring at the foundation level in the modeling software they have chosen (such as RSA and Revit), thereby ignoring the material and geometric non-linearities of the soil (which will be defined subsequently). Meanwhile, geotechnicians would tend to perform a simplistic and erroneous calculation that represents the loaded structure as a single-degree-of-freedom oscillator, in analogy with a single pendulum whose node weighs the entire mass of the structure. This is where the ingenuity of the macroelement technique comes into play, as it successfully reconciles the seemingly divergent perspectives of structural engineers and geotechnicians, and helps overpass, to the greatest extent possible, the challenges of calling on SSI in seismic design. [1]

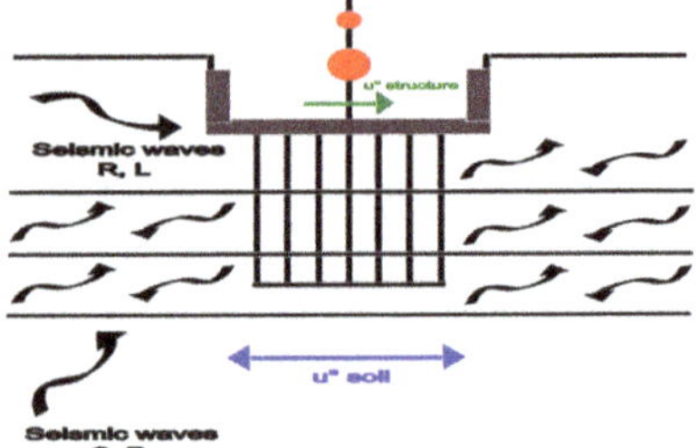

Fig. 2: Soil-structure interaction horizon.

On the logic of a macroelement

The basic principle of a macroelement is to split the soil-structure system into two domains separated by a fictive boundary, called the interaction zone. This disseveration aims to simplify and speed up seismic calculation by exploiting the well-known superposition theorem: the dynamic response of a given system equals the sum of the individual responses of each of its elements. This approach follows the divide-and-conquer paradigm. Nevertheless, while many theories, known as substructure methods, encouraged it by defining the SSI zone exactly at the physical borderline separating the foundations of the structure from the surrounding soil environment, the macroelement introduced a novel train of thought that locates this boundary further from the foundations. Therefore, considering SSI through a macroelement involves separating the soil into two domains: a near field (from structural foundations) and a far field. [2]

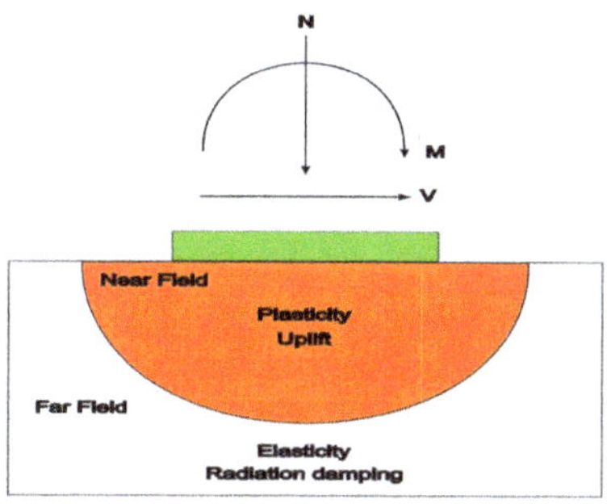

Fig. 3: Near and far fields of a macroelement.

The near field encompasses all nonlinearities in the vicinity of the soil-foundation system, which are of both material and geometric natures. The term linearity, in this context, refers to its mathematical meaning. Indeed, a function (or numerical application) is described as linear if the graph representing the output y as a function of the input x forms a straight line. This condition requires the slope (or rate of change) of the function to remain constant; otherwise, nonlinearity arises. In the context of dynamic calculation, nonlinearity corresponds to a nonlinear function of mechanical stress as a function of strain (or the ratio of displacement caused by the force to the initial length of the element before deformation). This nonlinearity is described as material when it is due to the intrinsic rheological properties of grains which, due to their irreversible destabilization during dynamic loading, push the soil-foundation system into a state of physical plasticity. Plasticity occurs when mechanical stress exceeds an allowable limit, which causes permanent deformations in the system even after the applied force comes to an end. The generated plasticity, which would have, for certain, led to failure had human intervention not occurred, induces a slope change on the stress-strain curve, hence bringing about material nonlinearity. Geometric nonlinearity, on the other hand, is caused by seismic (or dynamic in the general case) forces near the soil-structure interface, and so strong that the linear equilibrium, which previously characterized the perfect and lasting connection between the structure and the soil, is disrupted. A phenomenon called uplift occurs at the foundations which are partially lifted under its effect, and then rotate around their horizontal axis like a tree being uprooted. The combination of these two aspects of nonlinearity is handled within the near field of the macroelement using a dynamic calculation method suitable for bounded domains, such as FEM.

The far field, by contrast, is where linear soil response is taken into consideration, as it is distant enough from the soil-foundation system, impervious then to the dynamic effects of the tremor.

Similarly, radiation damping of the vibratory manifestation—i.e., the dissipation of mechanical wave energy over time due to Joule Effect—is integrated into the far-field soil modeling.

Thus, the macroelement method is worthy of its epithet "hybrid" geotechnicians use to describe it, since it combines the advantages of both direct and substructure methods, using the former to handle soil nonlinearities and the latter to define the near and far fields and the interaction horizon. Therefore, engineers should place importance on it, especially for seismic analysis.

A macroelement is also distinguished by its unique approach to determining the bearing capacity of the soil (its capacity to support the risk of settlement resulting from applied loads). In fact, instead of working with stresses (in N/m^2 or Pa) and dimensionless strains, which implies a microscopic view of SSI that continuum mechanics (a science that deals with the deformation of solids and fluid flow) impels, a macroelement typically analyzes the dynamic problem using dimensionless quantities or ratios over time, depending on the theory adopted to model the irreversible deformations experienced by the soil-foundation system during seismic vibrations. These parameters could be in the form of normalized forces (dimensionless) and normalized displacements (dimensionless) obtained by dividing the applied forces and displacements by their maximum admissible values, or in the shape of stress rates (in N/ (m^2.s) or Pa/s) and strain rates (in s^{-1} or Hz). This mathematical approach allows the macroelement to visualize SSI on a mesoscopic scale, dealing with microscopic phenomena (nonlinearities, damping, degrees of freedom, etc.) by manipulating macroscopic quantities (forces and displacements). Furthermore, if the model that reflects the soil plasticity includes the time factor, as is the case when stress and strain rates are introduced into the dynamic analysis, the macroelement becomes capable of retracing the loading history experienced by the soil. This allows for anticipating its future reactions to upcoming tremors and for better understanding the variability of its intrinsic properties over time. This second approach was chosen for our work with the goal of designing a new macroelement model that could faithfully represent the cyclic loading process (repeated or fluctuating application of mechanical loads, consisting of iterative loops of loading, unloading, and reloading cycles, as seen in earthquakes), a phenomenon highly time dependent. [3]

Macroelements through the years

The macroelement approach sparked interest and gained approval from geotechnical researchers as soon as it appeared in 1991, thanks to the works of Nova and Montrasio on estimating the bearing capacity of shallow foundations (those with moderate embedment in the soil, such as strip and isolated footings). This first macroelement model consisted of a two-dimensional plane study, and only regarded quasi-static loads (nearly constant loads) and the material nonlinearity of the soil. Nevertheless, it took a big step in the history of soil-structure interaction study.

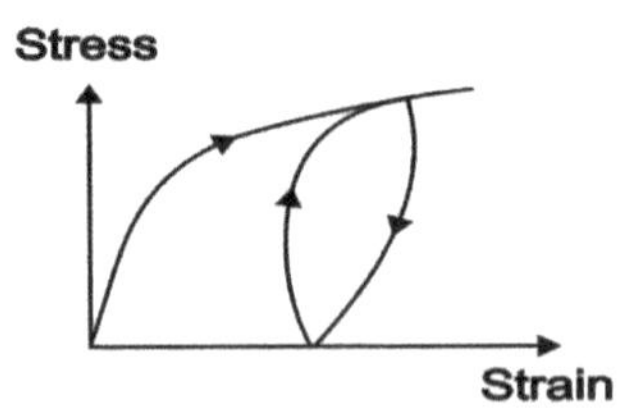

Fig. 4: Stress-strain curve for a cyclic loading.

Over the years, several researchers have recognized the need to improve the macroelements described in the literature. For instance, Paolucci adapted the Nova and Montrasio model to dynamic loads in 1997, Pedretti considered cyclic loads in 1998, Le Pape and Sieffert proposed a plasticity law inspired by thermodynamics in 2001, and Crémer incorporated geometric nonlinearity into her macroelement model, making history in the development of this hybrid method by theorizing on the uplift phenomenon in 2002. Besides, Cassidy designed a three-dimensional macroelement model, albeit for quasi-static loadings only, in 2004. Chatzigogos introduced stress and strain rates for performance-based designs in 2007, while Grange extended Crémer's macroelement to a three-dimensional study in 2008. [4, 5, 6, 7, 8, 9, 10, 11, 12]

Conclusion

In a nutshell, the macroelement concept has aligned with the guidelines of recent civil engineering standards, primarily Eurocode 7 which deals with geotechnical design, and Eurocode 8 which governs the seismic design of structures, leading to a comprehensive analysis of soil-structure interaction under seismic conditions. Indeed, according to Eurocode 7, a numerical design method for seizing geotechnical structures is only valid if it incorporates SSI among its input parameters, particularly when the stiffness of the constructed structure is significant. This standard also praises the optimization of settlement estimation and foundation sizing enabled by soil-structure interaction. Eurocode 8, for its part, mandates considering SSI within advanced calculation methods to thoroughly study foundations deformability and to accurately analyze their seismic response. Furthermore, this standard seeks to characterize the earthquake's effect on the natural frequencies (the inverse of the oscillation periods of the soil-foundation system) and on soil radiation damping, then attempts to account for the uplift a foundation might experience due to seismic forces. The stringentness of Eurocode 8 even requires engineers to provide appropriate justification if they decide to omit SSI from their designs.

This notwithstanding, even the best-developed models exhibit some limitations that need to be addressed. In this context, our research project aims to develop a new macroelement that will build upon Grange's work and will adopt a plasticity law based on stress and strain rates as in Chatzigogos macroelement (involving a class of plastic laws known as hypoplasticity), to accurately analyze cyclic loading. With this amendment, our work will foster a symbiosis between the methodologies used by structural engineers and geotechnicians for incorporating SSI into building design, hence will aid in introducing foundation calculations—a subtle geotechnical art—into the daily practice of construction professionals.

References

[1] Y. Abboud, Développement d'un macroélément pour l'étude des fondations superficielles sous charge sismique, Université Paris-Est, Ecole doctorale Sciences, Ingénierie et Environnement, Paris, 2017.

[2] C. Chatzigogos, Comportement sismique des fondations superficielles : vers la prise en compte d'un critère de performance dans la conception, Ecole Polytechnique, Paris, 2007.

[3] C. Chatzigogos, A. Pecker, J. Salençon, Macroelement modeling of shallow foundations, Soil Dynamics and Earthquake Engineering, Volume 29 (Issue 5): 765-781, 2008. https://doi.org/10.1016/j.soildyn.2008.08.009

[4] C. Crémer, L. Davenne, A. Pecker, Cyclic macro-element for soil-structure interaction: material and geometrical non-linearities, International Journal for Numerical and Analytical Methods in Geomechanics, Volume 25 (Issue 13): 1257-1284, 2001. https://doi.org/10.1002/nag.175

[5] S. Grange, Modélisation simplifiée 3D de l'interaction sol-structure : application au génie parasismique, Institut Polytechnique de Grenoble, Grenoble, 2008.

[6] S. Grange, P. Kotronis, Méthodes simplifiées pour le calcul non-linéaire de structures de génie civil, Les superstructures du bâtiment, Techniques de l'Ingénieur, 2016. https://doi.org/10.51257/a-v1-c6002

[7] D. Kolymbas, Introduction to Hypoplasticity, Advances in Geotechnical Engineering and Tunnelling 1, CRC Press, London, 1999.

[8] L. Montrasio, R. Nova, Settlements of shallow foundations on sand, Géotechnique, Volume 41 (Issue 2), 243-256, 1991. https://doi.org/10.1680/geot.1991.41.2.243

[9] A. Pecker, Dynamique des sols, Presses de l'école nationale des ponts et chaussées, Paris, 1984.

[10] A. Pecker, Dynamique des structures et des ouvrages. Paris : Presses de l'école nationale des ponts et des chaussées, 2006.

[11] J. Salençon, Théorie de la plasticité pour les applications à la mécanique des sols, Eyrolles, Paris, 1974.

[12] J.P. Wolf, Dynamic Soil-Structure Interaction, New Jersey, Prentice Hall, 1985.

Materials Research Proceedings 47 (2025) 309-318 https://doi.org/10.21741/9781644903391-36

Evaluation and modeling of a solar collector containing a thermal storage unit

E.H. SEBBAR[1,a*], N. ERAZA[1,b], N. LAAROUSSI[1,c], A. AIT MSAAD[2,d], M. MANSOUR[3,e]

[1]Mohammed V University in Rabat, Material, Energy and Acoustics Team (MEAT), EST Salé, Morocco

[2]Ecole Supérieure de Technologie de Fès, U.S.M.B.A, Route d'Imouzzer, BP 2427 Fès, Morocco

[3]National School of Architecture Rabat, Avenue Allal El Fassi, Medinet Al Irfane, B.P: 6372, Rabat, Morocco

[a]elhassan.sebbar@usmba.ac.ma, [b]n.eraza@research.emi.ac.ma, [c]najma.laaroussi@est.um5.ac.ma, [d]abdelouahad.aitmsaad@usmba.ac.ma, [e]m.mansour@enarabat.ac.ma

Keywords: Fusion/Solidification, Solar Collector, Thermal Energy Storage

Abstract. This study utilizes solar energy-saving methods, concentrating on thermal energy storage (TES) through the employment of a thermal storage unit and a solar collector. Using phase change materials (PCM), the integration of a thermal energy storage system in a solar collector is studied numerically. The behavior of the PCM during the three-day loading and unloading phase is the main focus of this investigation. The PCM in the solar collector was modeled using the enthalpy-porosity approach. This method faithfully replicates the PCM's melting and solidification procedures for heat transport. The location of the PCM in the collector and how to relate the mass flow rate with the temperature of the fluid were studied in detail to optimize the efficiency of thermal energy storage and heat production by the collector. The final results of the study are presented and discussed based on performance parameters including outlet temperature, liquid fraction, and melting and solidification process under different operating conditions.

Introduction

Globally, energy scarcity is now a significant issue, mainly due to rapid population growth, rising living standards, industrial development, and prosperity[1]. In actuality, fundamental energy sources or fossil fuels provide the majority of the energy needed by humans [2]. The excessive use of fossil fuels has seriously harmed the ecosystem and negatively impacted human health. Furthermore, these fuels' supplies are almost depleting, and their prices have increased recently and are expected to do so in the years to come [3]. Using renewable energies solves certain problems, but intermittency remains a major challenge. We therefore need to improve energy storage to ensure continuous, reliable use of these sources.

Latent heat storage and sensible heat storage are the three primary categories of thermal energy storage systems now in use [4]. Latent thermal energy storage (LTES) systems have found their way into various engineering applications and have been widely considered by researchers in this field. PCMs are fluids that are utilized to store thermal energy. To determine whether these systems are economically feasible, more research on a few specific technologies is required. Numerous dynamic processes have been created and evaluated during the previous 40 years utilizing numerical techniques.

By using these solar technologies in buildings, considerable energy savings can be realized, opening the door for their use in net-zero energy buildings [5]. Numerical integration study of a

PCM-based TES in a nuclear power plant. The balance between solidification and fusion is established in charge and discharge mode and the liquid fraction is constant over time [6]. For a study combining two granular PCMs, the bed wall's stored energy makes up 8–16% of the energy contained in the granular material with varying phase transition temperatures and a standard granular material [7]. Without accounting for tank heat losses, About 10% more heat is usable in a collector with a vertical square tank than in collectors using a cylindrical reservoir [8].

Studies on packed-bed latent storage systems show that a 96 m 2 home should have tanks that are 1.0 * 1.0 m in size and 0.05 kg/s^{-3} in flow rate [9]. A new thermal energy storage system with macro-encapsulation has been created. When the air inlet temperature increases from 425°C to 465°C the charging rate can be improved by 60.5% [10]. Extending the fins in a double-tube heat exchanger using PCM results in shorter melting times and deeper heat penetration. Additionally, it has been demonstrated that, during the phases of the charging procedure, the fin height of the heat exchanger determines its heat absorption capability [11]. The TES mechanism in concentrated solar power is limited by output threshold temperatures. The effective PCM consumption rate for PCM-TES without cascade is highly dependent on the melting temperature. The greatest rate of use is merely about 40% [12]. A numerical study is conducted on a solar-powered cooling system that includes a PCM tank and parabolic collectors. During non-solar hours, the absorption chiller investigated uses a PCM reservoir, which reduces cooling energy consumption by up to 30% [13]. Using two-dimensional rotation, a numerical investigation of the heat transfer properties during a phase transition of the paraffin-filled rectangular encapsulated vessel (RT -27) (PCM-REC) was conducted. As the PCM-REC spun in the direction of vertical heating, As radial height grew, The highest possible natural convection velocity increased to 0.0184 m/s from 0.0072 m/s [14].

This work constitutes an in-depth analysis of the behavior of a solar thermal collector integrated with a latent storage unit. The approach adopted is to use a PCM as a thermal storage medium to improve the solar collector's performance. The primary goal is to minimize temperature changes throughout the day and raise the water outflow temperature at night.

Numerical simulations were carried out to study the melting/solidification fronts of the PCM within the system. The choice of PCM was N-eicosane kerosene, selected for its compatibility with the operating temperature range of the solar collector. The results obtained were analyzed and discussed in depth.

Geometry description

Two PCM placement configurations, C1 and C2 (see Fig.1), were studied to determine the optimum configuration for the collector. Numerical simulations were carried out under specific meteorological conditions (see Fig.2 and Fig.3), with an initial temperature of 288 K and a mass flow rate of 0.0015 kg/s.

The second part of this study explores the impact of a variable mass flow rate over time, aimed at avoiding a drop in temperature to levels unacceptable for night-time use. The results obtained from CFD simulations were used to monitor temperature variations in the lower and upper volumes of the reservoir.

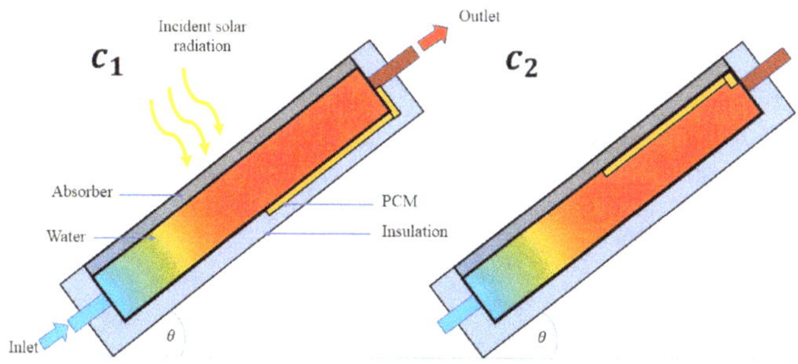

Fig.1: *The new geometry with PCM integration.*

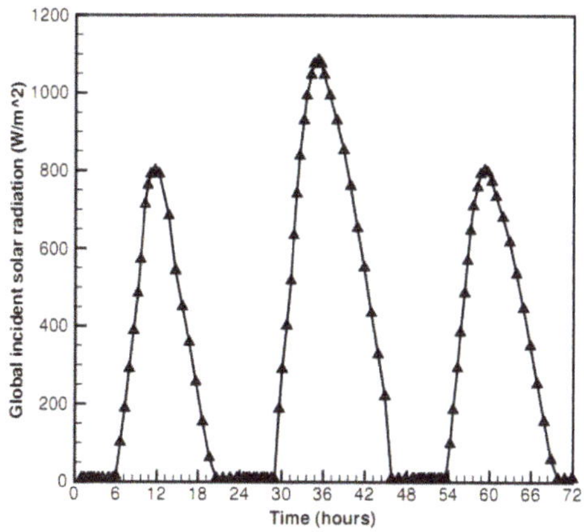

Fig.2: *The variation of the global incident flux imposed (W/m^2).*

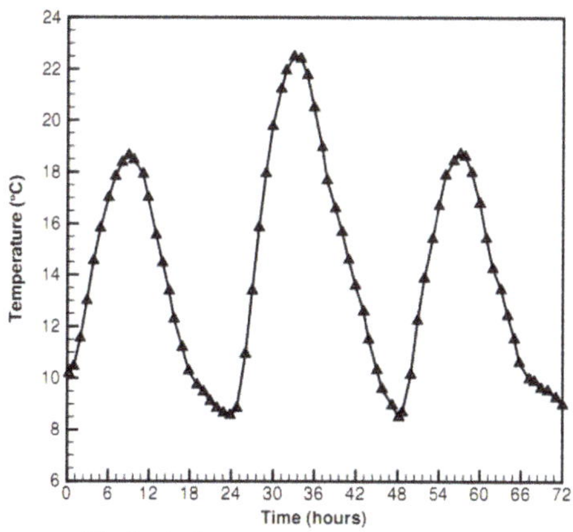

Fig.3: *Ambient temperature imposes.*

Physical model

The physical and numerical model is based on the following simplifying assumptions:

- The fluid is Newtonian and incompressible.
- Flow and heat transfer are bidirectional.
- With the exception of the density, which is thought to depend on temperature, all of the fluid's characteristics are taken to be constant.
- The viscous dissipation is neglected.

In Cartesian coordinates, the general CFD equations controlling the flow and heat exchanges are as follows:

- Equation of the continuity:

$$\frac{\partial \rho}{\partial t} + \vec{\nabla}\left(\rho \vec{U}\right) = 0 \tag{1}$$

- Equation of the moment :

$$\frac{\partial(\rho \overrightarrow{U})}{\partial t} + \vec{\nabla}\left(\rho \vec{U}.\vec{U}\right) = -\vec{\nabla}.P + \vec{\nabla}(\mu.\vec{\nabla}.\vec{U}) + \rho \vec{g} + \vec{S} \tag{2}$$

- Equation of the energy:

$$\frac{\partial}{\partial t}(\rho H) + \vec{\nabla}\left(\rho \vec{U} H\right) = \vec{\nabla}(k \vec{\nabla} T) \tag{3}$$

Where P is for pressure, H is for specific enthalpy, and V is the fluid's velocity. In the momentum equation, the source term S refers to the damping term in Darcy's Law. It is articulable in this way:

$$S = \frac{C_{st}(1-f)^2}{f^3 + \varepsilon} U \tag{4}$$

C_{st} is a constant that assesses the fusion zone's shape, and ε is intentionally set very low to prevent division by zero. Readers are advised to consult the earlier works for further information regarding the resolution process and the description of the phenomenon[15].

$$\dot{m} \begin{cases} 0.0015\ kg/s & if \quad T > 20°C \\ 0.0009\ kg/s & if \quad T \leq 20°C \end{cases} \tag{5}$$

The liquid-solid front is seen as a porous media and its porosity is taken into account while modeling the PCM melting process. Throughout the study period, the liquid fraction of material can range from zero to 1. The thermophysical characteristics of the PCM under investigation are displayed in **Table 2**.

Table 2: *Thermo physical properties of the PCM (n-eicosane ($C_{20}H_{42}$).*

Thermophysical properties	$C_{20}H_{42}$
Density (kg.m^{-3})	ρ_s =820, ρ_l = 780
Latent heat (kJ.kg^{-1})	237.4
Melting temperature (°C)	36.65
Specific heat capacity (kJ.kg^{-1}. K^{-1})	
Specific heat capacity (kJ.kg^{-1}. K^{-1})	$C_{p,s}$ =1.9, $C_{p,l}$ = 2.2
Thermal conductivity (W.m^{-1}. K^{-1})	k_s =0.212, k_l = 0.16

The liquid fraction is calculated based on an enthalpy balance of the PCM. For an isobaric process, the enthalpy variation of a PCM that does not change state can be written as:

$$dH = C_p dT \tag{6}$$

The specific enthalpy at temperature T of material is then:

$$H = h_{ref} + \int_{T_{ref}}^{T} C_p dT + fL \tag{7}$$

The liquid fraction f is defined as:

$$\begin{cases} f = 0 & if\ T < T_s \\ f = 1 & if\ T > T_L \\ f = \frac{T - T_s}{T_L - T_s} & if\ T_s < T < T_L \end{cases} \tag{8}$$

T_s is the temperature solidus and T_L is the temperature liquidus. The enthalpy variation is expressed in terms of the latent heat of the material L: $\Delta H = f.L$.

Simulation results and analysis

Fig.4 displays the average temperature at the solar collector outlet over 72 hours. This figure clearly shows that temperatures reach high values during the day, due to exposure to solar radiation. However, these temperatures can sometimes exceed the set or operating values. During the night, on the other hand, temperatures drop considerably due to the absence of sunlight.

This large temperature difference emphasizes how crucial it is to use PCM for a latent storage solution. The overall effectiveness of the solar collecting system is enhanced by PCMs, which make it possible to store excess heat during the day and expel it at night. This helps to maintain a more steady and higher temperature during the night.

- **Effect of PCM layer location**

Fig.5 and **Fig.6** show the temperature at the collector outlet and the average fraction in the PCM layer over 72 hours. Both figures demonstrate the effect of PCM integration. **Fig.5** shows that the temperature at the collector outlet, with PCM integration, is lower than the maximum values observed in the case without PCM (50°C instead of 70°C for the first day and 71°C instead of 96°C for the second day). This reduction in maximum temperature can be explained by the fact that during fusion, a sizable amount of heat is absorbed by the PCM layer.

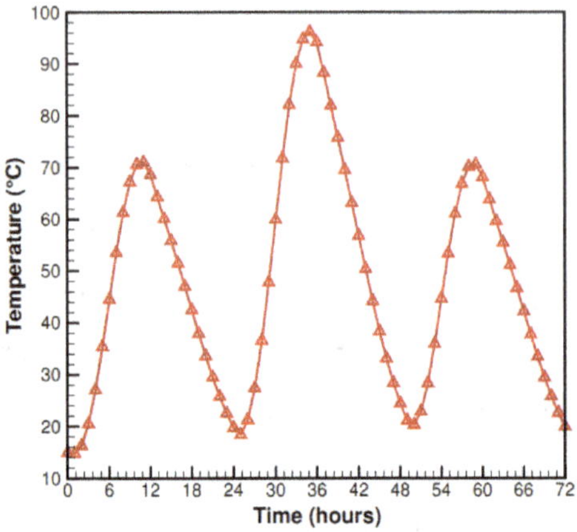

Fig.4: *Outlet temperature without PCMDuring the night, however, the temperature does not drop as drastically as in the case without PCM. This is due to the release of previously stored thermal energy, which contributes to maintaining a more stable and higher temperature.*

The impact of the location of the PCM layer on the outlet temperature is clearly illustrated in **Fig.5**. In configuration C1, where the PCM layer is positioned in the upper part of the collector, in direct contact with the absorber, it is observed that the temperature during the night drops to very low values. Although the PCM manages to store a significant amount of heat during the day, the water does not fully exploit the energy released by the PCM during the night.

In contrast, in configuration C2, where the PCM layer is placed in the lower part of the collector, in contact with the insulating wall, the outlet temperature during the night is maintained at acceptable levels, with a maximum difference of 10K compared to configuration C1. This better result is explained by the fact that the insulation prevents heat losses to the outside, thus allowing the energy released by the PCM to be fully recovered by the water. The inefficiency observed in configuration C1 is due to significant thermal losses to the outside, caused by convection between the absorber and the heat transfer fluid. As a result, the energy released by the PCM is not optimally transferred to the water, reducing the overall efficiency of the system. Conversely, configuration C2 shows that the insulation plays a crucial role in minimizing thermal losses, thus allowing a more efficient recovery of the energy stored by the PCM.

These observations highlight the critical importance of PCM location in the collector. Improper positioning can result in significant heat losses and reduce the benefits of latent storage, while optimal location, combined with effective insulation, maximizes the thermal performance of the system and improves its overall performance.

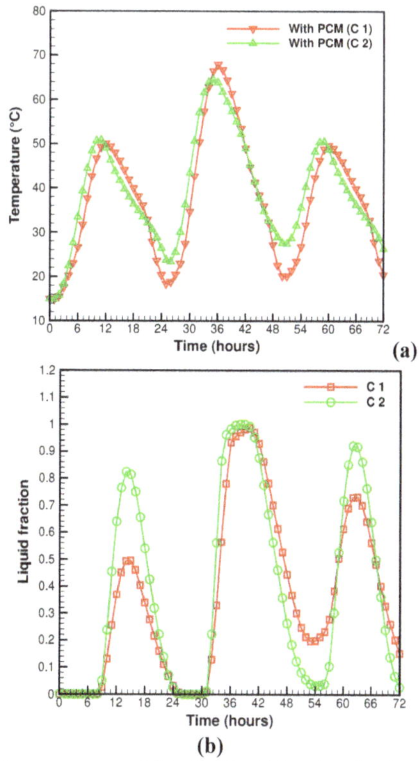

***Fig.5**: (a): Outlet temperature and (b): the liquid fraction for the two configurations*

- **Variation of mass flow**

The incorporation of PCM in the solar collector improves the outlet water temperature, making it closer to the desired values for practical use. However, this temperature can be further optimized by dynamically managing the heat transfer fluid flow rate. In general, a rise in the flow rate tends to reduce the outlet water temperature, because the thermal energy is distributed over a larger volume of water. To overcome this problem, it is proposed to decrease the flow rate when the outlet temperature starts to drop below the set values.

The idea is therefore to adopt a variable flow rate, adjusted according to the outlet temperature of the collector. When the temperature decreases, the flow rate is reduced, allowing the water to spend more time in the collector and absorb more heat, even if the energy input is lower. This flow modulation strategy ensures better recovery of the heat stored in the PCM, especially during the night.

The results shown in **Fig.6** show that this variable flow approach allows us to obtain a higher outlet temperature than that obtained with a constant flow, especially during night hours.

Adopting a variable flow, adjusted according to temperature variations, proves to be an effective strategy to maximize the performance of the solar collector equipped with PCM. This method not only improves the outlet temperature but also allows to exploitation of the energy stored by the PCM, thus optimizing the system's overall effectiveness.

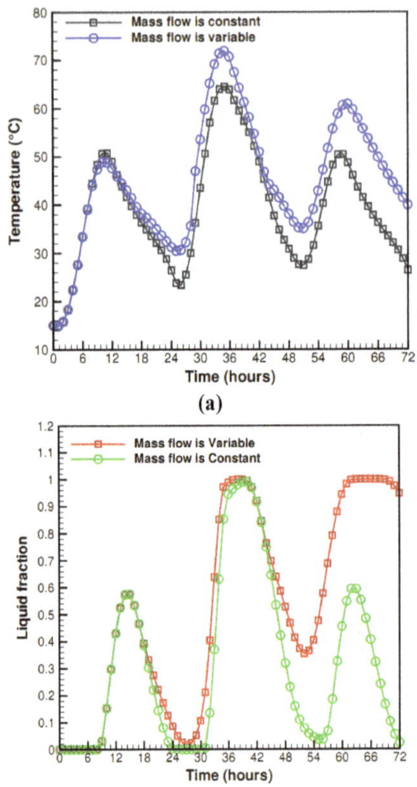

(a)

(b)

***Fig.6**: (a): Outlet temperature and (b): the liquid fraction at mass flow is variable and at mass flow is constant*

Conclusion

The obtained results highlight the crucial importance of integrating PCMs in solar collection systems to improve thermal efficiency, especially during nighttime periods. The study shows that the strategic location of the PCM in the collector plays a determining role in the thermal performance of the system. An optimized position of the PCM, combined with adequate insulation, allows for maximum heat recovery and minimizes thermal losses.

Furthermore, the adoption of a variable flow rate, adjusted according to the variations in the outlet temperature, proves to be a particularly effective strategy for maintaining high outlet temperatures even during the night. This approach allows us to fully exploit the energy stored in the PCM, thus offering a significant improvement compared to a constant flow rate system.

In conclusion, the combined use of PCMs and dynamic flow control offers a robust solution to improve the performance of solar thermal collectors. This approach not only optimizes the efficiency of thermal storage but also ensures a continuous supply of heat at acceptable levels over 24 hours, even in the absence of direct solar radiation. These results open the way to further optimizations when creating devices for thermal energy storage, to increase their efficiency and reliability under varied operating conditions.

References

[1] Koželj, R., Mlakar, U., Zavrl, E., Stritih, U., and Stropnik, R., 2021, "An Experimental and Numerical Analysis of an Improved Thermal Storage Tank with Encapsulated PCM for Use in Retrofitted Buildings for Heating," Energy Build., **248**. https://doi.org/10.1016/j.enbuild.2021.111196

[2] Goyal, P., Baredar, P., Mittal, A., and Siddiqui, A. R., 2016, "Adsorption Refrigeration Technology - An Overview of Theory and Its Solar Energy Applications," Renew. Sustain. Energy Rev., **53**, pp. 1389–1410. https://doi.org/10.1016/j.rser.2015.09.027

[3] Malekan, M., Khosravi, A., and Syri, S., 2019, "Heat Transfer Modeling of a Parabolic Trough Solar Collector with Working Fluid of Fe3O4 and CuO/Therminol 66 Nanofluids under Magnetic Field," Appl. Therm. Eng., **163**(March). https://doi.org/10.1016/j.applthermaleng.2019.114435

[4] Heier, J., Bales, C., and Martin, V., 2015, "Combining Thermal Energy Storage with Buildings - A Review," Renew. Sustain. Energy Rev., **42**, pp. 1305–1325. https://doi.org/10.1016/j.rser.2014.11.031

[5] Bosu, I., Mahmoud, H., Ookawara, S., and Hassan, H., 2023, "Applied Single and Hybrid Solar Energy Techniques for Building Energy Consumption and Thermal Comfort: A Comprehensive Review," Sol. Energy, **259**(January), pp. 188–228. https://doi.org/10.1016/j.solener.2023.05.006

[6] Ali, M., Alkaabi, A. K., and Lee, J. I., 2022, "CFD Simulation of an Integrated PCM-Based Thermal Energy Storage within a Nuclear Power Plant Connected to a Grid with Constant or Variable Power Demand," Nucl. Eng. Des., **394**(July), p. 111819. https://doi.org/10.1016/j.nucengdes.2022.111819

[7] Izquierdo-Barrientos, M. A., Sobrino, C., and Almendros-Ibáñez, J. A., 2016, "Modeling and Experiments of Energy Storage in a Packed Bed with PCM," Int. J. Multiph. Flow, **86**, pp. 1–9. https://doi.org/10.1016/j.ijmultiphaseflow.2016.02.004

[8] Sadeghi, G., Pisello, A. L., Safarzadeh, H., Poorhossein, M., and Jowzi, M., 2020, "On the Effect of Storage Tank Type on the Performance of Evacuated Tube Solar Collectors: Solar Radiation Prediction Analysis and Case Study," Energy, **198**, p. 117331. https://doi.org/10.1016/j.energy.2020.117331

[9] Liu, J., and Lu, S., 2024, "Thermal Performance of Packed-Bed Latent Heat Storage Tank Integrated with Flat-Plate Collectors under Intermittent Loads of Building Heating," Energy, **299**(December 2023), p. 131463. https://doi.org/10.1016/j.energy.2024.131463

[10] Li, M. J., Jin, B., Yan, J. J., Ma, Z., and Li, M. J., 2018, "Numerical and Experimental Study on the Performance of a New Two-Layered High-Temperature Packed-Bed Thermal Energy Storage System with Changed-Diameter Macro-Encapsulation Capsule," Appl. Therm. Eng., **142**(July), pp. 830–845. https://doi.org/10.1016/j.applthermaleng.2018.07.026

[11] Hosseini, M. J., Ranjbar, A. A., Rahimi, M., and Bahrampoury, R., 2015, "Experimental and Numerical Evaluation of Longitudinally Finned Latent Heat Thermal Storage Systems," Energy Build., **99**, pp. 263–272. https://doi.org/10.1016/j.enbuild.2015.04.045

[12] Li, M. J., Li, M. J., Tong, Z. X., and Li, D., 2021, "Optimization of the Packed-Bed Thermal Energy Storage with Cascaded PCM Capsules under the Constraint of Outlet Threshold Temperature," Appl. Therm. Eng., **186**(December 2020), p. 116473. https://doi.org/10.1016/j.applthermaleng.2020.116473

[13] Ajbar, A., Lamrani, B., and Ali, E., 2023, "Dynamic Investigation of a Coupled Parabolic Trough Collector–Phase Change Material Tank for Solar Cooling Process in Arid Climates," Energies, **16**(10). https://doi.org/10.3390/en16104235

[14] Liu, D., Xie, K., Zhang, H., Qiang, Y., Yang, D., Wang, Z., Zhu, L., Akkurt, N., Du, Y., Shen, M., Zhong, L., Yu, F., and Xu, Q., 2022, "Numerical Evaluation of Convective Heat Transfer Properties of Two-Dimensional Rotating PCM Melt in the Unilaterally Heated Rectangular Container," Renew. Energy, **193**, pp. 920–940. https://doi.org/10.1016/j.renene.2022.05.009

[15] Pirasaci, T., Wickramaratne, C., Moloney, F., Yogi Goswami, D., and Stefanakos, E., 2017, "Dynamics of Phase Change in a Vertical PCM Capsule in the Presence of Radiation at High Temperatures," Appl. Energy, **206**(January), pp. 498–506. https://doi.org/10.1016/j.apenergy.2017.08.187

Vernacular Architecture: Support for Territorial Development | Materials Research Forum LLC
Materials Research Proceedings 47 (2025) 319-326 | https://doi.org/10.21741/9781644903391-37

Energy efficiency applied to vernacular architecture, case of the rammed earth buildings: A literature review

CHEIKHI Wiçal[1,a *], BABA Khadija[1,b]

[1]LGCE Laboratory, Higher School of Technology, Salé, Morocco

[a]wical.cheikhi@est.um5.ac.ma, [b]Khadija.baba@est.um5.ac.ma

Keywords: Vernacular Architecture, Rammed Earth, Raw Material, Clay, Mud Wall, Energy Efficiency

Abstract. Rammed earth architecture is a widely spread vernacular architecture on a global scale and represents an iconic construction model in our country. Indeed, this architecture is an integral component of Moroccan architectural heritage. Long neglected in Favor of modern construction techniques, this method is nowadays experiencing a renaissance due to its energy, environmental, socio-economic, and aesthetic qualities. However, although its aesthetic and historical value is well-established, its energy, ecological, and environmental aspects deserve further exploration. This article aims to shed light on these aspects by placing them within the current energy trend, through the study of the thermal characteristics of the construction material and the analysis of the conceptual choices of rammed earth architecture in Morocco.

Introduction

Earth architecture is undoubtedly one of the oldest and most widespread vernacular architectures, intimately linked to the history of humanity. Although there is no consensus on when humans began building with earth, many authors [1–3] trace the beginning of its use in construction to between 12,000 and 7,500 BC. According to the United Nations Centre for Human Settlements, 40% of current habitats worldwide are made of raw earth. This material was favored by ancient civilizations for both simple constructions and monumental works. Indeed, 57% of the sites on UNESCO's "World Heritage in Danger" list are built with earth. The pyramids of ancient Egypt, the ruins of the Pre-Columbian city of Chan-Chan in Peru, the Great Wall of China, and the Kasbahs of southern Morocco are striking examples.

Fig. 1: Kasbah Ait Benhaddou- Ouarzazate- Morocco [4]

Since ancient times, humanity has relied on this material to build across the planet. However, after World War II, the use of rammed earth construction significantly declined. Replaced by industrial materials, it lost prominence due to the general trend in architectural evolution: a search for quicker implementation, the rise of ready-to-use industrial materials, construction delegated to professionals, and laws imposing inflexible standards. After the oil crises of the 1970s, a revival of earth architecture in the Southern regions. In Northern countries, efforts to find alternatives to non-renewable resources also began around this time. Europe then became a leader in the resurgence of earth architecture, first in France, then in Germany and Italy, and today in Austria and Portugal.

Indeed, rammed earth architecture proves to be unbeatable in a challenging and urgent area: environment-friendly architecture. In the quest for architecture with genuine environmental qualities, energy-efficient, using unlimited resources, and producing no waste, rammed earth architecture ranks at the forefront.

Moreover, the earth material requires very little energy to produce: while the production of one cubic meter of concrete requires 400 to 800 kWh, the same amount of earth only requires 5 to 10 kWh per cubic meter [5].

This work focuses on highlighting the energetic aspects of rammed earth architecture through the study of the characteristics and benefits of the construction material, especially in a world context focused on sustainable development and energy efficiency.

Socio-economic benefits:

The first benefit of building with earth would be economic. Local, natural, and renewable, the cost of this material is lower than other ones, more expensive but yet more commonly used today. This characteristic can be a real advantage, especially in developing countries where labor costs are still accessible. Thus, in addition to its attractive price, earth construction has the benefit of creating local jobs, as it primarily relies on manual work.

Earth architecture has demonstrated its adaptability not only in poor countries where people lack the means to build contemporary houses but also in poor areas of wealthy countries where housing for the poor is often precarious. Likewise, the aesthetic qualities of the material have also shown adaptability to both rural and urban construction contexts.

Environment benefits:

Transport, Manufacturing, and End-of-Life Phase:

One of the environmental advantages of this material is its very low energy consumption and minimal pollution during manufacturing compared to conventional materials. Indeed, before being implemented, these materials go through various phases such as material extraction, transport, processing, and manufacturing, resulting in significant energy consumption and a considerable environmental impact. In contrast, locally sourced and manually produced materials (such as rammed earth) limit transportation processes as they are available locally, and consume very little energy during manufacturing, leading to a very low environmental impact since they are made manually outside of factories.

Another environmental advantage lies in the demolition phase of buildings. Traditional earth materials are easily recyclable and do not create waste problems like conventional materials. It can be easily recycled and reused for constructing new buildings. Additionally, it can simply be returned to nature without the need for treatment or storage, resulting in zero environmental impact. The issue of waste production from the deconstruction of buildings at the end of their life and its management is not as pressing as it is for other composite materials.

Construction Phase:

The low energy consumption during the construction phase of rammed-earth buildings is well established. In this context, several authors have conducted comparative studies on the energy consumed during the construction of buildings using either local materials or industrial materials.

The study by reference [6] supports this view. The authors compared energy consumption during the construction of houses in Grasse, in the south of France, between prototypes using local materials (stone masonry + earth mortar, and rammed earth) and others using concrete. The results of this study were quite revealing. They demonstrated that the use of local materials (stone and rammed earth in this case) significantly reduces energy consumption during the construction phase compared to buildings that use concrete.

Table 1 below summarizes the results of the aforementioned study, showing that among the three building variants studied, the rammed earth option is the least energy-consuming.

Table 1: Energy demand during the construction of stone, rammed earth, and concrete houses [6]

	Stone masonry and earth mortar	Rammed Earth	Concrete
Energy (GJ)	97	70	239
Transport (t.Km)	1390	1041	6707

Living Phase:

The energy performance of rammed earth buildings during the Living phase remains a significant topic of debate. In this regard, several authors have focused on studying the thermal and energy performance of rammed-earth buildings compared to other conventional construction methods. The results obtained from these studies are as varied as they are numerous: Reference [7] conducted a comparative study of the energy performance of two buildings with the same layout and spatial composition but using two different construction materials: rammed earth on one side and concrete with a masonry envelope on the other. The results of this study showed a reduction of approximately 32% in heating requirements for the building constructed with rammed earth compared to the masonry-built variant. Similarly, the study indicated a decrease in thermal losses through the envelope of the rammed earth building compared to that of the concrete building. Table 2 below summarizes the key findings of these results.

Table 2: Heat losses and heating needs for concrete and rammed earth buildings [7]

	Masonry and concrete	Rammed earth
Heat losses (Kw)	10,48	6,05
Heating needs (kWh/m^2)	49,94	33,96

In the same vein, the study conducted by reference [8] can also be cited. In his study, a comparison between the comfort of occupants in three buildings has been made: one is a "green building," entirely built using rammed earth techniques, while the other two are "conventional." By comparing the energy bills (electricity and gas) of the buildings, it is noted that the green building has a lower energy consumption compared to the other two.

The results from these two studies provide a positive assessment of the energy performance of rammed earth buildings compared to their concrete counterparts during the habitation and usage phases.

A comparative study between two buildings in Mexico, one made of hollow concrete blocks and the other of adobe, was conducted by reference [9]. The study showed that the energy consumed to maintain a comfortable temperature in an adobe building is five times lower than that consumed in a conventional building.

Thermal characteristics:
Rammed earth buildings have always been known for the level of thermal comfort they provide. In both summer and winter, rammed earth houses, for example, maintain stable and pleasant indoor temperatures day and night, significantly minimizing temperature fluctuations. This characteristic is due to the use of an exterior envelope made of fairly thick walls (around 50 cm for rammed earth constructions in southern Morocco), ensuring good thermal inertia and timely heat retention (capturing heat during the day and releasing it at night within the dwelling).

These thermal qualities observed historically in rammed earth constructions across various climatic contexts (Morocco, France, Australia, etc.) suggest a promising synergy between the use of this ancient technique and the application of bioclimatic architecture measures in construction.

Thermal resistance of rammed earth:

Thermal resistance (R, expressed in $K \cdot m^2/W$) is a characteristic of the material that indicates its ability to resist cold and heat. Thus, the higher the R-value, the more insulating the material is. The study of the thermal resistance of rammed earth has been the subject of several investigations. In this context, reference [10] conducted a study to determine the R-value of samples from rammed earth house walls in Australia. The samples measured 1 meter in length, 1 meter in width, and 20 cm in thickness, with an average mass of around 400 kg. The results showed that most of the R-values obtained from the measurements were below 0.4, while local regulations require a minimum R threshold of about 1.3 for walls. Therefore, these walls exhibit poor thermal resistance despite the good level of thermal comfort provided by these houses.

Consequently, the authors suggested that the low resistance of a rammed earth wall could be compensated for by its good thermal mass and concluded that this factor should be considered in regulations to accurately assess the thermal capacity of the walls.

Thermal Mass of Rammed Earth

Thermal mass is a characteristic that helps reduce the energy consumed for heating and cooling in a building [11]. Indeed, thermal mass allows for the stabilization of indoor temperature regardless of external climatic variations. Fig. 2 illustrates this phenomenon [12].

In addition to its ability to stabilize indoor temperature, a wall with a high thermal mass has a good phase shift capacity: it absorbs heat from solar radiation during the day and only releases it back into the building when the outdoor temperature drops, specifically at night.

To determine the role of thermal mass in ensuring a good level of comfort in rammed-earth constructions, several studies have been conducted. References [13,14] conducted a study leading them to believe that the high thermal mass of rammed earth can compensate for its low thermal resistance value. Like reference [9], they confirm that relying solely on resistance criteria to evaluate the thermal comfort of earth buildings is an incomplete approach and that the criterion of thermal mass should now be taken into consideration.

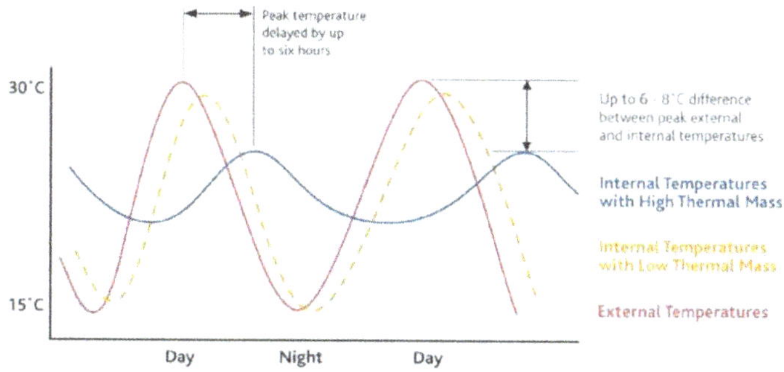

Fig. 2: Effect of the thermal mass on the Stabilisation of the inner temperature [12]

Other authors who have worked on the thermal characteristics of rammed-earth constructions have highlighted the importance of mastering solar design measures and tools to maximize the benefits of the material's thermal mass. Reference [15] demonstrated that good solar design can enhance the thermal performance of rammed earth buildings by significantly reducing energy consumption. The principal outputs of this study are summarized in Table 3 below.

Table 3: thermal performance of earth buildings with bioclimatic design [15]

Building	Ambient temp (° C)	Inner temp (° C)
Mudgee	-8	14
Gulf of Carpentaria	45	26/27
Hunter Valley	2-42	17-30
Mona Vale	7-42	18-27

Still in Australia, in a cold and humid climate in winter and hot and dry in summer, reference [16] conducted measurements on a rammed earth office building. The results showed that the high thermal mass of rammed earth can enhance its thermal performance. Therefore, the assessment of this performance cannot be based solely on its resistance value.

The aforementioned studies all tend to reach the same conclusion: thermal mass plays a crucial role in determining the thermal performance of rammed earth constructions. However, to this day, there are no tools available to accurately calculate the performance of thermal mass [17] and integrate it into the thermal simulation of buildings.

Effect of adding an insulation on a rammed earth building:

The integration of thermal mass into standards, thermal regulations, and simulation software is challenging; as a result, they currently only use thermal resistance as the criterion to evaluate the energy performance of the material.

To ensure that rammed earth constructions comply with thermal regulations, studies have been conducted. References [18–21] indicate that the strategic placement of thermal insulation layers on rammed earth walls could meet regulatory requirements, even though it increases the energy consumed during construction. The studies mentioned above show that this increase does not significantly impact total energy consumption, which remains 40 to 60% lower compared to buildings using conventional materials. Thus, a rammed earth wall remains highly efficient from an energy perspective, even though its thermal inertia is not taken into account in the evaluation of its performance.

Materials Research Proceedings 47 (2025) 319-326 https://doi.org/10.21741/9781644903391-37

The recent revival of the rammed earth technique is being paired with adaptations to modern construction methods. For instance, many earthen structures now incorporate wall insulation to enhance their thermal inertia and insulation performance. In this context, reference [22] performed thermal conductivity tests on rammed earth walls equipped with extruded polystyrene insulation. These tests indicated that combining a high-mass wall with low-conductivity insulation yields a wall with lower thermal conductivity than an uninsulated rammed earth wall. The resulting composite walls demonstrated excellent thermal properties suitable for various heating climates. Similarly, reference [23] conducted a comparative study of the indoor performance of two rammed earth houses in Willunga, South Australia, one featuring traditional rammed earth walls and the other with insulated walls. The findings revealed that the insulated house could be up to 5 degrees warmer than its uninsulated counterpart, potentially resulting in lower heating energy consumption.

In the same way, reference [24] conducted a comparative study on the energy performances of a rammed earth building located in Marrakech – Morocco, before and after adding insulation to the walls. The results show an improvement in the thermal comfort inside the building as well as a slight decrease in the heat losses after the addition of the insulation, as respectively shown in Fig. 3 and Fig. 4 Below.

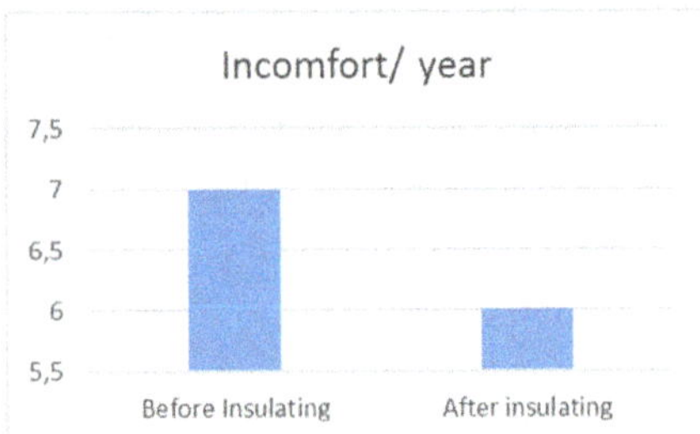

Fig. 3: Hours of discomfort before and after adding the insulation [24]

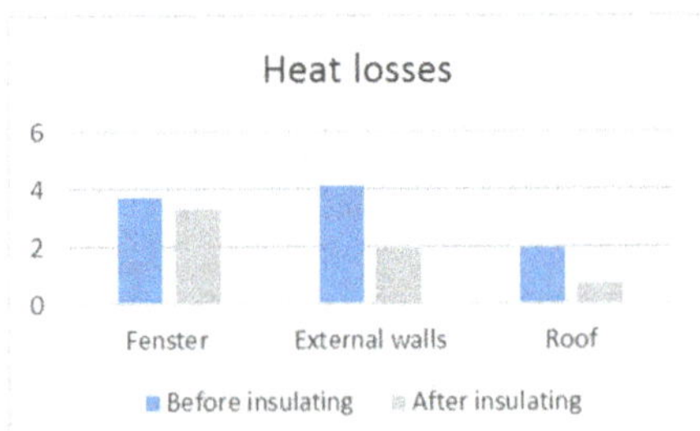

Fig. 4: Heat losses before and after adding the insulation [24]

Summary

In summary, it should be emphasized that rammed earth construction offers countless energy, environmental, socio-economic, and sustainability benefits. This material, along with its various applications in architecture, can align perfectly with the contemporary goals of resource conservation, sustainable development, and energy efficiency. Furthermore, the wide range of methods for its implementation, its availability in many parts of the world, and its adaptability to

Materials Research Proceedings 47 (2025) 319-326
https://doi.org/10.21741/9781644903391-37

both hot and cold climates make it a universally relevant material that deserves to be brought back to the forefront. In conclusion, rammed earth architecture may be vernacular, ancient, and affordable, but it still finds its place in the current world context.

References

[1] G. Minke, Building with earth: Design and technology of a sustainable architecture, Birkhäuser, Germany, 2006.

[2] Bjørn Berge, Chris Butters, Filip Henley, The ecology of building materials, Architectural Press, 2009.

[3] F. Pacheco-Torgal, S. Jalali, Earth construction: Lessons from the past for future eco-efficient construction, Construction and Building Materials 29 (2012) 512–519. https://doi.org/10.1016/j.conbuildmat.2011.10.054

[4] L.A.-M.W. News, Ouarzazate's Ait Ben Haddou Kasbah 'Most Distinguished' African UNESCO Heritage Site: Afrika News, (n.d.). https://www.moroccoworldnews.com/2017/12/235653/ouarzazate-ait-ben-haddou-kasbah-african-unesco-heritage-site-afrika-news (accessed October 11, 2024).

[5] M. Lamrani, Étude du comportement thermique des matériaux locaux (Caractérisation et modélisation), PHD thesis, Mohammed V university, 2018.

[6] J.C Morel, A Mesbah, M Oggero, P Walker, Building houses with local materials: means to drastically reduce the environmental impact of construction, Building and Environment 36 (2001) 1119–1126.

[7] W. Cheikhi, K. Baba, Study of indoor performances of a building using Rammed earth, MATEC Web of Conferences 149 (2018).

[8] W.L. Paul, P.A. Taylor, A comparison of occupant comfort and satisfaction between a green building and a conventional building, Building and Environment 43 (2008) 1858–1870. https://doi.org/10.1016/j.buildenv.2007.11.006

[9] M.Á. Porta-Gándara, E. Rubio, J.L. Fernández, Economic feasibility of passive ambient comfort in Baja California dwellings, Building and Environment 37 (2002) 993–1001. https://doi.org/10.1016/S0360-1323(01)00085-3

[10] 28 April 2000 Anna Salleh-ABC Science Online Friday, CSIRO gives mud walls a thumbs down, (2000). https://www.abc.net.au/science/articles/2000/04/28/121761.htm (accessed October 11, 2024).

[11] D. Baggs, N. Mortensen, Thermal Mass in Building Design, (2006).

[12] D. Allinson, M. Hall, Investigating the optimisation of stabilised rammed earth materials for passive air conditioning in buildings, in: Bengalore, India, 2007.

[13] S. Fotios, A Revised Kruithof Graph Based on Empirical Data, LEUKOS 13 (2017) 3–17. https://doi.org/10.1080/15502724.2016.1159137

[14] L. Soudani, M. Woloszyn, A. Fabbri, J.-C. Morel, A.-C. Grillet, Energy evaluation of rammed earth walls using long term in-situ measurements, Solar Energy 141 (2017) 70–80. https://doi.org/10.1016/j.solener.2016.11.002

[15] N. Mortensen, The Naturally Air Conditioned House, Oom2.Forumotion.Com (n.d.). https://oom2.forumotion.com/t36617-the-naturally-air-conditioned-house (accessed October 11, 2024).

[16] P. Taylor, M.B. Luther, Evaluating rammed earth walls: a case study, Solar Energy 76 (2004) 79–84. https://doi.org/10.1016/j.solener.2003.08.026

[17] A. Wilson, Thermal Mass and R-value: Making Sense of a Confusing Issue, BuildingGreen (1998). https://www.buildinggreen.com/feature/thermal-mass-and-r-value-making-sense-confusing-issue (accessed October 11, 2024).

[18] S. Goodhew, R. Griffiths, Sustainable earth walls to meet the building regulations, Energy and Buildings 37 (2005) 451–459. https://doi.org/10.1016/j.enbuild.2004.08.005.

[19] M. Hall, D. Allinson, Assessing the effects of soil grading on the moisture content-dependent thermal conductivity of stabilised rammed earth materials, Applied Thermal Engineering 29 (2009) 740–747. https://doi.org/10.1016/j.applthermaleng.2008.03.051

[20] P. Mendonça, Non-structural adobe walls in housing buildings : environmental performance, International Symposium on Earthen Structures, Bengalore, India (2007).

[21] M.L. Parra-Saldivar, W. Batty, Thermal behaviour of adobe constructions, Building and Environment 41 (2006) 1892–1904. https://doi.org/10.1016/j.buildenv.2005.07.021

[22] V. Soebarto, Analysis of indoor performance of houses using rammed earth walls, in: Glasgow, Scotland, 2009: pp. 1530–1537.

[23] M. Hall, D. Allinson, Assessing the moisture-content-dependent parameters of stabilised earth materials using the cyclic-response admittance method, Energy and Buildings 40 (2008) 2044–2051. https://doi.org/10.1016/j.enbuild.2008.05.009

[24] W. Cheikhi, K. Baba, A. Nounah, Effect of adding insulation on the energy performances of rammed earth buildings in hot and arid climates, E3S Web of Conferences 150 (2020).

Materials Research Proceedings 47 (2025) 327-342 | https://doi.org/10.21741/9781644903391-38

Towards resilient architecture in technological innovation and AI

Mocerino CONSIGLIA[1,a*], Abderrahim LAHMAR[2,c], Mohamed AZROUR[2,b], Abdelilah LAHMAR[3,d]

[1]Former adjunct professor, PhD Urban Recovery and Regeneration, PDTA-Faculty of Architecture, Sapienza University of Rome, Italy

[2]Materials Sciences for Energy and Sustainable Development Team, Department of Chemistry, Faculty of Sciences and Techniques, Moulay Ismail University, Errachidia, Morocco

[3]Laboratory of Condensed Matter Physics, University of Picardie JulesVerne, Amiens, 33 rue Saint Leu, Amiens 80039, France

[a]l.mocerino12@gmail.com, [b]m.azrour@umi.ac.ma, [c]abderecte@yahoo.fr, [d]lahmar2@yahoo.fr

Keywords: Technological Innovation, Resilience, Passive Systems,Nanotechnologies, Energy Efficiency

Abstract. The new trends in construction aim at a process innovation in which the new construction, organizational, and production methods are integrated with qualitative techniques and efficient technologies, on principles that tend more towards resilience in architecture. In fact, from a strongly eco-sustainable approach, there is a tendency towards a greater verification of the quality and efficiency in construction that highlights a new trend towards resilient design, in line with a desired vernacular design in architecture. Technological innovation follows the great impact in socio-economic and industrial developmentindicating new housing performances, in efficient and resilient building envelopes that can avoid major climatic and environmental risks. Above all, building according to principles of resilience to growing global warming, meteorological disasters, and air pollution, in line with COP28, the UNFCCC, the Intergovernmental Panel on Climate Change (IPCC), the Paris Agreement, WMO,UNEP, etc. The main objectives are user well-being in the habitat of efficient contextualized envelopes with the use of vernacular building techniques integrated with the technological innovation of intelligent and biocompatible materials, nanotechnologies and innovative nanomaterials,sustainable bio-inspired systems, and natural engineering. So integration criteria of high-performance and adaptive building envelopes for new construction and redevelopment, integratedwith sustainability strategies with passive heating and cooling, increased waste disposal, reductionof energy costs, reduction of water dependence with increased use of wastewater, for the reductionof environmental pollution and optimization of local renewable resources. The methodologies areabove all the application of resilient design with political and passive strategies with innovative glass systems, shading, innovative, recyclable techniques and materials of green building, high efficiency, and low cost, from renewable sources and energy saving, bioswales for the collection and management of polluted rainwater, green roofs, critical systems to resist flooding, global warming, and above all meteorological risks and active strategies with intelligent digital systems and AI. The challenge is that of decentralization and decarbonization with durable energy-efficientbuildings for user well-being with structures in urban and territorial contexts that increase the fabric of the social community.

Introduction

Vernacular architecture, which has distinguished itself mainly as a global building typology by adopting, from the genius loci [1], efficient local materials, techniques, and technological systems, especially to respond to relative climatic needs and local cultures, has highlighted in the new building processes, innovative aspects that aim at resilience in architecture. They are distinguished

Materials Research Proceedings 47 (2025) 327-342 https://doi.org/10.21741/9781644903391-38

by characterizing aspects that invest construction and technological systems, in a growing development of resilience principles [2] that increase the approach of new design and architectural planning. Therefore, from a strongly eco-sustainable approach, there is a tendency togreater checks for quality and efficiency in construction that aim at a new trend of resilient designin the vernacular architecture project. The main objectives are to build or redevelop contextualizedarchitectural envelopes for user well-being, through the use of efficient techniques and technologies, use of sustainable natural and bio-inspired engineering systems, intelligent and biocompatible materials, innovative nanotechnologies and nanomaterials and high-performance systems. and AI, for user well-being. Like infrastructure and community projects, buildings, in particular, must be resistant to extreme climate and weather changes, maintaining their habitabilityover time both due to the vulnerability of their natural aging and in the event of flood risk, seismicvulnerability, various flood disasters, loss of renewable energy supply, air pollution, increasing global warming, fires, etc.

Such phenomena for further strategic objectives could be prevented in a preventive manner, with a notable proactive function of control and anticipation, through resilience, to any reactivity. It is within an international and global debate of socio-environmental policies and scientific research, that technological innovation stands out for these resilience objectives, indicating a process innovation in terms of implementation, organization and production, and requalification of the pre-existing in terms of efficiency. Therefore also in terms of quality, costs, and times in line with the principles against climate change of COP 28.

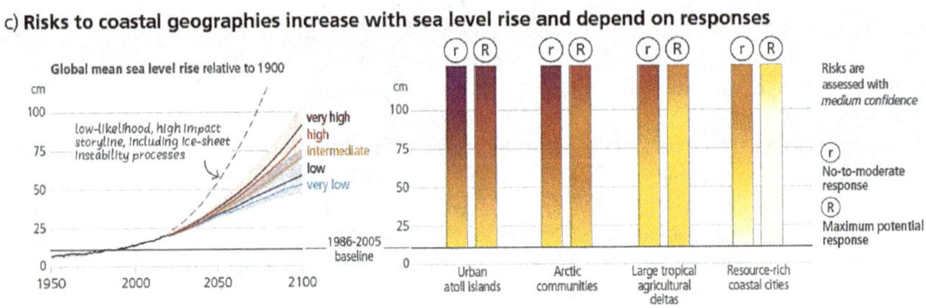

Fig.1. Synthesis Report (SYR) [3]

COP 28 - Conference of the Parties to the United Nations Framework Convention on Climate Change The United Nations Framework Convention on Climate Change (UNFCCC) was held in Dubai in 2023 with a fundamental agreement on decarbonization and the need for transition from fossil fuels. As in the Synthesis Report (SYR) (Fig.1) of the sixth assessment report(AR6) of 2023 of the Intergovernmental Panel on Climate Change (IPCC), based on technical scientific and socio-economic assessment analyses the information about the risks of theimpact of climate change, and how the interaction between mitigation and related adaptation is the climate challenge and identifying effective and efficient system transitions. In the perspectiveof sustainable development and climate resilience also for the growing urbanization, aiming at theconservation of biodiversity with the adoption of equity solutions for climate change, tracing the sources of global emissions, and being able to follow the limit of 1.5° C.

On the same line the challenge of climate change to remain at a global level, below 2° C, remains the same Paris Agreement, since temperatures are recorded at a global level, above the average of about 2.3° C compared to 1.45° C with a very hot September in Europe in 2023 according to European State of Climate 2023.

Principles and analysis are also in agreement with the World Meteorological Congress (WMC) for resilience, exchange of data and observations, and climate adaptation with objectives for 2027

Materials Research Proceedings 47 (2025) 327-342 https://doi.org/10.21741/9781644903391-38

of protection for all, through life-saving alert systems and with the Global Greenhouse Gas Watch to strengthen the monitoring of gases that store heat.

The same objectives of reducing climate criticalities and resilience with sustainable development are highlighted in UNEP with aid for countries to reduce greenhouse gases and climate adaptation, protect human health and the planet, to also end poverty, build sustainable and resilient communities, etc. The Buildings Performance Institute Europe (BPIE) identifies 10 political priorities [4], strategies from 2024-2029, to ensure that zero-emission resilient buildings are the key to social well-being and economic gain, energy security. Among these are increased financing,adoption of AI for data collection, increased financing, adapting the built environment for climate resilience, successful implementation of the EPBD, "Green Homes" Directive, promoting cooperation, and strengthening industrial competitiveness and sustainable construction (from the New European Bauhaus to the High-Level Forum on Construction), with the Heating & Cooling Strategy, improving energy security and resilience. Furthermore, green buildings and public procurement by public authorities for the construction or renovation and redevelopment of buildings, in terms of sustainability and energy efficiency, policies to ensure that climate change is sustained and equitable through monitoring and benchmarking

In this perspective, vernacular architecture is therefore at the center of the debate for the sustainabledevelopment of a new building process in which criteria for the integration of new pre-existing high-performance and adaptive architectural envelopes of redevelopment, sustainable recovery, and resilience are outlined for the challenges of new climate changes. Therefore, technological design of new building processes in the innovation of technological/construction systems, with materials and components in which nanotechnologies predominate with passive strategies in passive heating and cooling systems, with sustainable natural and bio-inspired engineering systems, intelligent materials, for the reduction of dependence on water and increase of wastewater with waste disposal. In a trend for residential housing units for user well-being with comfortand functionality, the industrial and tertiary sectors envelope an international and global debate of socio-economic and industrial development with the use of intelligent systems of active strategieswith HVAC, and optimization of local resources with genius loci.

The methodologies are those for resilient design with buildings able to resist seismic vulnerabilities, floods, overflows, fires, and global warming, with systems of vernacular architecture typologies, high-performance systems of insulation, and natural ventilation, for the livability of homes in cases of reduction of energy load, and water supplies in case of optimization of the use of renewable energy, efficient closures resistant to hurricanes, etc. So, from a sustainable approach tothe more adaptive one of the resilient architectures of energy efficiency against environmental stress, but also with results of economic advantages for costs and efficiency, with an increase ininvestments for owners and businesses, increased demand and real estate value and with greater opportunities for resale and rental. The use of resilient architecture [5], on the principles of greenbuilding, with green roofs, bioswales for the collection and management of rainwater, constructionof compostable systems, passive cooling and cooling systems, strategies with alternative renewable energy systems in cases of energy losses due to hurricanes, energy efficiency accordingto new European directives such as Dir. (EU) 2024/1275 of "green houses" etc. and according to the International Finance Corporation (IFC) green buildings represent an investment opportunity of 24.7 trillion dollars by 2030. Also, according to new European directives, to last longer and to better resist the effects of time, with consequent lower financial loss.

The challenges are decarbonization and decentralization with resilient design and vernacular architecture practices, creating resistant, durable new or redevelopment buildings resistant to storms, hurricanes, forest fires, seismic risks, etc. with the identification of durable and global risk-resistant construction and technological systems in urban-territorial and landscape contexts, etc.

This involves, in particular for buildings, long-term savings in maintenance, recovery, and redevelopment costs and above all environmental and planetary protection. Resilience at a territorial level with the enhancement of ecosystem services: control of erosion of coasts, rivers, canals, etc. Challenges with resilient architecture to strengthen the social community with parks, for example with the Japanese "Koban" model, community mailboxes and noticeboards, creation of resilience hubs such as emergency meeting places during service interruptions, etc. Many resilient design practices are highlighted globally with quite comprehensive results with good practices of applying vernacular architecture typology.

Resilient design in architecture: building process innovation

In the broad international debate to combat climate change and which invests the greatest criticality in the built and newly built environment, there are regulations and laws in force, with proposals and programs of management policies for the protection of man and the planet. According to the principles of the COP28 conference - United Nations Framework Convention on Climate Change (UNFCCC), the Intergovernmental Panel on Climate Change (IPCC), the Paris Agreement, WMO, UNEP, they aim at the conscious management of the planet and human well-being. This occurs mainly through decarbonization, in the reduction of CO2 also to the potential of BECCS (Bio-Energy and Carbon Capture and Storage) technology with biomass practices that absorb CO2 from the air through the process of photosynthesis and also through Plug-in strategies [6], etc. counteracting climate change that causes floods, hurricanes, droughts, and more, with the greatest global warming highlighting temperatures of about 2.3°C, above average compared to 1.45°C, a high index of the seas including the Mediterranean considered a hotspot together with the Antarctic, as reported by the World Meteorological Organization (WMO) and the European Copernicus Service, an initiative of the European Space Agency (ESA) and the European Commission, in the European State of Climate in 2023 Report.

In the USA, the Resilient America Act (Resilient America Act) bill involves an increase in disaster-predictive funding through the FEMA Building Resilient Infrastructure and Communities (BRIC)program and with the adoption of building codes and standards by municipalities and citizens to support resilience in the built environment with a view to economic and environmental sustainability. In fact, for every $1.00 invested for pre-disaster purposes, approximately $11.00 is saved through the adoption of codes and standards as reported by the NIBS (National Institute of Building Science). Further strategies are reported in the UNEP (United Nations Environment Programme) publication, Beating the Heat: A Sustainable Cooling Handbook for Cities [7], for the sustainable cooling of urban spaces that have become "heat islands" with temperatures above 5° Celsius due to materials such as asphalt and concrete, etc. Therefore, in the building sector, the resilient architectural design comes first with the fundamental objective of countering climate change, with all the related risks and consequent damages that it entails for man and the planet, and external events, to then return to an original condition. It requires an integrated project with a multidisciplinary approach of architects, energy managers, engineers, scientists, urban planners, landscapers, local administrators, communities, etc. The proactive action of resilient architectural design comes before the active action of the criticality of climate change and externalevents, to warn and solve the damages that are highlighted, in the buildingsector and infrastructure with the additional advantage of improving the safety and quality of life of the user, with cooling and heating through natural ventilation integrated with intelligent BMS systems. It widely promotes sustainable development and through process innovation interacts, atdifferent scales, both on the new and on the built in the building sector, on the rigor of scientific research, at the urban level with the expansion of the connective with open grids and at the peri- urban level, on the territory, on the communities and on the ecosystem. Therefore, the perspectiveof a resilient architecture is that of an innovative design in which the increase in the value of the property and the rental demand of about 6% according to the Urban Land Institute and resale at high costs is also highlighted, with greater savings for maintenance, restoration and reconstruction works resisting environmental challenges.

For this purpose, the principles of vernacular architecture that are highlighted in the process innovation at the production, construction, and organizational level on operational and organizational models with traditional techniques and adoption of local materials, according to local traditional cultures, are predisposed to the integrative approach of intelligent technological systems towards AI.

A new building process in which a sequence of organizational, instrumental, and activity aspects is highlighted according to a process of operational phases, programming with localization of interventions and technical-financial, architectural design with the assignment of works and related execution, testing, management, and maintenance. To achieve results, technological innovation can be characterized by the introduction of new methodologies, tools, and new figures/subjects or by the innovation of activities, procedures, norms, and rules.

The Ecofin Council of the EU, among the strategies for safeguarding the collective good, sustainability, resilience, and equity for all, approved on 13 July 2021 the National Recovery and Resilience Plan (PNRR), including the Italian one, of the European Recovery Plan, Next Generation EU, with grants and loan agreements of up to 13% of the total amount of the plan for the Member States. Therefore, for the strategic purposes of resilience design in architecture, the aim is, according to incentive policies and international standards, such as the new European Directive EPBD (Energy performance of building directive, EPBD) 2024/1275 "Green House Directive", to resilience in the building sector, both new and pre-existing, with the redevelopment of the built environment, for energy efficiency (43% of buildings with a 55% reduction in energy consumption), among the macro objectives of reducing harmful emissions with zero emissions in 2050. In particular, for the renovation of buildings in the residential sector, the reduction of energy by 16% by 2030, 20/22% by 2035, while for non-residential buildings, 26% by 2033. Instead, by 2028, public buildings will be zero emissions (ZEmB-zero emission building), and by2033, new residential buildings. "Solar-ready" buildings for the installation of photovoltaic or solar thermal systems. Instead, from 2027, solar energy will gradually have to be installed in existing public and non-residential buildings.

Strategies with increased automation in intelligent buildings can reach approximately 5-25% of energy costs with smart readiness with investments in efficiency and energy savings, through the installation of technical management systems of the building TBM (Technical Home and Building Management) and for automatic control with BACS (Building Automation and Control Systems) multiple intelligent technologies are installed with the additional aspect of the ability to prepare and install intelligent technologies integrated with management systems with IT (Information Technology) and IoT (Internet of Things) technologies and BIM and cloud technologies that also contribute to the management of the resilience of water infrastructure systems, both for drinking water and wastewater capable of adapting to weather changes.

They are aimed at predictive and resilient action with low environmental impact and to detect, first of all, anomalies of possible damage and then intervening withintelligent devices and restoring the initial conditions. Such as gas or smoke detectors, through sensors to activate evacuation and alarm systems. Furthermore, they create internal comfort with high-performance levels, according to the user's needs, with automatic adaptation capabilities andensuring internal microclimatic conditions, there will be an indicator and a methodology with an optional stand of the European Union.

In this perspective, Artificial Intelligence (AI) [8], through the development of advanced analysisand virtual models connecting all systems contributes to efficiency and energy savings in buildingmanagement, through predictive monitoring with direct control of security systems,

Materials Research Proceedings 47 (2025) 327-342 https://doi.org/10.21741/9781644903391-38

lighting, ventilation, and all plant management of the building with control of any faults or anomalies maximizing efficiency and reducing costs.

Another strategic aspect of resilient design is the reduction of financial losses through the creationof reserve generator sets with reinforced building structures. Furthermore, the creation of adaptabletechnological spaces, with flexible use of spaces leads to continuous productivity during any interruptions of damage from outside.

For smart readiness which represents the ability to prepare and install intelligent technologies forinternal comfort with high performance levels, according to the user's needs, with automatic adaptation capabilities and ensuring internal microclimatic conditions, there will be an indicator and a methodology with an optional stand of the European Union. Furthermore, for safety and security, indoor air quality, flexibility with control capacity for energy saving and to operate on “demand response” for the adaptation of the amount of energy consumed also according to the needs of the occupant (user) in which the inputs of the electrical network are useful and also for thermal comfort with active systems with the adoption of intelligent technologies and HVAC (Heating, Ventilation, and Air Conditioning) systems for internal microclimate with passive overrides and adopting natural ventilation systems and with smart and green and eco-sustainable materials.

For example, HVAC fans guarantee the extraction of smoke by introducing flows of fresh air, in the event of fires, ensuring that they are not supplied with additional oxygen, but using pressurization and compartmentalization according to the structures of the buildings and vertical connections. In fact, in sports facilities, where there are no fire doors, the intelligent HVAC system can automatically detect the intake of smoke and the introduction of air, so even in hospitalsthe effectiveness of the detection is mainly to leave the passages free or pressurize the rooms, preventing the spread of bacteria in the air. Even in the presence of possible fires, the override function is similarly to that of the firefighters' maneuver, to control the function of the engine rotation and guarantee the internal temperature, even in extreme situations (source) and constituting the safety of the inhabitants, allowing the inhabitants to evacuate by containing the damage and the costs of restoration or recovery caused by the fire.

Parametric design methodologies [9], for the resilient architecture of a new building process withBIM systems and adoption of the BMS (Building Management System) of MEP systems, and BEMS (Building Management Energy Systems) systems for the reduction of C02, user comfort and well-being with greater cost reduction.

Building design that aims at an architecture made with durable and maintainable materials that improve resilience, as for the same infrastructures and ecosystems, this principle is followed by connectivity to external communication networks, with possible infrastructure enabling a possible innovative service.

It also follows the interoperability of all BACS (Building & Automation Control System) UNI EN15232-1 systems, for the intelligent management of all building systems connected to IoT. In thisscenario of adoption of digital technologies, AI is among the fundamental parameters for resilientdesign which, in addition to the purpose of energy savings for highly polluting buildings (about 40% of CO2 emissions come from the construction and building sector in general), there is a community-wide prospect of reducing emissions from the building sector by 6% from 2020 to 2030, including net-zero objectives, including resistance to seismic risks and catastrophic and anthropic events. Therefore, economic advantages are highlighted with a reduction in insurance premiums and further financial advantages from the results of an invoked process innovation in which the technological system of resilient architectures indicates various systems, both passive and digital, for the management of the building's systems in a resilient architectural design and development of structural typologies that aim to protect the building in the event of environmental and hydrogeological damage.

Even green building practices, for example, low environmental impact and sustainable, with greenbuildings including wood with high seismic resistance, fire resistance, fireproof, durable, with highenergy performance with the application of different technologies and innovative, sustainable, andsmart materials, nanotechnologies vernacular architecture techniques, bio-inspired materials withthe use of concrete, wood, polymers. Within this field of investigation, new techniques are soughtto systematically mix materials (concrete, wood, polymers), with some specific properties including porosity, opacity, and stiffness, and with the help of new parametric and associative design tools, etc.

Many researches and schools of Architecture such as ETH, ILEK, UCL explore the use of graded and anisotropic multi-materials, sustainable and bio-inspired, distant from the conventional isotropic ones. The new types of resilient buildings offer economic advantages with lower environmental impact results and an increase in real estate value of about 4%, according to the GBC (Green Building Council) of the USA. Therefore, wooden buildings are also built for skyscrapers that almost as challenges show off their heights and resilience, and among these the 2023 project of the tallest skyscraper in the world, the C6 by Fraser & Partners in Perth, Australia, with an elegant and monolithic diagrid grid facade, 50 floors and 191 meters high, with a hybrid solid wood structure used for the beams, windows, etc. constituting 42% of the materials adopted in the project. The wood is used with Cross Laminated Timber (CLT) technologies, Laminated Veneer Lumber (LVL), which is a new type of engineered wood, and Glulam with a basement and access to a fleet of 80 shared Model 3 Testa, with EV charging ports for each car. The building contains 245 apartments spread over 48 floors with a rooftop of approximately 500 square meters with restaurants, edible gardens and gardens, and entertainment spaces. A sustainable and resilient skyscraper designed to be zero carbon for its entire life cycle, it adopts an integrated network with renewable energy (Fig.2).

Fig.2. Wooden skyscraper (project), Perth, Australia. C6 by Fraser & Partners[10]

For example, it is essential, in the design of resilient architecture, to integrate green spaces that contribute to high standards of decor, compositional quality, and liveability in communities, especially in urban areas, promoting not only biodiversity but also the reduction of heat islands that are created with the overheating of materials and the torrid climate in cities.

The criteria for integrating new pre-existing high-performance and adaptive architectural envelopes of redevelopment, sustainable recovery, and resilience for the challenges of climate change. Therefore the resilient design of new building processes in the innovation of technological/construction systems, with materials and components in which nanotechnologies stand out in passive systems such as heating and cooling, with sustainable naturalistic engineering systems (develop) and bio-inspired (develop), intelligent materials, for the reduction of dependence on water and increase of wastewater with disposal of waste of green building and eco-sustainable materials.

It contributes to sustainable development by improving the resilience of the community whose quality can be improved by businesses, in attracting greater investment within resilient urban planning with flexible and adaptable infrastructures to recover quickly.

And therefore, in addition to buildings, resilient design involves the entire urban planning with cities with low environmental impact and minimizing the damage resulting from climate change, seismic and climatic events, and hydrogeological disasters. In fact, against such risks, even the integration of equipped green spaces, according to naturalistic engineering with green parks, simultaneously become devices and green infrastructures for the absorption of waste water with the improvement of the quantity of water, as happens in some cities including Philadelphia, with the Green City program, Clean Waters, as in Hamburg in Germany with resilient infrastructures, Riode Janeiro, etc.

Performance of systems and materials

Resilient design in architecture, with the use of BIM, considers some fundamental parameters including the shape of the building, with exposure of the internal residential units to the cardinal points, the color, concerning the context, the position of the building for the suitablesolar exposure and that of the prevailing winds, presence of vegetation and water and with materials resistant to floods, seismic events, adaptive with high performance, natural of green building and sustainability and possibly local of the genius loci and with renewable energy, etc. As in many vernacular and resilient architectures including the house at the foot of Etna in Sicilyin the city of Catania, Felcerossa house (Fig.3) made with a load-bearing wooden structure and straw bales (a) and a pitched roof, with the use of natural materials, adopted throughout the house,with internal walls and floors treated with a clay and lime mixture finished with a spray compound (b) of 50° skimmed milk and 50% water, to avoid the crumbling of the clay. Instead, the externalwalls were treated with natural waxes and linseed oil for the consolidation of the mortar, while forthe colored walls, a mix of oxides, water, and milk was applied.

Materials Research Proceedings 47 (2025) 327-342 https://doi.org/10.21741/9781644903391-38

a b

Fig.3. Resilient Home, Felcerossa, Catania, Sicilia: (a) load-bearing structure in wood and strawbales, (b) clay and lime mixture finishes finished with a compound spray [11]

Generally, it is necessary to recover their functionality after natural disasters or those caused by anthropogenic factors for buildings. To this point, it requires an integrated building/systems and participatory project, which protects above all the building and the user from the dangers of damages coming from natural phenomena and climate change by resisting shocks and stresses such as floods, environmental disasters, floods, strong winds, hurricanes, forest fires, etc. It is mainly based on green infrastructures at the territorial and urban levels, integrated water management(wastewater collection and greywater recycling), and smart technology. To avoid the rise in sea levels, floods, load of strong winds, and climate change with all the consequent damages, aresilient architectural design is carried out within a mega project in which participatory urban planning is important with a series of inputs. Residents collaborate, with a series of inputs, on the plan, through mechanisms such as community workshops and social media, as well as online surveys, and with the possibility of voting for projects in their neighborhood, as in the city of Medellin in Colombia.

In fact, in America, both states and localities provide grants for pre-disaster mitigation efforts on FEMA (Federal Emergency Management Agency) programs of the BRIC (Building Resilient Infrastructure and Communities)[12], strengthened by the NAHB (National Association Home Builders) based on the Resilient AMERICA Act - Resilient Assistance for Mitigation for Environmentally Resilient Infrastructure and Construction by Americans Act, which goes from 6% to 15% of the funding calculation in adopting updated building codes and building standards, to incentivize resilience in the built environment with a relative saving on investments of about 11 dollars compared to 1 dollar of investment for recovery and redevelopment after damages as reported by the NIBS (National Institute of Building Science). Even in New York City with the Rooftop Garden Project, communities are participating in urban planning for resilience, through green infrastructure projects. Likewise in Canada in the city of Vancouver, citizens through awareness-raising communication initiatives on social media, online surveys, and community workshops transmit input to local authorities to reflect on their needs for the plan, as well as in many other cities.

Therefore, within the large urban planning for resilient design, process innovation is manifested, geological surveys are carried out in the territory with careful reading of the maps, especially for the control of the areas at seismic and hydrogeological risk, before any intervention, both residential and infrastructural, which are followed by the adoption of regulations and laws in force with application and implementation of programs. Among the most feasible forms of intervention to restore the building from a flood are those of isolating and waterproofing it, as well as raising the ground and raising the floor.

For territorial planning, a checklist of buildings that are aging and on which to operate with mandatory retrofits with urban planning is drawn up together with a series of works of adaptation to the climate and with zoning that can guarantee areas not vulnerable to the seismic risk of possible interventions.

The main features of this resilient design are represented by high-performance systems and sustainable and innovative materials with the creation of redundant systems, passive solar energy collection systems, mainly as storage control components in heavy structures - floors and walls, storage masses, solar collectors, greenhouse effect, green infrastructures that guarantee the functioning of the service, as an alternative to damaged systems, use of naturalistic engineering, integrated water management, with wastewater collection and grey water recycling. In particular, some types of passive systems, from the perspective of resilience, are also identified in bioclimatic architecture for direct solar gain with radiative heat exchange which consists of the installation of transparent vertical closures on the southern facades of the building, for the best solar collection with the greenhouse effect, as well as in the already announced innovative wintergarten, of German architecture, having the dual function of thermal buffer and "Gemütlichkeit" inside the homes. To absorb internal heat, all vertical closures, exposed to the north of the building, require thermal insulation which is mainly accumulated by floors, attics, walls, roofs, etc. since their mass, with high thermal inertia, becomes a heat accumulator with a time delay and attenuationof the thermal wave, through insulating systems to minimize heat dispersion.

Natural materials also work together, such as wood, local stone, glass, etc. porcelain stoneware, which is distinguished by a high percentage of recycled material, making it singularly resilient, highlighting its unalterable and very waterproof, fireproof function with a long life cycle. Also natural insulators (especially suitable for the construction of wooden houses), including mushroom mycelium, whose roots, drying out after being sprinkled around the wood, transform into a fireproof insulating barrier, replacing the insulating foams. Like other natural insulators such as rock wool, cork, or sheep's wool.

Also, insulating concrete blocks (Fig.4), indicate durability and speed of construction, high thermal insulation, and energy savings by creating very resistant walls through a system with a dual function of internal and external insulation with $R = 8.70\ m^2K/W$

For solar gain, solar collector systems are used, photovoltaic panels on the roofs and facades.

Fig.4. Insulating formwork block: 2 polystyrene walls, connected with metal spacers, in which concrete is poured up to the height of one floor. This type of wall guarantees resistance and stability. [13]

Instead, in indirect solar gain systems the heat exchange is connective, and finally, isolated gain in which the capture surface is separated from the thermal accumulation, not heating the surroundingindoor environment. At the level of resilient communities, even pedestrian lanes separated from cycle paths, and bodies of water, constitute an advantage for new design in the field of resilience in architecture, etc. In particular, green infrastructures including green walls and urban reforestationprojects, with green roofs reduce the urban heat island effect (UHI- Urban Heat Island), a phenomenon that occurs through a microclimate with an increase in temperature, which can reachover 5° in urban areas, compared to peripheral and rural areas and caused by the cities themselvesdue to the density of buildings, motor vehicle traffic, that emitted by construction materials including cement and road asphalt, etc. Furthermore, socioeconomic and demographic factors alsointeract to cause heat islands, with a high environmental impact and risks in the urban and connective space. Therefore, green interventions are very decisive for the best air quality and resilience, also promoting the flow of wastewater, in streets, alleys, urban spaces with squares, equipped green areas, parks, and roadsides, etc., and in which the infrastructures themselves are designed for greater adaptability and flexibility against extreme flood events, seismic risks and vulnerability. In particular, for resilience in both new and pre-existing or historic buildings, innovative, minimally invasive systems are adopted, with low impact and guaranteeing broad structural safety and resilience of the building, throughout the technological system of the building, including the load-bearing structures, the vertical partition system, the infill closures, the fire-fighting systems, the anchoring system, etc. Smart technologies are applied to them, in the various typologies for anti-seismic strategies, according to current regulations such as UNI 15129:2008, including the same hysteretic, re-centering, passive seismic isolators (elastomeric, sliding) including sliding pendulum ones (Friction pendulum): in rubber/steel, rubber/steel with lead core, with different reactions in the two horizontal directions) and viscous dissipators, which through discharge valves using fluids such as lead or silicone, dissipate the seismic energy which, together with the ISAAC technologies with the Active Mass Damper, represents a safety for the resilience and anti-seismic resistance of the building.

The function of the tuned mass damper by Fip Mec, an Italian company, on the project is to reduce the amplitude of the oscillations with safety, avoiding the structural collapse of the building

Materials Research Forum LLC
https://doi.org/10.21741/9781644903391-38

due to any seismic actions counteracted by accelerations generated by the ground towardsthe bottom of the building, by the action of the wind and those of normal management of the building, ensuring comfort for users. In the case of the 101-story skyscraper, with a height of about 580 meters, hence the name Taipei 101 in the city of Taipei in Taiwan (Fig. 5), a building of Taiwanese culture consisting of eight truncated pyramid-shaped modules with features of bamboo trunk, a symbolic plant of Taiwan, the anti-seismic device was installed between the 87thand 92nd floor of the 508-meter-high building on 101 floors. It consists of one of the largest steelspheres in the world with a diameter of 5.5 meters, suspended from steel cables and connected to the load-bearing structure of the building with a weight of approximately 660 tons (equal to the weight of two Boeing 747 airplanes) and formed by 41 overlapping disks. The function of the tuned mass damper (TMD), a passive device as a harmonic absorber, is to reduce the amplitude of the oscillations safely avoiding the structural collapse of the building due to possible seismic actions, counteracted by accelerations generated by the ground towards the bottom of the building and by those of the wind. In fact, in Taipei, vertical oscillations are generated by strong winds that counteract the action of the wind, which can reach speeds of up to200 km/h at a height of 500 meters, through the action of hydraulic connecting pistons that can dissipate the energy produced by the earthquake and the wind. During the design phase, accelerometer sensors are defined and then installed in various parts of the building that indicate, through real-time reading, the force generated by the AMD. At the top of the skyscraper, the massof the sphere, which acts as a suspended damper, due to its inertia, oscillates in counter phase and stimulated by an actuator that makes it vibrate by exerting a force on the steel mass, concerning the trend of the structure, counteracts the oscillation of the building both due to seismic actionsand wind actions, through connecting pistons that, by compressing and stretching, dissipate the energy produced by the oscillations. In reality, the active version of the TMD is called Active MassDamper (AMD, an active system developed by ISAAC.

e f

Fig.5. Taipei 101 Skyscraper, Taipei a Taiwan- tuned mass damper (TMD): (e) sphere suspendedfrom steel cables [14], (f) detail of steel cables connected to the sphere [15]

Instead, for hysteretic seismic isolators, resilient design becomes optimal also to obtain calibratedvalues of horizontal stiffness and equivalent damping, with their installation at the base of the buildings between the foundations and pillars, managing to dissociate the superstructure of the building from the seismic action and reducing the transmission of seismic energy as well as to obtain calibrated values of horizontal stiffness and equivalent damping and aimed at the suitable isolation system for the building structure.

Fig.16 Seismic isolators that guarantee the control of the deflections andaccelerations of a building [16]

Within an anti-seismic design for greater safety, in the event of earthquakes with a probability of exceeding about 10% in 50 years according to the ULS (Ultimate Limit State), especially in the most vulnerable and exposed countries to earthquakes. Among these dissipators, the hysteretic ones stand out, which can absorb the deformation of shaped steel elements placed in a restraint system or coupled to a mobile support, in some cases, managing to dampen the seismic movement actions (Fig.16).

Other different technological systems that are adopted in buildings are those for passive cooling, resilience, and combating climate change with the use of vernacular architecture with some examples around the world including India. The plan of the housesis often distributed on a central courtyard with the use of dry stone walls, raw earth bricks, doublewall closures, green roofs, solar panels for solar thermal energy, double glazed window systems for high energy efficiency, pitched and cantilevered roofs to collect rainwater and convey it to thewater table and also to a closed circuit as in some cases in India etc. in verandas and trellises, drystone walls, double walls with external and internal waterproofing in urban and rural centers, in aninterface of local associations and administrations for the management of domestic water treatment. Even for the structural typologies, new foundation systems are inserted such as standardplinths at higher heights from from the support plane, in case of floods or inundations. This house in Alibaug designed by Architecture Brio (Fig.17) features a system of metal chains that directrainwater from a hidden gutter to a catch base.

Materials Research Proceedings 47 (2025) 327-342
https://doi.org/10.21741/9781644903391-38

Fig. 17. Alibaug House, designed bay Architecture Brio: wastewater collection system from gutter to catch basin [17]

Conclusions

Resilience in architecture in technological innovation that is highlighted in line with sustainability in construction is acclaimed with a strong attitude, in new construction and pre-existing construction, since it is aimed at the efficiency and safety of living in safe contexts against natural and anthropic vulnerability. For this purpose, an appropriate resilient design is constantly invoked that can in architecture give the best results for an efficient future in global construction supported by both regulations and laws that govern the various States. COP28, the UNFCCC, the Intergovernmental Panel on Climate Change (IPCC), the Paris Agreement, WMO, UNEP, etc. aim at the conscious management of the planet and human well-being against climate change.

Since such is the challenge of remaining globally below 2 ° C and also supported by the World Meteorological Congress (WMC) for resilience, with life-saving alert systems, and the Global Greenhouse Gas Watch to strengthen the monitoring of gases that store heat. Furthermore,the Buildings Performance Institute Europe (BPIE) also aims to identify 10 policy priorities with strategies that ensure zero-emission resilient buildings, for social well-being, economic gain and energy security, including AI for data collection, increased funding, promoting cooperation with the implementation of the EPBD, New European Bauhaus at the High-Level Forum on Construction), with the Heating & Cooling Strategy, to improve energy security and resilience. Furthermore, programs of FEMA (Federal Emergency Management Agency) of BRIC (Building Resilient Infrastructure and Communities) strengthened by NAHB (National Association Home Builders) based on the Resilient AMERICA Act - Resilient Assistance for Mitigation for Environmentally Resilient Infrastructure and Construction by Americans Act, NAHB (National Association Home Builders), New York City with the Rooftop Garden Project, etc.

Therefore, the main objectives are to build and redevelop architectural envelopes contextualized for user well-being, with the use of efficient techniques and technologies, natural engineering systems and intelligent materials, nanotechnologies and nanomaterials, bio-inspired and sustainable materials, biocompatible, nanotechnologies, high-performance systems. Also, strategic objectives that could prevent, in a preventive manner and with a proactive function, of control and anticipation, through resilience, any natural and anthropic reactivity, in an innovation of the building process in terms of construction, organization, and production. Adoption of redundant systems, accumulation components in heavy structures - floors and walls, accumulation masses, solar collectors, greenhouse effect, green infrastructures, use of naturalistic engineering, integrated water management, etc.

The following are integration criteria of resilient architectural envelopes with sustainability principles and passive systems for innovative heating and cooling, minimally invasive, low-impact systems, ensuring broad structural safety and resilience of the building, with optimization of local renewable resources and the genius loci and resilient design methodologies with political strategies and adoption of AI and intelligent BEMS, IoT, and IT (Information of Technology) systems.

The challenge is to aim for new resilient design through decentralization and decarbonization withdurable and efficient buildings for user well-being, with a strongly eco-sustainable approach towards vernacular architecture technologies.

References

[1] C.N.Schulz, Genius Loci: Genius loci. Paesaggio ambiente architettura, Milano, Electa 2023

[2] A. Ilki, D. Çavunt, Y.Selim Çavunt, Building for the Future: Durable, Sustainable, Resilient,2023 , Volume 2, Springer. https://doi.org/10.1007/978-3-031-32511-3

[3] Climate Change 2023 Synthesis Report https://www.ipcc.ch/report/ar6/syr/downloads/report/IPCC_AR6_SYR_FullVolume.pdf

[4] BPIE, Building Prosperity 10 Policy Priorities To Achieve Growth And Social Value ThroughResilient, Healthy And Affordable Zero-Emission Buildings, 2024, https://www.bpie.eu/wp- content/uploads/2024/07/Building-prosperity_10-policy-priorities-for-the-new-COM_Final.pdf

[5] S.Walker, Design for Resilience-Making the Future We Leave Behind, Publisher: The MIT Press Ltd, 2023. https://doi.org/10.7551/mitpress/14924.001.0001

[6] Beltracchi, C. (2024). Resilience and the Metaverse: A Toolkit Approach. In: Ruttico, P. (eds)Coding Architecture. Digital Innovations in Architecture, Engineering and Construction. Springer, Cham. https://doi.org/10.1007/978-3-031-47913-7_5

[7] Beating the Heat: A Sustainable Cooling Handbook for Cities https://www.unep.org/resources/report/beating-heat-sustainable-cooling-handbook-cities

[8] S. Dehghan, H. Norouzi, H. Gholami, Artificial Intelligence Methods for the Construction ofBuildings, Ed.Lambert Academic Publishing 2024

[9] F. Di Paola , A. Mercurio, Parametric Experiments in Architecture-A Connection Joint Design for Sustainable Structures in Bamboo. Springer 2023. https://doi.org/10.1007/978-3-030-96276-0

[10] C6 by Fraser & Partners and Grange Development, https://thelocalproject.com.au/articles/c6-by-fraser-partners-and-grange-development-issue-11-commercial-feature-the-local-project/?srsltid=AfmBOoryEjedstb2d4XGiWMaQcfB8dLCAjzCY8V-Xe2L94vNqDJukoku

[11] La Famiglia Che Vive in una Casa di Paglia alle Falde dell'Etna https://www.houzz.it/magazine/la-famiglia-che-vive-in-una-casa-di-paglia-alle-falde-dell-etna-stsetivw-vs~112694875

[12] FEMA, Building Resilient Infrastructure and Communities, 2023,

https://www.fema.gov/grants/mitigation/learn/building-resilient-infrastructure-communities

[13] Insulating concrete blocks: a system for resilient buildings, https://www.euromac2.com/en/les-blocs-coffrants-isolants-un-systeme-qui-permet-la-construction-de-batiments-resilients/

[14] Si chiama Tuned Mass Damper, la maxisfera italiana ha … http://www.blueplanetheart.it/2024/04/si-chiama-tuned-mass-damper-la-maxisfera-italiana-ha-salvato-la-torre-di-taipei

[15] Tecnologia italiana evita il possibile crollo del grattacielo ... https://www.meccanicanews.com/2024/04/06/tecnologia-italiana-evita-il-possibile-crollo-del-grattacielo-taipei-101/

[16] Architecture and construction: structural seismic isolators https://amusementlogic.com/general-news/architecture-and-construction-structural-seismic-isolators/

[17] Cos'è l'Architettura Resiliente che Risponde al Cambio del Clima https://www.houzz.it/magazine/cos-e-l-architettura-resiliente-che-risponde-al-cambio-del-clima-stsetivw-vs~155675738

Optimization and enhancement of building envelope characteristics for implementing energy-efficient buildings

Omar BOURASS[1,a]*, Khalid EL HARROUNI[1,b] and Aziz ET-TAHIR[2,c]

[1]National School of Architecture of Rabat, Morocco

[2]Mohammed V University of Rabat, Morocco

[a]omar.bourass@hotmail.com, [b]k.elharrouni@enarabat.ac.com, [c]ettahiraziz@hotmail.com

Keywords: Energy-Efficient Buildings, Thermal Comfort, Sustainable Architecture, Building Envelope, Thermal Performance, Thermal Dynamic Simulation, Energy Needs Gains and Loss

Abstract. The purpose of this scientific paper is to study the impact of optimizing the characteristics of the envelope of a house localized in a desert climate, on the annual energy needs in terms of heating and cooling, solar energy, and internal gains and losses. In this regard, simulations were carried out, first on a basic house which took into consideration the characteristics of the usual construction method in this specified geographical area, and secondly, on the same basic house but improved in terms of its envelope characteristics and taking in consideration some of the strongest point of vernacular architecture in this area, namely, the harmony between environment and buildings. Based on our analysis, it turns out that by simply optimizing the characteristics of the envelope, we can save a considerable amount of energy throughout the life cycle of a house and we manage to save three times less energy, more energy gains, and less energy losses compared to the basic method of construction.

Introduction

The buildings and construction sectors combined are responsible for 36% of global final energy consumption and nearly 40% of total direct and indirect CO2 emissions [1].

In Morocco, the building sector is the largest consumer of final energy, with a share of 25% of the country's total energy consumption, and this energy consumption is expected to increase rapidly in the coming years [2]. One of the major priorities of Morocco's new energy strategy is to increase the contribution of renewable energies to electricity production by 52% by 2030. Developing energy efficiency in buildings is one of the aims of this strategy.

In this paper, we tried to see if an efficient envelope of a house can contribute to reducing energy consumption in the arid climate of South Morocco.

Aims

We studied the impact of optimizing the characteristics of the envelope on the annual energy needs in terms of heating and cooling, and also on the internal and solar energy gains and finally on the energy losses.

Our simulations were carried out on two study cases in the city of Errachidia in the South-East of Morocco: firstly, on a house that we named "basic" since it takes into consideration the characteristics of the new-usual construction method in these areas compared to old vernacular architecture, and secondly, on the same basic house but, improved in terms of its envelope characteristics.

The goal is to determine the degree of impact of the optimization of the characteristics of the envelope of the buildings of this geographical area in comparison with the usual mode of construction, to be able to show the degree of energy savings that can be achieved while guaranteeing the same thermal comfort.

Materials Research Proceedings 47 (2025) 343-353 https://doi.org/10.21741/9781644903391-39

Previous Studies

Several studies have been interested in studying the impact of different parts of the building envelope in a desert climate similar to our case study, indeed, (Bolattürk, 2006) calculated the optimal thickness of the insulation as well as the energy savings relating to the winter period in four climatic zones of Turkey [3], (Fezzioui et al, 2008) studied the influence of the building envelope on its energy demand during the overheating period, in the cities of southern Algeria [4]. (Mia Mefteh et al, 2008) have shown that the reduction of losses in energy rehabilitation consisted of improving the various components of materials and reducing thermal conductivity [5].

(Bekkouche et al, 2009) studied the influence and impact of the thermal inertia of stone on the efficiency of the interior and exterior thermal insulation of a building in Ghardaia in the south of Algeria and found that the insulation acts as a thermal barrier, slowing down energy losses during the cold seasons [6]. (Daouas et al, 2010) have shown that the most profitable cases for energy savings are the use of stone, brick sandwich panels, and expanded polystyrene for insulation, with an optimum thickness of 5.7 cm [7]. (Guechchati et al, 2010) used TRNSYS 16, using the multizone model (type 56), to study the effect of thermal insulation on the energy consumption of a Psycho-pedagogical Center in Morocco and showed that the economy in the energy was obtained by using the full insulation of the center and thus opted for the insulation of the roof coupled with the external insulation of the walls with 6 cm of expanded polystyrene [8].

Methodology

In our case study, we worked with "TRNSYS 16" a complete and extensible simulation environment for the transient simulation of systems [9].

To generate meteorological data for our area, we used Meteonorm, a database containing comprehensive climatological data for solar engineering applications at every location on the globe, it's a standardization tool permitting developers and users of engineering design programs access to a comprehensive, uniform meteorological data basis [10].

The city in which we developed our model is situated in the south-east of Morocco characterized by its desert climate. Indeed, the city of Errachidia has a significant temperature difference, between very high temperatures in the summer (39°C on average in July) and very low temperatures in the winter (3°C on average in January). The temperature reaches 42°C in July and falls below 0°C in January;

In this study, we took into account the results of the works that were carried out for the study of the impact of orientation (Bourass and Ettahir, 2019) [11], and exterior walls (Bourass, Ettahir and Kettani, 2019) [12] on energy consumption.

The house object of our thermal simulations has 70 m^2 as gross floor area, 57 m^2 as a net floor area, and 171 m^3 as a net volume. The length is 10,60 m and the width is 6,60 m. The type of glazing used in our first case is simple glazing with a thermal transmission of 5.74 $W/m^2.K$ while it's triple glazing in the improved case with a thermal transmission of 2 $W/m^2.K$. The windows of the main façade in the basic model are oriented to the North, while in the ameliorated model, they are oriented to the South.

The set indoor temperature is 22 °C for heating, and 26 °C, for cooling. The ventilation rate is set at 0.6V/h (the house is ventilated between 6 and 7 AM and between 8 and 9 PM), while the infiltration is 0.6 V/h.

Materials Research Proceedings 47 (2025) 343-353 https://doi.org/10.21741/9781644903391-39

Construction Details

Table 1: Construction details and thermal characteristics used in (basic house)

Constitution (from Inside to Outside)		Thickness (m)	Conductivity (Kj/Hmk)	Capacity (Kj/Kg.K)	Density (Kg/m³)	Thermal Transmission (W/m².K)
Floor	Floor Tiling	0,01	1,3	0.801	790	1.623
	Cement Screed	0,05	2,52	1	1300	
	concrete slab	0,15	4.068	1	1400	
	hardcore	0,15	2,52	1	1800	
		Tot : 0,36				
Exter Walls	Plaster Mortar	0,02	1,26	1	1200	2.179
	Concrete block	0.20	3.46	0.65	1300	
	Cement Mortar	0,03	4,5	0.84	2000	
		Tot :0,25				
Roof	Plaster Mortar	0,02	1,26	1	1200	1.702
	Concrete Slab	0.25	4.068	1	1400	
	Bitumen Roofing	0.02	1.805	1	1700	
	Cement Screed	0,05	2,52	1	1300	
	Roof Tiling	0,01	1,3	0.801	790	
		Tot : 0,35				

Table 2: Construction details and thermal characteristics used in the "improved house"

Constitution (From Inside to Outside)		Thickness (m)	Conductivity (Kj/Hmk)	Capacity (Kj/Kg.K)	Density (Kg/m³)	Thermal Transmission (W/m².K)
Floor	Floor Tiling	0,01	1,3	0.80	790	0.134
	Cement Screed	0,05	2,52	1	1300	
	XPS	0.20	0.105	1.18	35	
	concrete slab	0,15	4.068	1	1400	
	hardcore	0,15	2,52	1	1800	
		Tot :0,56				
Exter Walls	Plaster Mortar	0,02	1,26	1	1200	0.133
	Hollow Brick	0,05	1,69	0.794	720	
	XPS XTRUD	0.20	0.105	1.18	35	
	Hollow Brick	0,15	1,69	0.794	720	
	Cement Mortar	0,03	4,5	0.84	2000	
		Tot :0,45				
Roof	Plaster Mortar	0,02	1,26	1	1200	0.134
	Concrete Slab	0.25	4.068	1	1400	
	XPS	0.20	0.105	1.18	35	
	Bitumen Roofing	0.02	1.805	1	1700	
	Cement Screed	0,05	2,52	1	1300	
	Roof Tiling	0,01	1,3	0.801	790	
		Tot :0,55				

Results and discussion

Based on the results of the annual energy needs, gains, and losses (Fig. 1), we can observe a big difference between the two studied cases. Indeed, in the improved case, we managed to save three times less energy needs, more energy gains, and less energy losses compared to the basic case.

In terms of the distribution of annual energy needs, gains, and losses (Fig. 2), we note that:

- The energy needs represent 40% in the basic model, while it is only 22% in the improved case;
- The energy losses represent 34% in the basic model, while it is only 28% in the improved case;
- For energy gains, the improved model marks an important quantity of energy gain, which is equivalent to the sum of its energy losses and needs (50%), while in the basic model, they represent only 26%.

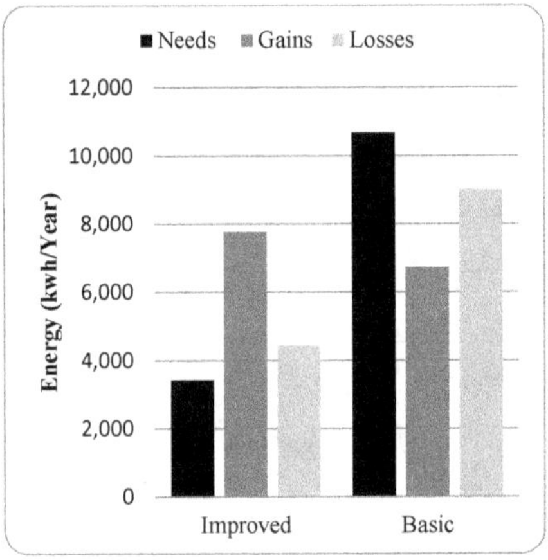

Fig. 1 Annual energy needs, gains, and losses

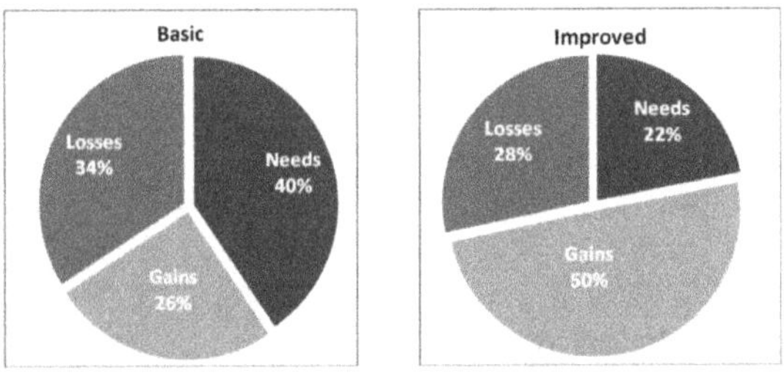

Fig. 2 Distribution of annual energy needs, gains and losses

- **At the level of energy needs:**

In general, for the annual energy needs for heating and cooling, we can see that significant energy savings are achieved in the improved model compared to the basic model. Indeed, the improved model saves three times less energy on the annual energy needs of the heated floor area (57 kWh/m²/year), compared to the basic model (178 kWh/m²/year) (Fig. 3). We observe also, that most of the energy savings recorded in the improved case are achieved in the energy consumption for heating (68 kWh/year in the improved model, compared to 6502 kWh/year in the basic model)

(Fig. 4). Those concerning cooling are less important (3363 kWh/year in the improved model, compared to 4167 kWh/year in the basic model.

We can see these differences in terms of the repartitions of annual heating and cooling needs (Fig. 5). Indeed, in the basic model, heating represents a large part of the needs (61%), while cooling represents 39%. In the improved model, the major part of the annual energy needs is mainly due to cooling (98%), while the heating ones are insignificant 2%.

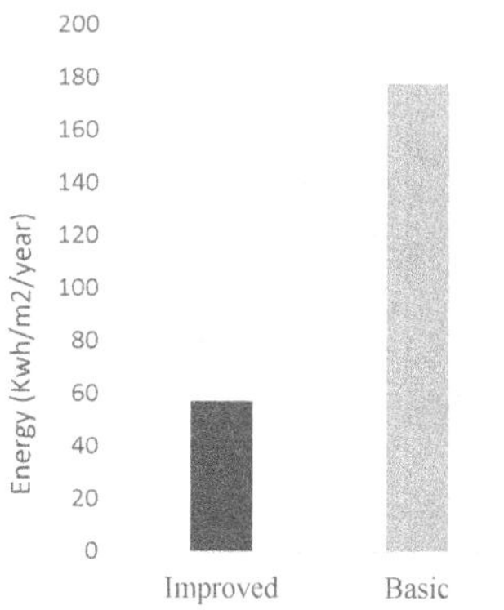

Fig. 3 Annual energy needs per m² (heating and cooling)

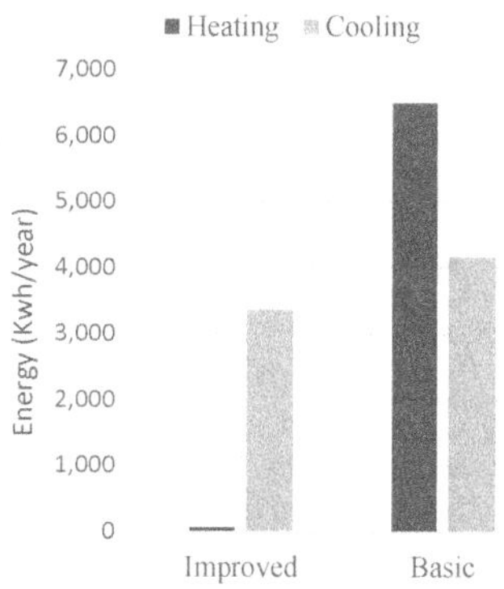

Fig. 4 Annual energy needs (heating and cooling)

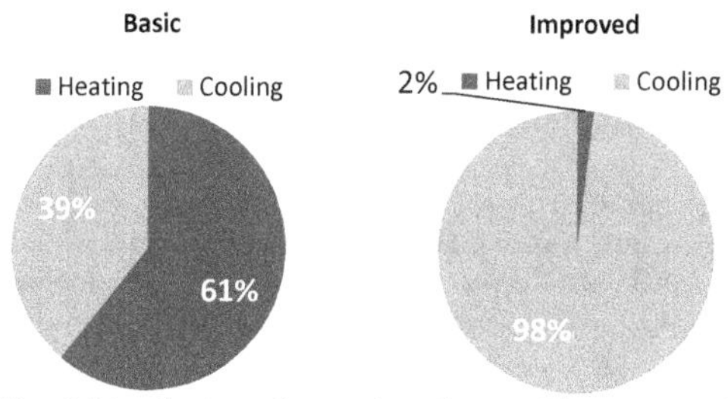

Fig. 5 Distribution of annual cooling and heating needs.

By analyzing the graphs representing the monthly energy needs for heating and cooling (Fig. 6) and (Fig. 7), we can get an idea about our deducted results in terms of their repartitions per month and see the difference between the two models.

Indeed, in the basic model, energy needs for heating, are spread over 09 months starting from September with insignificant consumption (5.29 kWh) and ending in May with a consumption of (54 kWh). The most important consumptions for heating are carried out over 04 months (November, December, January, and February). December (1671 kWh) and January (1886 kWh) remain the two months where there are much higher energy consumption rates for heating.

The energy needs for cooling, is spread over 09 months starting from April to September, with insignificant energy consumption over 04 months (April, May, September, and October). The most important energy consumption rates for cooling are made over 03 months, June (737 kWh), July

Materials Research Forum LLC
https://doi.org/10.21741/9781644903391-39

(1594 kWh), and August (1273 kWh). July and August remain the two months where there are much higher energy consumption rates for cooling.

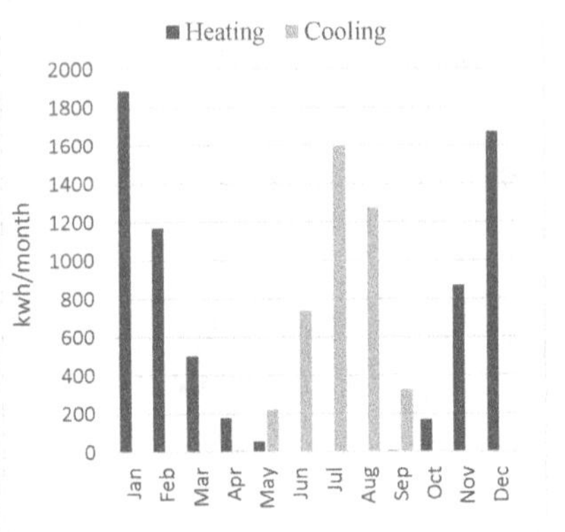

Fig. 6 Monthly cooling and heating energy needs (Basic case)

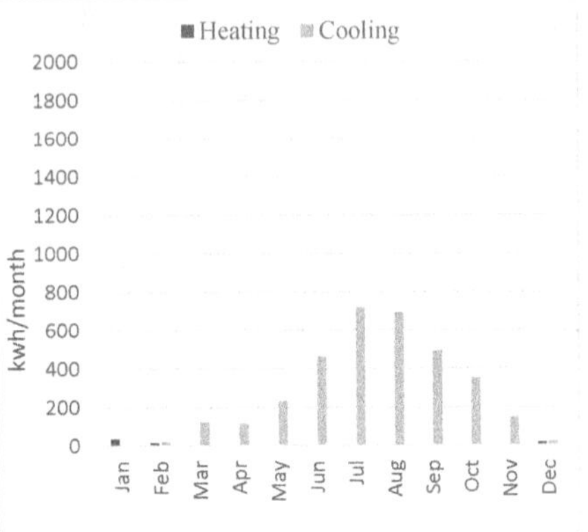

Fig. 7 Monthly cooling and heating energy needs (Improved case)

In the improved case, the energy needs for heating are spread over a period of 09 months from April to September, with very insignificant consumption throughout the whole year not exceeding 37 kWh in January.

The energy needs for cooling are spread over the whole year, with less consumption over 07 months (from November to May). The most significant consumptions are made over five months (ranging from June to September. The two months that mark high rates of consumption for cooling needs remain July (717 kWh) and August (691 kWh).

- **At the level of energy gains**

For the annual energy gains, we have already shown at the level of the general balance that, there is no big difference between the two models, indeed the basic model registers gains of the order of 6718 kWh/year and the improved model registers slightly significant energy savings of around 7758 kWh/year (Fig. 1).

The gains due to infiltration, ventilation as well as internal ones are identical in the two models (Fig. 8). However, the difference is observed at the level of solar gains (4422 kWh/year for the improved case and 1724 kWh/year for the basic case), as well as at the level of gains due to transmission (228 kWh/year for the improved case and 1288 kWh/year for the basic case). (Fig. 8)

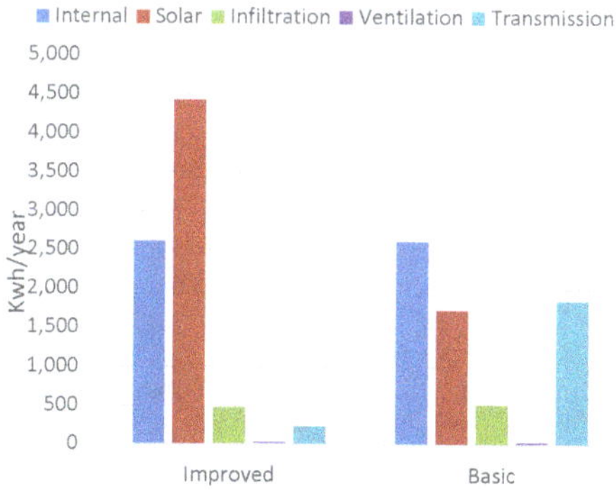

Fig. 8 Annual energy gains

By analyzing the graphs representing the monthly internal and solar energy gains (Fig. 9) and (Fig. 10), we notice that the improved model manages to gain enough solar energy compared to the basic case, especially during the cold months when the outdoor temperature begins to drop. Indeed, the more the temperature drops, the more the improved building gains solar energy, and the more the outside temperature increases the less it uses this energy since during the summer months we need to protect ourselves from the sun's rays than to benefit from it.

The basic model acts as the opposite of the improved model it gains more solar energy during the summer months, in other words, during the months when it is not needed, and marks less solar gains, during the cold months, when it is needed to benefit from it as much as possible.

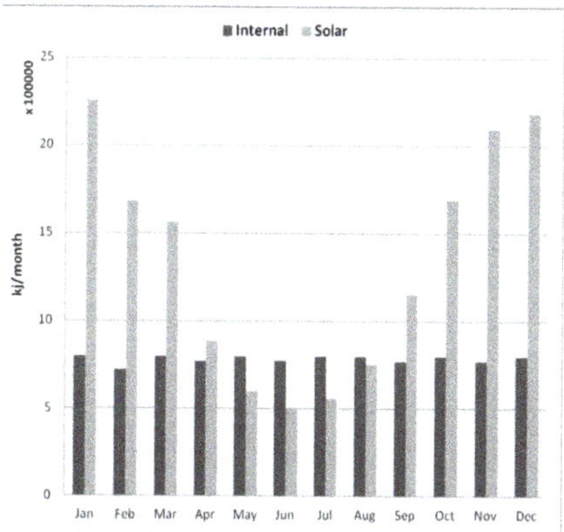

Fig. 9 Monthly solar and internal energy gains (Improved case)

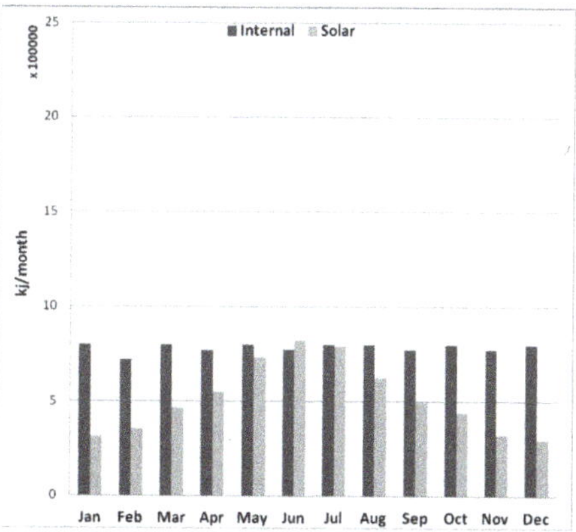

Fig. 10 Monthly solar and internal energy gains (Basic case)

Materials Research Proceedings 47 (2025) 343-353
https://doi.org/10.21741/9781644903391-39

- **At the level of energy losses**

In terms of energy losses, we notice that as in the energy needs, the results (Fig I) differ radically from the improved case to the basic one. Indeed, the latter marks the double of energy losses (9007 kWh/year) than the improved case (4429 kWh/year).

Regarding the distribution of energy losses (Fig. 11) and (Fig. 12), for the case of the basic model, the major part of the energy losses is achieved mainly by transmission (88%), the losses by infiltration constitute only 11%, while the losses by ventilation are insignificant (1%). In the case of the improved model, the energy losses are achieved mainly by transmission (52%), infiltration (44%), and ventilation; those are unimportant (4%).

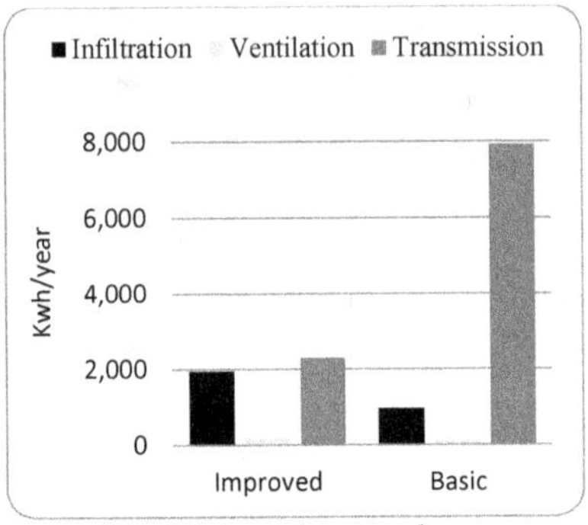

Fig. 11 Annual energy losses

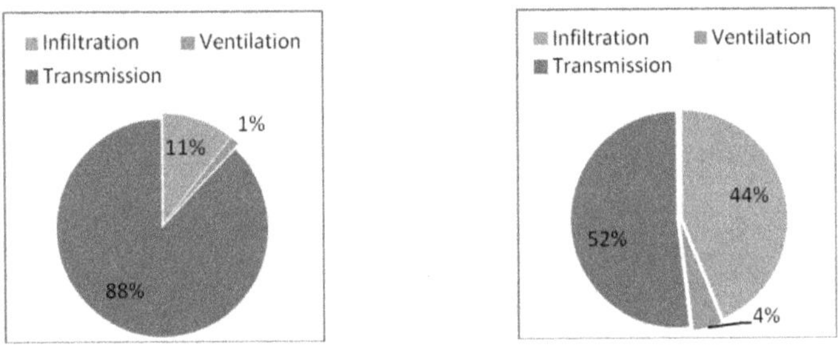

Fig. 12 Repartition of annual energy losses in the basic case (Left) and improved case (Right)

The energy losses occur between October and May, the most significant are between November and February. December and January remain the months when there are more losses, considering that the outside temperature marks significant drops in these months.

By analyzing the graphs (Fig. 13 and Fig. 14), we can see clearly that there is an important difference between the two models. Indeed, the basic case loses huge amounts of energy compared to the improved one especially (from January to May and from October to December). The most important energy losses are realized in autumn, and winter exactly in the months of November, December, and January.

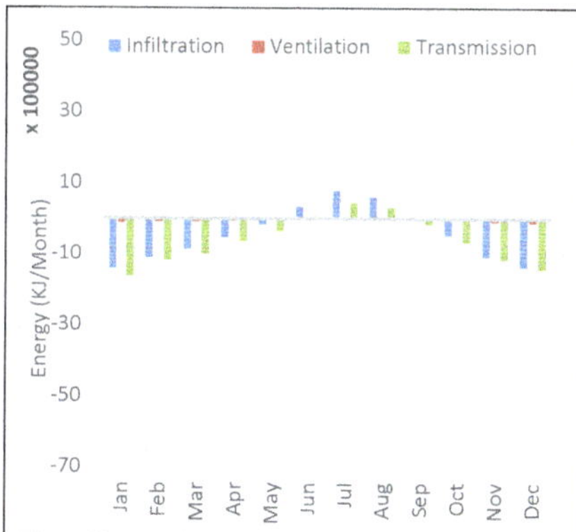

Fig. 13 Monthly energy losses and gains due to infiltration, ventilation, and transmission (Improved case)

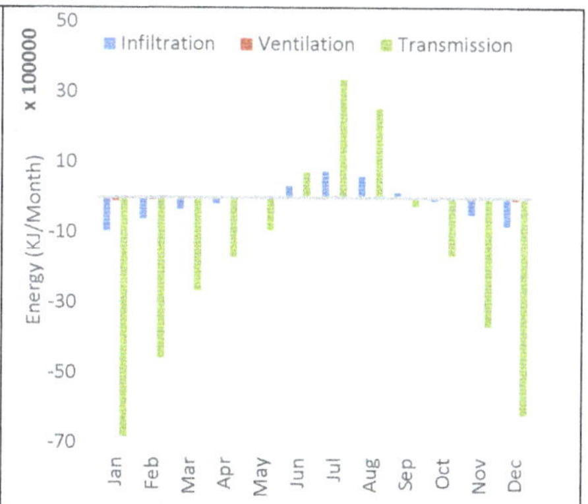

Fig. 14 Monthly energy losses and gains due to infiltration, ventilation, and transmission (Basic case)

Explication

The difference in results can be explained by the nature and characteristics and the types of materials used in the two cases:

- **Type of materials and composition of the floor:** In the basic case, we used an assembly of hardcore, a concrete slab, a cement screed, and floor tiling, with a thickness of 0.36 m and thermal transmittance of 1.62 W/m^2.K, while in the improved case, we used a same composition but we add a 0.20 m of XPS insulation (extruded polystyrene), the thickness of the whole floor, in this case, is 0.56 m and the thermal resistance becomes 0.13 W/m^2.K, which means that the floor, in this case, will participate in the insulation of the house more than in the first case.
- **Type of materials and composition of the exterior walls:** in the basic case we used an assembly of a concrete block, plaster, and cement mortar, with a thickness of 0.25 m and a thermal transmission of 2.18 W/m^2.K. In The improved case, we opted for exterior walls with a thickness of 0.45 m composed of two walls of hollow brick separated by a 0.20 m layer of XPS insulation (extruded polystyrene). In this case, the thermal transmission is 0.13 W/m^2.K, the exterior walls become then so resistant and the house more insulated.
- **Type of materials and composition of the roof:** in the basic case we used a composite of concrete slab, bitumen, cement screed, and roof tiling with a thermal transmission of 1.70 W/m^2.K, while in the improved case, we used the same composition but we added a 0.20 m layer of XPS insulation (extruded polystyrene), with a thermal transmission of 0.13 W/m^2.K, the roof becomes then so performant than in the basic case.
- **The type of glazing:** used in our basic case is a simple glazing with a thermal transmission of 5.74 W/m^2.K, while it's a triple glazing in the improved case with a thermal transmission of 2W/m^2.K. Indeed, the third pane of glass located halfway between the inner and outer panes of double glazing creates two volumes of air locks which improves the energy performance of glazing
- **The orientation of the main façade:** The windows of the main façade in the basic model are oriented to the North, while in the ameliorated model, they are oriented to the south. Indeed, we opted for this choice based on the study of the impact of the orientation of a house in this geographical area on energy consumption (Bourass and Ettahir, 2019) [11], in which it was demonstrated that the orientation has a significant impact on the energy consumption of

heating and cooling and that the orientations on the south side (south, south-east south-west) are the least energy-consuming, while the orientations on the north side (north, north-east, and northwest) are more energy consuming, and in consequence, the South remains the best choice in terms of energy saving since it makes the best balance between energy gains and losses.

Thus, the type and characteristics of materials used in the improved case in the exterior walls, roof, and floor, especially their coefficient of thermal conductivity, their thermal transmission, and their thickness, but also the type of glazing, and orientation, allow the house to be more insulated during both cold and hot months. It protects indeed against the rising temperatures of summer, and also stores the heat during winter and autumn, thanks to the good thermal inertia efficiency, of the envelope.

Thanks to the thermal characteristics of materials chosen in the case of the improved house, and to the good thermal inertia efficiency of the envelope, the improved house marks the least energy needs but also the least energy losses.

Conclusion

Based on our results, optimizing the characteristics of the building envelope can lead to significant energy savings. Specifically, in the improved scenario, we achieved energy savings that were three times greater, with more energy gains and reduced energy losses compared to the basic case.

This study demonstrates that the optimal selection of materials for the exterior walls, roof, and floor—particularly concerning their thermal conductivity, thermal transmission, and thickness—as well as the type of glazing and orientation, can substantially reduce the energy needs of residential buildings for heating in winter and cooling in summer. Our research, which builds on previous studies of envelope components, has highlighted the energy weaknesses of construction methods in this specific region and climate. Ultimately, we can affirm that certain considerations can be addressed during the architectural design and conception of a building, while others can be corrected in existing structures. Therefore, it is important to note that the findings of this study can be directly utilized by architects, engineers, and real estate developers to design energy-efficient buildings in this region and beyond, provided they consider the thermal characteristics of the zones where their projects are located.

References

[1] International Energy Agency, The United Nations Environment Programme (2018), The Global Alliance for Buildings and Construction (GlobalABC). “2018 Global Status Report: Towards a zero-emission, efficient and resilient buildings and construction sector”, 2018, ISBN No: 978-92-807-3729-5, Copyright © United Nations Environment Programme, 2018.

[2] Agence Nationale Pour le Développement des Energies Renouvelables Et De L’efficacité Energétique « Les bonnes pratiques de l’Efficacité Énergétique dans le bâtiment», 2010

[3] Bolattürk A. « Determination of optimum insulation thickness for building walls with respect to various fuels and climate zones in Turkey. » Applied Thermal Engineering 26, Pp 1301-1309, 2006. https://doi.org/10.1016/j.applthermaleng.2005.10.019

[4] Fezzioui N. & Draoui B. & M. Benyamine & Larbi S. « Influence des Caractéristiques Dynamiques de l’Enveloppe d’un Bâtiment sur le Confort Thermique au Sud Algérien. » Revue des Energies Renouvelables, Vol. 11, N°1, pp. 25 - 34, 2008. https://doi.org/10.54966/jreen.v11i1.53

[5] Mia Meftah M. , Benmanssour.M.B « Etude de faisabilité de l’utilisation de l’énergie solaire en vue de la réhabilitation thermique du bâti existant.» Colloque National : Pathologie des Constructions : Du Diagnostic à la Réparation Département de Génie Civil Université Mentouri Constantine - 25 et 26 Novembre 2008.

[6] Bekkouche S.M.A. & Benouaz T. & Cheknane A. « Influence de l'isolation thermique intérieure et extérieure d'un batiment en pierre situé à Ghardaia » SBEIDCO – 1st International Conference on Sustainable Built Environment Infrastructures in Developing Countries ENSET Oran (Algeria) - October 12-14, 2009

[7] Daouas N. & Hassen Z. & Ben Aissia H. « Analytical periodic solution for the study of thermal performance and optimum insulation thickness of building walls in Tunisia.» Applied Thermal Engineering 30, 319–326, 2010. https://doi.org/10.1016/j.applthermaleng.2009.09.009

[8] Guechchati R. & Moussaoui M.A.& Mezrhab Ahm. & Mezrhab Abd. « Simulation de l'effet de l'isolation thermique des bâtiments Cas du centre psychopédagogique SAFAA à Oujda ». Revue des Energies Renouvelables Vol. 13 N°2 ; pp 223 – 232, 2010. https://doi.org/10.54966/jreen.v13i2.191

[9] The Solar Energy Laboratory, University of Wisconsin-Madison "Trnsys 16 a Transient System Simulation Program, Volume 1 Getting Started", 2007.

[10] Meteotest: Jan Remund, Stefan Müller, Stefan Kunz and Christoph Schilter "Meteonorm: Global Meteorological Database" Handbook Part I, Software Version 7/ May 2012"©Meteotest.

[11] Omar Bourass, Aziz Et-Tahir "The Impact of Building Orientation on Energy Consumption in a Domestic House in Desert Climate" International Journal of Innovative Technology and Exploring Engineering (IJITEE) ISSN: 2278-3075, Volume-9 Issue-1, November 2019. https://doi.org/10.35940/ijitee.A4347.119119

[12] Omar Bourass, Aziz Et-Tahir, Kamal Kettani, "The Impact of Exterior Walls Materials on Energy Consumption in a Domestic House in Desert Climate" International Journal of Innovative Technology and Exploring Engineering (IJITEE) ISSN: 2278-3075, Volume-9 Issue-2, December 2019. https://doi.org/10.35940/ijitee.B6210.129219

Materials Research Proceedings 47 (2025) 354-368
https://doi.org/10.21741/9781644903391-40

Hydraulic networks and vernacular architecture: Ingenuity and sustainable management of a vital resource

Ouafa MESSOUS[1*], Khadija KARIBI[1], Karima BERDOUZ[1], and Nezha TLEMÇANI MEKAOUI[1]

[1]National School of Architecture of Rabat, Rabat Institute, Morocco

*o.messous@enrabat.ac.ma

Keywords: Hydraulic Networks, Vernacular Architecture, Water Scarcity, Sustainable Management, Resilience

Abstract. The scarcity of water resources, exacerbated by demographic pressures and climate change, presents a significant challenge to contemporary societies. Ancient hydraulic networks and vernacular architecture, developed by communities in response to arid environments, offer valuable insights into resilient and sustainable water resource management. Rooted in an intimate understanding of local ecosystems, these systems reflect the ingenuity of past societies in achieving a hydrological balance within fragile environments. This paper examines these historical water management strategies, exploring how the principles of modularity, simplicity, and adaptability embedded in these systems can inspire contemporary sustainable water management solutions. By analyzing traditional hydraulic networks, particularly the systems in Moroccan cities like Figuig, Fez, and Tetouan, this study reveals how ancient water distribution, governance, and social engineering models continue to offer relevant and low-cost solutions adaptable to modern water scarcity challenges. Through case studies and primary source analysis, the resilience of these networks underscores their potential as a model for contemporary sustainable water strategies, emphasizing the importance of integrating ancestral knowledge with modern technologies to address pressing environmental demands. This research advocates for a hybrid approach that combines heritage preservation with innovation, ensuring resource sustainability for future generations.

Introduction

Insights from Ancient Water Systems for Modern Sustainability

The scarcity of water resources, intensified by demographic pressures and climate change, poses a major challenge for contemporary societies. Traditional hydraulic networks and vernacular architecture, developed by ancient communities as responses to arid environments and local constraints, stand as fascinating examples of resilient and sustainable water resource management [7]. Rooted in an intimate understanding of the natural environment and adaptive local techniques, these systems reflect the ingenuity of past societies in their ability to structure, organize, and maintain a delicate hydrological balance within fragile ecosystems.

In the face of current challenges, it is both timely and relevant to examine these ancient practices and explore a key question: How can studying these networks, which embody simplicity and resilience, guide us in designing modern strategies for sustainable water management?

1. Study of Primary Sources and Understanding Urban Structures: Ancient hydraulic networks, integrated within specific social and technical systems, invite a deep analysis of primary sources, such as archival deeds. These documents, which reveal historical water management and distribution practices, offer essential insights into the organization and evolution of urban

Materials Research Forum LLC
https://doi.org/10.21741/9781644903391-40

layouts. The often-underutilized information they contain helps clarify the influence of hydraulic systems on vernacular architecture and the structuring of inhabited spaces.

2. Architectural Ingenuity and Optimized Management of Scarcity: Traditional hydraulic networks demonstrate the technical ingenuity and organizational efficiency of vernacular architecture in managing limited resources. Through carefully designed systems, these networks enable balanced water management rooted in a highly developed spatial and social organization. This capacity to maximize the use of scarce resources reflects an expertise that ensures the sustainability and resilience of the water supply.
3. Resilience in the Face of Urban and Social Transformations: The durability of these systems, despite profound urban, social, and administrative changes, reveals a remarkable structural resilience. Functional and efficient over centuries, these water networks offer highly relevant, low-cost solutions adaptable to various contexts. Their enduring presence highlights the effectiveness of a minimalist and integrated model of water management.
4. Perspectives for Sustainable Management of Contemporary Water Resources: The analysis of these vernacular networks sheds light on inspiring models of hydraulic engineering for contemporary water management. The modularity, simplicity, and adaptability of these ancient systems provide strategic insights for addressing modern water scarcity challenges. By drawing on these models, it becomes possible to design integrative and sustainable approaches to resource management, leveraging historical ingenuity to meet the pressing needs of our time.

Case Study: Traditional Hydraulic Networks in Morocco

To illustrate how vernacular hydraulic networks address the challenges of water scarcity, three emblematic cases will be presented: the hydraulic systems of Figuig, Fez, and Tetouan. These examples showcase the diversity and adaptability of these systems, which embody a remarkable historical expertise that integrates architecture, engineering, and sustainable resource management.

1. Figuig: Ingenuity and Community Sharing in an Oasis Environment
   In Figuig, an exemplary hydraulic network demonstrates local governance and optimized sharing practices to meet both agricultural and domestic needs. This system of equitable management, despite modern pressures, remains a clever response to the limitations of water availability [11].
2. Fez: A Model of Historical Complexity and Resilience
   The Fez network, one of Morocco's oldest, reflects the evolution and resilience of an infrastructure that has supported the city's growth since the 9th century. The millennia-old water management in Fez demonstrates how a network can adapt to the increasing demands of a dense urban environment.
3. Tetouan: The Durability of the Scando Network at the Heart of Urban Heritage
   Tetouan's Scando network, still operational today, represents a unique example of sustainability. Designed to address the specific topography and local uses, this autonomous network harmoniously integrates into the architecture of the medina, exemplifying a living heritage that continues to meet contemporary needs.

1. The Hydraulic Network of Figuig: Water Management and Distribution Systems

The hydraulic network of Figuig is built upon two key pillars: water management and distribution systems. These components, grounded in both social and technical engineering, ensure the efficient use of water resources within an arid environment.

1.1 Water Management: A System of Equitable Sharing

Water management in Figuig is governed by principles of equitable sharing, based on customary rules that prioritize need over the ability to own or purchase water. This approach ensures that water is distributed according to real needs rather than financial means. The system is administered by *aiguadiers* (water custodians), who oversee distribution according to community-defined schedules and volumes.

The customary rules, validated by the *j'maâ* (the community assembly), provide a flexible and collectively accepted structure, allowing the system to adapt continually to environmental changes and community needs [11]. This social engineering is deeply rooted within the community fabric, demonstrating a sustainable governance model that has enabled Figuig's hydraulic system to remain resilient despite occasional tensions.

Through successive adaptations, this governance model has proven robust, and essential in the context of water scarcity and increasing pressures on the resource [11]. The resilience of this approach is a testament to the enduring balance between collective governance and resource sustainability, critical for water management in arid regions.

1.2 Distribution Systems: Collection, Storage, and Allocation

The distribution of water, from collection to consumption, follows a structured process to optimize every drop [9]. This journey begins with capturing water sources and directing it to storage basins and channels (*cheneaux* and *seguia*) leading to distribution points. Key elements in this system include:

Foggaras and Seguia: These underground and surface channels ensure the transportation of water [9]

Aqueducts: Used to overcome natural obstacles and maintain network continuity.

Retention Systems and Cisterns (*ahfir*): Built-in depressions to collect both rainwater and groundwater [9], an ingenious feature,

The behbouha, consists of subterranean chambers designed to maintain stable temperatures suitable for changing climates [9]. In summer, when outdoor temperatures reach 40-45°C, water in the *behbouha* remains cool at around 20-25°C. In winter, as outdoor temperatures drop, the water stays warmer, providing a pleasant environment. These spaces serve as refreshing areas in summer and bathing places in winter.

Figure 1. Ahfir : Land depression fitted with underground cisterns, source: Messous, 2019

Figure 2 : Bahbouha nearly 20 meters below ground level, source: Messous, 2019

Figure 3 Water storage basin for distributing irrigation shares by time and volume, with shaded cultivation under palm trees. Source: Messous, 2019

1.3 Prioritized Water Management: Usage, Quality, and Crop Stratification

Water is allocated according to a strict priority order and cleanliness level. This hierarchy ensures that essential uses—such as crop irrigation, drinking water, and domestic needs—are given precedence. After domestic use, water is repurposed for secondary needs like washing or cleaning [11].

This tiered approach allows for progressive water use, maximizing every drop until it is no longer viable. Furthermore, agricultural practices are organized to limit water evaporation. Crops are stratified, with trees and palms forming a protective upper layer that shields against direct sunlight. This arrangement creates a cooler microclimate for lower crops, which benefit from the shade provided by the upper canopy, reducing water loss through evaporation.

2. Example of Fez: The Hydraulic Network as a Key to Understanding the Urban Form of the Medina

The study of Fez's hydraulic network offers an exceptional framework for understanding the organization and structure of the medina—a system with roots dating back to the Idrisid and Almoravid eras. These irrigation systems play a fundamental role in shaping the urban form of Fez, organized around traditional principles of precedence, and regulating benefits and harm in water usage [8]. These principles, deeply embedded in local social practices, directly influence water distribution and the way urban space is organized around this vital resource.

Figure 4 : The Jnane Sbile Garden in Fez is a large recreational space but, first and foremost, a major component of the city's hydraulic system, designed for the storage and distribution of river water; source: Messous, 2018

2.1 Principles of Precedence and Water Usage

Fez's hydraulic network is governed by carefully established management rules to minimize conflict and regulate water usage, based on two key principles:

a. Principle of Harm Prevention: This principle nullifies any action that might harm a downstream user, ensuring that water usage does not adversely affect others. It aims to uphold peaceful and equitable resource management, safeguarding the rights of all users [1]
b. Principle of Benefit: Conversely, water usage is permitted when it provides a benefit without causing harm. This principle allows for flexibility in usage as long as it yields a collective advantage or remains neutral to others' rights [3].

These foundational principles not only prevent conflicts but also foster a community-based, adaptive management system [3].

Figure 5 : Regulation and distribution system at the entrance to the major sectors of the Fez medina, located at Jnane Sbile, source: Messous, 2018

2.2 Regulation of Water Usage: Rules of Priority and Allocation

The rules governing precedence and allocation in water use are applied rigorously to ensure fair management, particularly within the densely populated urban context of the medina. Key rules include:

Upstream Priority**:** Users located upstream of a water source have priority access to the water based on their geographical position.

Downstream Priority and Planting Precedence**:** Downstream users retain priority rights when they have cultivated their land before upstream users, highlighting the historical importance of tenure in resource use [8].

Irrigation Priority**:** Agricultural needs take precedence over other activities, underscoring the central role of irrigation for local crops in water usage [8].

Figure 6 : Jnane Sbile in Fez, source: Messous, 2024

Reference to Canonical Texts and Jurisprudence

In resolving water-related conflicts, the community first refers to canonical texts, which serve as foundational legal guides for water use decisions. In cases where such texts are silent, local jurisprudence is consulted. Judges and scholars adapt these rules to meet the specific needs of Fez, thus providing a coherent legal foundation for water management practices [8].

Materials Research Proceedings 47 (2025) 354-368 https://doi.org/10.21741/9781644903391-40

Local Expertise and the Role of Water Custodians

In unforeseen circumstances, the expertise of local specialists (ذوي الخبره والبصر) becomes essential. These experts, typically seasoned *aiguadiers* (water custodians) or experienced community leaders, possess deep knowledge of the hydraulic network and its unique features. Their expertise enables well-informed decision-making in the absence of formal regulations; ensuring water management aligns with practical needs and local realities [8].

2.3 Community Management and the Tradition of Shared Water Resources in Fez: Balancing Private Ownership and Public Use

The hydraulic system in Fez is based on a balance between private ownership and public use of water, demonstrating how natural resources, though initially privately owned, are progressively transformed for collective benefit [3].

Fez's primary water sources—such as Aïn el-Baghal, Aïn el-Khayl, Sidi F'raj, Aïn Azliten, Aïn Aâllou, and Aïn Al Khadoune—are renowned for their high flow rates and stand as symbols of this shared, community-based water management [3].

Private Assets Serving the Public: Water as a Community Gift

Water sources that supply mosques and public fountains in Fez often began as private properties. Driven by a spirit of communal generosity, the owners of these resources chose to transform them into public assets by donating them to religious institutions or making them available for communal use. This generosity extended beyond humans, with certain fountains, such as the Chrabliyines fountain, specifically designed to provide water for animals, including dogs and cats [8], [3].

Figure 7 : View of the Sahne, the central space of the Bouanania Madrasa in Fez, an example of optimized water use, source: Messous, 2024

The Tradition of Sharing: An Informal Practice of Solidarity

In addition to official donations, access to water is often governed by a longstanding tradition of sharing, rooted in goodwill and solidarity. This custom requires well or water source owners to grant access to neighbors, especially during times of drought. Known as the tradition of sharing, this practice prevents owners from denying water access to other residents of the medina, ensuring equitable access even in times of scarcity [8].

This practice is further reinforced by legal principles that prohibit denying vital resources, such as drinking water, particularly for human consumption. Consequently, this tradition of sharing could be grounded in legal values affirming the inalienable nature of water for essential needs [6]; [3].

 Materials Research Forum LLC
https://doi.org/10.21741/9781644903391-40

Regulated Access: A Compromise for Resource Management
To prevent conflicts and regulate the use of these precious resources, a compromise system has been adopted: water is accessible to all but at specific times. This rule provides collective access to water resources in the morning, meeting residents' needs for household tasks while preserving the resource for other times of the day. This system of time-based sharing reflects a community organization that balances resource conservation with the daily needs of residents [8].

Hydraulic Network Devices: Distribution and Protection
The hydraulic network of Fez includes various devices that contribute to the distribution, management, and protection of water within the medina:
Distributors and Dividers: Strategically placed, these devices channel water into different canals to meet the needs of each neighborhood [3].
Protective Dams: Located upstream, these structures shield the medina from floods and regulate surface water flow.
The Grand Distributor of Bab Boujeloud: This main distributor manages water flow evenly, adapting to the city's topographical variations and demographic pressures [3].

This blend of technical and aesthetic considerations underscores a thoughtful approach to water management that is integral to Fez's urban landscape.

Figure 8 Oued Fez at the entrance to Jnane Sbile in Fez, source: Messous, 2018

2.4 The Watercourse in Fez: A Social and Technical Transformation of the Hydraulic Network
The example of Fez provides a perspective on how a natural watercourse evolves into an organized and socially integrated hydraulic network within the medina. This transformation involves human intervention and sophisticated engineering, revealing two key aspects of water management in Fez: the technical structuring of the watercourse and its aesthetic and functional integration into the city.

Figure 9 : Oued Fez violently flowing through the Fez medina after a storm, highlighting the importance of developments made to prevent damage, source: Messous, 2019

From Natural Water to Urban Watercourse

As it flows into the city of Fez, the watercourse traversing the region is initially an unstructured natural element, simply following the land's topography. Outside the city, this waterway draws the attention of residents without yet undergoing extensive technical organization; it flows freely, its path shaped by the natural terrain.

However, upon entering the heart of the medina, this watercourse transforms into a significant social and technical phenomenon. The residents' interventions convert this natural flow into a sophisticated network, vital to the urban fabric and daily life. Water becomes inseparable from human activity; neglect or lack of maintenance would lead to severe disruptions within the medina, affecting its organization and functionality.

Water Engineering and Management: Technical and Aesthetic Aspects

The hydraulic network management in Fez relies on advanced engineering that combines technical efficiency with aesthetic considerations. This engineering approach unfolds in two primary aspects:

- **Technical Aspect:** The water flow is controlled through a precise system of distributors and dividers, particularly at the entrance to the medina, where a large arched distributor channels water to different neighborhoods. This technical organization ensures a steady and regulated water flow throughout the city [3].
- **Aesthetic and Functional Aspect:** Beyond distribution, water serves an aesthetic role, enriching public spaces with fountains and water wheels (*norias*). These features, while functional, also beautify the urban environment, creating pleasant settings for residents [8].

Materials Research Proceedings 47 (2025) 354-368 https://doi.org/10.21741/9781644903391-40

Figure 10 : Nouria at Jnane Sbile, source: Messous, 2018

Figure 11: View of the large water retention basin at Jnane Sbile, source: Messous, 2024

3. Case of Tetouan: The Scando Hydraulic Network

The Scando hydraulic network of Tetouan is a unique example in Morocco whose etymology is thought to be an iteration of the word 'Escondido,' meaning 'hidden'", as the only traditional urban hydraulic network still in operation [8]. This system, which began expanding as early as the 15th century, is based on a complex and autonomous underground infrastructure, showcasing ingenuity in urban water management and distribution [3]. Comprising 17 main sources located in the northern part of the city, between Bab Nuader and Bab Mqabar, the network demonstrates remarkable continuity over time. Today, its management is shared between the municipality of Tetouan and the company Amendis.

Figure 12 : The side door in the photo provides access to the Scando distributor; source: Messous, 2024

3.1 Underground Architecture and Defensive Secrecy

The architecture of Tetouan's hydraulic network, known as "Scando," was long kept secret, shrouded in legends and protective measures. This secrecy served a strategic purpose: in the event of an attack, it preserved the integrity of the network to prevent any disruption in water supply, ensuring the city's resilience against external threats [1] This underground design not only makes the network self-sufficient but also separates it from surface infrastructure, allowing it to pass through private properties and public roads without direct overlap [3]. Built according to customary laws and rules, the network unfolds deep below the surface, safeguarded from external threats.

3.2 The Complexity of the Scando Network: Autonomy and Unconventional Organization

The Scando network consists of autonomous underground channels that do not overlap with public roadways. It follows the natural topography and sometimes passes beneath built structures, demonstrating a unique architectural freedom informed by specific governing rules. This independence, both in organization and distribution, allows for precise adaptation to the city's water needs. Its autonomous functioning reflects a profound understanding of local resources and unique geographical conditions.

Role of the Habous Institution in Water Management and Public Services

Beyond managing places of worship, the *habous* (religious endowments) institution plays a multifaceted role in the management of public resources and services in Tetouan, including water and the maintenance of communal infrastructure [3]. Key functions of the *habous* institution in Tetouan include:

- Public Services and Infrastructure: The *habous* funds the upkeep of essential city infrastructure, particularly public fountains and defensive structures [8].
- Support for Vulnerable Populations: This institution aids the most vulnerable social groups by providing access to basic resources, including water [3].
- Management of Property and Water Resources: The *habous* oversees the maintenance of water channels and reservoirs, applying the principle of *ripta* (monetary contributions) to fund network upkeep, under the guidance of *Arbab al-Bassar*—experts responsible for regulating consumption based on needs and technical expertise [1]
- Water Management Expertise: The *Arbab al-Bassar*, hydraulic advisors, are essential to maintaining the network, resolving disputes, and adapting rules to meet local needs. Their expertise ensures the network's resilience and effective functioning within the community.

Private Water Management: Contracts and Responsibilities

Water management among private individuals within the Scando system relies on formal contracts and a defined set of responsibilities [1] These contracts govern the terms for selling and sharing surplus water and include clauses on the maintenance and sustainability of hydraulic infrastructure.

- Sale and Sharing of Surplus Water: The sale of surplus water between individuals is strictly regulated. Participants are required to contribute to the maintenance costs of pipelines and other hydraulic infrastructure. Before any contract is established, the *Arbab al-Bassar* (water management experts) conduct a technical assessment to evaluate the feasibility of new uses or divisions within the network [3].
- Maintenance and Creation of New Divisions: Preserving the integrity of the network is paramount. No individual is permitted to resell or divert water without a contract validated by experts. This validation ensures that new divisions do not compromise the overall water flow or the system's durability. Any water diversion is documented in an official record, creating a formal trace of resource usage and maintenance.

Social Management of the Scando System: Water Access Based on Social Status

Water distribution within the Scando system is organized to consider social criteria and the specific needs of residents, who are categorized according to their socio-economic status [1]

- Water Access for Affluent Residents: Individuals from affluent families benefit from privileged water access, known as *faïd* or *ma'ae jari*. This status not only allows them sufficient water for domestic use but also grants them the right to manage any surplus as they see fit, whether by redistributing, selling, or using it to irrigate gardens and plants on their properties [3].

- Restricted Use for Medium and Small Households: Residents of medium or small households have more limited access rights, receiving only enough water to cover basic domestic needs without the option to manage or sell any potential surplus. This restriction, based on social status, ensures that resources are distributed equitably, aligning with real needs and the capacity of the infrastructure [7].

Figure 13 : Erzini P Erzini Palace, a resilience space within the Scando system; source: O. Messous, 2024

3.3 The Architecture of Public and Social Fountains in Tetouan: A Tool of Power and Social Cohesion

Tetouan's network of public fountains, integrated into the Scando hydraulic system, is distinguished by an architectural and spatial organization that addresses both the social needs and power dynamics of the medina. The placement of fountains follows a zoning logic based on public and artisanal spaces, as well as the social standing of the patrons who financed them [8].

Fountains in Artisanal and Public Areas: Simplicity and Functionality

In the medina's artisanal and public zones, fountains are numerous but feature a simple, functional design. These fountains are intended for collective use, meeting the daily needs of artisans and residents by providing accessible water for artisanal activities and refreshments for passersby. Their modest architecture reflects their purely utilitarian function, aimed at efficiently serving a large population without an emphasis on prestige [3].

Monumental Fountains at the Medina's Gates: Expressions of Power by Local Elites

In contrast, the fountains at the main gates of the medina are often monumental structures, richly decorated and funded by local elites. These fountains are more than mere water points; they symbolize prestige and influence for the patrons who commissioned them. These patrons, often competing for local power, used the architecture of the fountains to display their social status and community role [1] These fountains serve as prestige monuments, marking the city's principal entrances. They convey a political message of benevolence and commitment to society while reinforcing the authority of the elites. Through these fountains, patrons demonstrate their ability to meet community needs, thereby strengthening their legitimacy and local power.

Materials Research Proceedings 47 (2025) 354-368 https://doi.org/10.21741/9781644903391-40

Figure 14 : Operational public fountain in Tétouan, part of the Scando system; source: O. Messous, 2024

Water Distribution in Public Spaces: Baths and Public Toilets

Water was not only provided for domestic or artisanal use; it also played an essential role in public services, supplying baths and toilets. These spaces, crucial for hygiene and well-being, contributed to the quality of life in Tetouan's medina. The provision of water to these public facilities underscores the hydraulic system's complexity and its role in fostering social cohesion and public health [7].

The Political and Social Role of Water Provision

The network of fountains and public baths reveals the political dimension of water provision: local elites used water management as a way to consolidate their authority and reinforce their local standing. By supporting these infrastructures, they demonstrated their commitment to the population's welfare, while also solidifying their social status and influence [3] ; [1]. This dynamic of power elevates water to a symbol of prestige and political legitimization, beyond its basic public utility function.

3.4 The Scando Hydraulic System of Tetouan: Components and Functionality

The Scando hydraulic system in Tetouan is a sophisticated, sustainable water management model characterized by diverse components and ingenious devices. Key components and their functions are as follows [2]:

Autonomous Pipelines**:** The Scando network is based on autonomous pipelines that transport water from each source to specific distribution points. These often-underground pipelines deliver water independently, without interference with other urban networks such as roadways.

Reservoirs: Located just after the water sources, reservoirs provide temporary storage. They serve as accumulation points, enabling efficient flow management before the water is directed into main channels. These reservoirs are essential for stabilizing the water supply.

Division Devices (Qisma) : The division device, or *Qisma*, is designed to split water into multiple channels. It consists of an inlet pipe and several outlet pipes, known as "masharib." This device allows precise water division to distribute it to different neighborhoods or uses, optimizing the water flow.

Materials Research Proceedings 47 (2025) 354-368 | https://doi.org/10.21741/9781644903391-40

Figure 15: Scando: A system for dividing water from the main pipeline into secondary branches based on the principle of communicating vessels; source: O. Messous, 2024

Evacuation Tubes and Maintenance Devices (Tafjira) : The evacuation tubes, or *tafjira*, temporarily divert water during maintenance or repair work. These devices ensure flexible and secure network management, minimizing water supply interruptions during technical interventions.
Maïda Al Faïd (معدة الفيض) : This component allows property owners to access surplus water. Those with access to the Maïda Al Faïd have an additional water volume they can use for various purposes, such as selling, sharing, or personal use.
Maïda Al 'Adl (معدة العدل) : Unlike the Maïda Al Faïd, the Maïda Al 'Adl restricts water use to an exclusive point. Owners of this component may only use the water for personal consumption, with no option to redistribute it.

Figure 16 : Scando: Example of Maïda Al Adl; source: O. Messous, 2024

Rukba (الركبة) : The *rukba* is a connector that allows adjustments in water distribution direction. It functions as a bend or junction within the pipeline system, enabling the redirection of water flow as needed.
Regulation System (Taâlaâ - الطالع) : The *Taâlaâ* (or *Tawali*) is an essential regulation system within the Scando network. Based on the principle of communicating vessels, it regulates water levels and maintains consistent pressure throughout the network. This device ensures equitable and stable distribution, adapting to variations in height and flow rate.

Conclusion: Lessons and Challenges for Water Resource Management

The study of traditional hydraulic systems offers key insights not only into the architecture and management of this vital resource but also into sustainable and resilient strategies for addressing contemporary challenges.

The Value of Primary Sources and Waqf Documents

Primary sources, such as waqf documents, are invaluable for an interdisciplinary approach, providing essential insights into:

- The structure and organization of hydraulic networks;
- The formation and evolution of urban fabrics;
- The management systems implemented to regulate tensions and resolve water-use conflicts.

Analyzing these documents reveals not only the technical dimensions of these systems but also offers a broader view of city management, characterized by a complex balance of engineering, customary law, and social needs.

Ingenuity and Purpose: Results-Oriented Resource Management

The ingenuity of these systems lies in a purpose-driven approach to management, focused on outcomes rather than resources. By optimizing limited resources, these systems highlight the importance of:

- Water Conservation: Prioritizing water preservation and efficient distribution in response to community needs;
- Resource Optimization: Reducing waste and maximizing the impact of every drop of water;
- Conflict Management and Anticipation: Establishing rules and sharing mechanisms to prevent tensions and ensure fair distribution.
- These examples demonstrate that water management aims not only to supply water but to foresee, adapt, and innovate through a needs-based approach.

Resilience and Adaptability: Balancing Tradition and Modernity

Resilience in traditional hydraulic systems extends beyond mere endurance; it represents the ability to evolve and integrate new paradigms. This requires:

- Balancing Tradition with Modernity: Traditional water management systems now intersect with modern technologies like digitization and artificial intelligence, which open new possibilities;
- System Hybridization: Integrating modern, interactive tools, such as intelligent water management systems, while respecting traditional methods and expertise;
- Heritage Continuity: Preserving these ancient systems while adapting their practices to meet the growing challenges of resource scarcity and current demands.

Future Perspectives

- In the face of increasing water scarcity, especially in arid regions, and the challenges posed by climate change, integrating ancestral knowledge with contemporary technological advancements becomes essential. This hybrid approach could provide more resilient and sustainable solutions to meet future needs. It also raises a crucial question for policymakers:

Materials Research Proceedings 47 (2025) 354-368 https://doi.org/10.21741/9781644903391-40

how can heritage preservation and innovation be combined in natural resource management to ensure a more stable and equitable future?

- These traditional systems, symbols of ingenuity and resilience, offer valuable lessons for rethinking water management. They remind us that, often, true efficiency lies in the simplicity of methods and the clarity of purpose.

References

[1]Abboud, H.A. New Issues in the History of Tetouan: Water, Urban Development, and Society. Tetouan: Series of Publications by the Mohammed Ahmed bin Abboud Foundation, 2023.

[2]ERRAMI Khalid. Water Resource Management in Tetouan through Habous, Family, and State Resources. Rabat: Ministry of Endowments and Islamic Affairs Publications, n.d.

[3]ERRAMI Khalid. The Original System of Water Distribution in the City of Tetouan (1862-1913). Tetouan: Publications of the Asmir Association of Tetouan, 2008.

[4]El-Fassi, M. Figuig: Resilient Oasis on the Desert's Edge. Casablanca: Éditions du Maghreb, 2015.

[5]El Hassan, A. Traditional Water Management Systems in Moroccan Oases. In: F. Zohra (Ed.), *Oases of the Maghreb: Environmental and Social Challenges*, Rabat: University Press, 2018, pp. 123-145.

[6]Fazza, M.A., Aftit, M., Jamal. The Oasis and Societal Transformations: Human, Space, and Society. Rabat: Mohammed V University, Faculty of Letters and Human Sciences, 2023.

[7]Hamouda, H. Water and the Making of the Sacred: An Anthropological Study of Oasis Society Structures in Morocco. Rabat: Mohammed V University, Faculty of Letters and Human Sciences, 2024.

[8]Mellouli Idrissi, A. The Water System of Fez: A Historical, Field, and Documentary Study, 2023.

[9]Ministry of Agriculture and Maritime Fisheries. Report on Water Management in Moroccan Oases, Ministry of Agriculture, last modified March 15, 2022.

[10] https://www.agriculture.gov.ma/rapport-oasis.

[11]Ministry of Agriculture and Maritime Fisheries. Report on Water Management in Moroccan Oases. Rabat: Ministry of Agriculture, 2020.

[12]Vallat, J.P., et al. Moroccan Heritage: An Oasis at the Heart of Cultures. Paris: L'Harmattan, 2014.

Keyword Index

About the Editors

Hassane KHARMICH is a Full Professor of Cities and Urban Planning. He serves as the Publishing Assistant for the African and Mediterranean Journal of Architecture and Urbanism (AMJAU) https://revues.imist.ma/index.php?journal=AMJAU. Additionally, he is in charge of the Research Laboratory titled "Housing, Architecture, and Urbanization of Territories" (LabHAUT), as well as the Reception Structure for doctoral students at the Doctoral Studies Center: "Architecture and Associated Disciplines, National School of Architecture, Rabat". He has authored numerous publications and oversees various research programs focusing on the medina, architectural heritage, metropolization, and urban planning.

ORCID ID: 0000-0003-4511-0628

https://hkharmich.wixsite.com/my-site-1

https://scholar.google.com/scholar?hl=fr&as_sdt=0%2C5&q=kharmich+hassane&btnG=&oq=k harmich

Khalid EL HARROUNI is a full professor of civil engineering and computational mechanics and director of UNESCO chair *"Education and Research on Sustainable and Bicoclimatic Urban Planning and Architecture"*. As the deputy director of research at Ecole Nationale d'Architecture in Rabat, Morocco, he is responsible for the Doctoral Studies Centre 'Architecture and related Disciplines' and serves as scientific coordinator for the research theme 'Sustainability in Architecture and Urbanism'. He is also the editor-in-chief of *African and Mediterranean Journal of Architecture and Urbanism (ISSN 2665 - 7953)* https://revues.imist.ma/index.php?journal=AMJAU and an expert member of the two International Council on Monuments and Sites (ICOMOS) International Scientific Committees on the Analysis and Restoration of Structures of Architectural Heritage, and Energy and Sustainability. He has more than thirty-five years of research and professional experience in civil engineering, urban planning, building energy efficiency, advanced computational techniques, and heritage architecture.

ORCID ID: 0000-0002-3391-6889

https://www.scopus.com/authid/detail.uri?authorId=6603628901

https://scholar.google.com/citations?hl=fr&user=KTS0D9YAAAAJ&view_op=list_works

Scopus Author ID: 6603628901, Scopus Author ID: 57205495322, ResearcherID: AAK-6890-2020

Khadija BABA is a civil engineer from Mohammadia School of Engineering Rabat in Morocco (EMI Rabat), she holds a PhD in Civil Engineering from Mohammadia School of Engineers and is a Professor at the Higher School of Technology of Sale, Mohammed V University of Rabat in Morocco.

She is the director of the soil mechanics, structures and materials team and member of the Civil and Environmental Engineering Laboratory, Mohammadia School of Engineers, Mohammed V University in Rabat (Morocco).

Her research activities are mainly focused on the valorization of local construction materials. she has published more than eighty articles and book chapters in indexed journals and co-edited over 4 books.

SCOPUS: 57803630600 and ORCID: 0000-0003-4603-4321

Khadija KARIBI, Co-responsible for the UNESCO Chair "Education and Research in Sustainable Urban Planning and Bioclimatic Architecture" at the National School of Architecture in Rabat, is an architect-urban planner and a doctor in urban geography. She is also a Full Professor at the National School of Architecture in Rabat. Her expertise lies in the fields of urban planning, urban development, and public space practices. She oversees the "Urban Project" architecture workshop, teaches a lecture course on Urban Planning Currents, and conducts a workshop on Urban Planning tutorials. Additionally, she serves as the Scientific Coordinator of the doctoral program "Architecture and associated disciplines." She has published a book, collective works, and articles, and has actively participated in various research projects and urban studies. She is also a co-editor of the African and Mediterranean Journal of Architecture and Urbanism (ISSN 2665-7953)

https://revues.imist.ma/index.php?journal=AMJAU
ORCID ID: 0009-0000-0880-7317

www.ingramcontent.com/pod-product-compliance
Lightning Source LLC
LaVergne TN
LVHW052224200926
848481LV00001B/208